ADVANCES IN RICE BLAST RESEARCH

Developments in Plant Pathology

VOLUME 15

The titles published in this series are listed at the end of this volume.

Advances in Rice Blast Research

Proceedings of the 2nd International Rice Blast Conference
4–8 August 1998, Montpellier, France

Edited by

D. THARREAU

M.H. LEBRUN

N.J. TALBOT
and
J.L. NOTTEGHEM

KLUWER ACADEMIC PUBLISHERS

DORDRECHT / BOSTON / LONDON

A C.I.P. Catalogue record for this book is available from the Library of Congress.

ISBN 0-7923-6257-8

Published by Kluwer Academic Publishers,
P.O. Box 17, 3300 AA Dordrecht, The Netherlands.

Sold and distributed in North, Central and South America
by Kluwer Academic Publishers, ̣
101 Philip Drive, Norwell, MA 02061, U.S.A.

In all other countries, sold and distributed
by Kluwer Academic Publishers,
P.O. Box 322, 3300 AH Dordrecht, The Netherlands.

Printed on acid-free paper

TABLE OF CONTENTS

EPIDEMIOLOGY AND INTEGRATED DISEASE MANAGEMENT

PATHOGEN POPULATION STUDIES

PATHOGEN GENETICS AND MOLECULAR BIOLOGY

PREFACE

The idea of a 2nd IRBC was born in 1993 in Madison when participants to the 1st IRBC concluded with optimism that another IRBC will be organised 5 years later, linked to the 7th ICPP. When specialised meetings are organised very frequently, the interest of the IRBC meetings, which bring together researchers with the rice blast disease as common and sometime unique interest, was questionable. But since the beginning it was an evidence that IRBCs are a unique opportunity to examine with a wide scope the results in different areas, and to evaluate when the results obtained by specialists will provide useful tools to control the disease. We also knew that they are important opportunities to accelerate exchange of information and establish new collaborations.

Several meetings about blast were previously organised, the first one was held at IRRI in 1963 and provided the state of knowledge about *Pyricularia oryzae* and the disease it causes. At this period one of the main return of this conference was the definition of an "international set of differential varieties" and the beginning by IRRI of the IRBN nurserie's network. At that time the genetic characterisation of the resistance of this set of cultivars was incomplete except for the Japanese cultivars. The origin of the variability of the fungus was a real mystery. When we organised in Montpellier with some colleagues a symposium on the rice blast in 1981, several ideas were discussed: the importance and the polygenic support of the partial resistance, the existence of complex and diverse populations of blast. The first IRBC was the 1st Rice Blast meeting of the "molecular time" and those who had not followed carefully the investment of Universities in the rice blast as a model were surprised to discover that rice and blast had become "the model" for studying cereals-fungi interactions. The methods for studying populations of *Magnaporthe grisea* were developed, the genetics of *M. grisea* was manageable, mapping and cloning resistance genes and corresponding avirulence genes were in progress, strategies for obtaining resistant transgenic cultivars were proposed. In 1998, participants may have observed the presentations with at least two points of view. Those who are in the world of fundamental research had a rare opportunity to discuss together, to note the progress obtained and to analyse the fantastic progress in the knowledge of the interactions at the molecular level as well as at the population level. But for those who work at the level of farmer fields, they may consider that the progress are still too slow, when many of these results are far from applications. The knowledge of populations structure and diversity may help for improving strategies of resistance gene management but the methods are not totally satisfying. No efficient artificially designed resistance is available. Very efficient new fungicides sometimes environmentally safe are developed by Companies but unfortunately they are often too costly for the low input agriculture which is the main rice producer. The management of breeding for resistance remains a complex task as breeders have to integrate many objectives together. To make efficient breeding and screening methods for cultivars with improved resistance, it will be necessary to develop simple and reliable methods for resistance characterisation. Only simple strategies for resistance durable enough to be

ix

manageable will have an impact. Of course progress of science are always too slow for those who need the benefits but we have many reasons to be optimistic: progresses were never so fast, and projects in development are many.

The particular atmosphere of Montpellier summers may had a special effect, but it seem that most participants have appreciated this 2nd IRBC and a very wide majority asked for a 3rd IRBC. The objective of IRBC scientific committee is also to organise the meetings in place and time which make possible the exchange of information between many people. So the next IRBC should be linked to another rice meeting, most of us would like to have it in 2000 or in 2001.

Jean Loup Nottéghem

ACKNOWLEDGEMENTS

We would like to thank our sponsors who have made this 2nd IRBC possible:

the AGRO Montpellier and CIRAD kindly provided facilities, human and financial ressources but also encouragement to the organizing committee,

the Région Languedoc-Roussillon, Rhône Poulenc Agrochimie and Novartis provided financial support necessary to welcome all IRBC participants,

the French Embassy in Vietnam and the French Embassy in Russia contributed to travel expenses of our colleagues.

We are very grateful to our sponsors for facilitating exchange of information and collaborations between our colleagues who all contribute in progress to the control of the rice blast disease.

GENETIC ANALYSIS OF BLAST RESISTANCE IN IR VARIETIES AND RESISTANT BREEDING STRATEGY

T. IMBE, H. TSUNEMATSU, H. KATO, AND G.S. KHUSH
International Rice Research Institute
P.O. Box 933, 1099 Manila, Philippines

1. Introduction

To develop rice cultivars with durable resistance to blast, the following breeding strategies have been proposed: accumulation of minor genes (field partial or quantitative resistance), pyramiding of major genes (true, complete or qualitative resistance), and a combination of major and minor genes, multilines of major genes, etc. Knowledge of the genetic constitutions of major genes in rice cultivars and the pathogenicity of blast isolates are prerequisite for all of these strategies.

We have limited information on blast resistance in *indica* varieties including IRRI bred IR cultivars, despite the fact that many resistance genes have already been identified in Japan even in *indica* cultivars and their *japonica* derivatives. Therefore, we need to analyze the genetic constitutions of resistant gene of important *indica* cultivars. We are establishing simple differential systems for the identification of pathogenicity of blast isolates of each country. By using such identified isolates, the resistance genes of different varieties in the world will be easily identified.

2. Pathogenicity of IRRI blast isolates

Blast resistance genes in most of the *indica* varieties including IRRI bred IR cultivars have not been identified due to their complexities. Among Japanese varieties, 14 alleles on 8 loci have been identified. Except for *Pi a*, *Pi i*, *Pi k-s*, and *Pi sh*, the other 10 alleles have been introduced into Japanese varieties from foreign *indica* donors. Therefore, those genes may exist also in *indica* varieties.

We have tested the pathogenicity of a few hundred blast isolates by inoculating them to Japanese differential varieties and IRRI near-isogenic lines (NILs) having known resistance genes. These blast isolates had been collected from all over the Philippines or were found in the IRRI blast nursery. They were classified into around 30 different pathotypes according to their pathogenicities.

The pathogenicities of the blast isolates are usually used for the analysis of resistant genes of cultivars. Based on the results of inoculation, the pathogenicities of

D. Tharreau et al. (eds.), Advances in Rice Blast Research, 1–8.

the standard blast isolates are shown in Table 1. The pathogenicity of the isolates had been analyzed by using Japanese differentials and IRRI NILs, but virulence to some resistance genes was not clear because of the extra gene(s) in the differentials or NILs.

All the isolates were compatible to the *Pi t* gene but moderately incompatible to *Pi sh*. Other blast isolates tested so far also showed the same reactions to the two genes. These results are contrary to that obtained with Japanese blast isolates, most of those were compatible to *Pi sh* and incompatible to *Pi t*. The *Pi 19(t)* gene was also not effective against all isolates. It was identified in Aichi Asahi and other Japanese differentials against a blast isolate from Yunnan Province, China (Hayashi *et al.*, 1998). It was allelic or closely linked to the *Pi ta* locus. So far, we have not found any Philippines blast isolates that are incompatible to *Pi t* and *Pi 19(t)*.

All of the *Pi k* alleles (except for *Pi k-s*), *Pi 1*, and *Pi 7(t)* have shown the same reactions to the isolates. Other isolates not listed in the table also showed the same reactions to the genes, and we have no isolate to distinguish these genes from each other at IRRI. *Pi 1* gene was closely linked to the *Pi k* locus (Inukai *et al.*, 1994), and *Pi 7(t)* was found to be allelic or closely linked to *Pi 1* (Wang *et al.*, 1994; Inukai *et al.*, 1996). Therefore, the origin of these genes was presumed to be the same. Till distinguishable isolates are identified, these genes are designated *Pi k** gene in this report.

The *Pi i* and *Pi 3* genes showed exactly the same reaction and were closely linked to each other (Inukai *et al.*, 1994). Therefore, the two genes should also be treated as one. The *Pi 5(t)* gene showed a tendency similar to that of *Pi i* and *Pi 3* genes in terms of reaction to the isolates (Wang *et al.*, 1994; Inukai *et al.*, 1996). However, the line with *Pi 5(t)* showed a higher resistance level than those with *Pi i* and *Pi 3*. We have to confirm the resistance pattern of the *Pi 5(t)* gene by inoculating the NILs with *Pi 5(t)* because we could not know whether the line possesses extra gene(s) for moderate reaction or whether *Pi 5(t)* itself showed higher resistance.

3. Classification of IR cultivars

All the IR cultivars and some of their ancestors were inoculated with 14 isolates of different pathotypes. In the IR cultivars, the existence of *Pi a*, *Pi b*, *Pi k-s*, *Pi k**, *Pi ta*, *Pi z-t* genes and a new gene, which was registered as *Pi 20* has been either confirmed or presumed. Three genes among them, *Pi k**, *Pi ta*, and *Pi 20*, showed a relatively broad spectrum of resistance and were regarded to be the key genes responsible for resistance to natural blast populations.

The cultivars were classified roughly into eight groups with distinctly different reaction patterns to the isolates (Table 2). This grouping was characterized by *Pi 20*, *Pi ta* and *Pi k**. The IR30 group has none of them and the IR24 group has *Pi 20*. The IR36 group has *Pi ta* and the PSBRc1 group has *Pi k**. The IR74 group, IR56 group, and IR64 group have *Pi k** + *Pi 20*, *Pi k** + *Pi ta*, and *Pi ta* + *Pi 20*, respectively. Three cultivars, PSBRc10, PSBRc18, and PSBRc20 were classified as the PSBRc10 group. These three

Table 1. Pathogenicity of IRRI blast isolates.

Blast isolate	Blast resistance gene (Pi)																				
	k-s	a	b	i	3	5(t)	k	k-h	k-m	k-p	1	7(t)	z	z-5 (=2(t))	z-t	ta (=4(t))	ta-2	20	sh	t	19(t)
B90002	S	R	R	S	S	M	R	R	R	R	R	R	M	M/R	R	R	R	-	M	S	S
C923-49	S	R	M	S	S	M	R	R	R	R	R	R	R	M/S	R	S	R	-	M	S	S
BN111	S	S	S	M	M	R	R	R	R	R	R	R	M	R	S	S	R	R	M	S	S
BN209	S/M	S	R	S	S	M	R	R	R	R	R	R	M/S	M	S	S	R	-	M	S	S
JMB840610	S	S	R	S	S	M	R	R	R	R	R	R	M/R	R	R	S-	R	-	M/R	S	S
V86010	R	S	R	S	S	M	R	R	R	R	R	R	M/S	M	R	S	R	-	M/R	S	S
V850196	R	S	S	M	M	-	R	R	R	R	R	R	-	M	S	R	R	S	M	S	S
IK81-3	S	S	S	S	S	M	R	R	R	R	R	R	-	M	S	R	R	S	M	S	S
IK81-25	S	S	S	R	R	R	S	S	S	S	S	S	M	R	S	R	R	S	M	S	S
PO6-6	S	S	S	M	M	R	R	R	R	R	R	R	M	M	M	S	S	S	M	S	S
Ca89	S	S	S	S	S	M	R	R	R	R	R	R	M	R	M	S	S-	S	M	S	S
V850256	S	S	(S)	S	S	M	R	R	R	R	R	R	M	R	M	M	R	-	M	S	S
M36-1-3-10-1	S	S	S	M	M	M	S	S	S	S	S	S	M	M	S	S	R	R	M	S	S
M64-1-3-9 1	S	S	S	S	S	S	S	S	S	S	S	S	M	R	R	R	S	S	M	S	S
M39-1-3-8-1	S	S	S	M/S	M/S	M	S	S	S	S	S	3	M	R	R	R	R	S	M	S	S
M101-1-2-9-1	S	S	R	S	S	M	R	R	R	R	R	R	M	S	R	S	R	-	M	S	S

cultivars were susceptible to the isolate Ca89, of which pathogenicity was similar to that of PO6-6 except for the virulence to *Pi i* and *Pi 3*. Therefore, the genotypes of the three cultivars were inferred to be *Pi ta*, *Pi 20*, and *Pi i* (or *Pi 3*). This indicates that some additional blast isolates are necessary to identify other resistance genes.

Table 2. Classification of IR cultivars and presumed blast resistance genes.

Group	Cultivar	Presumed gene
IR30	IR20, IR28, IR30, IR45, IR66	*(Pi b, Pi k-s)*
	IR29, IR34	*(Pi b, Pi k-s, Pi z-t)*
IR24	IR8, IR22, IR24, IR26, PSRc30	*Pi 20 (Pi b, Pi k-s)*
	IR43, Peta	*Pi 20 (Pi b)*
	PSBRc 2	*Pi 20 (Pi b, Pi k-s, Pi z-t)*
IR36	IR5, IR42, IR44	*Pi ta (no Pi b)*
	IR32, IR36, IR38, IR40, IR50, IR52, IR54, IR58, IR60, IR62, IR65, IR68, IR72, PSBRc4	*Pi ta (Pi b)*
PSBRc1	PSBRc1	*Pi k**[a]
IR74	IR74	*Pi k*, Pi 20*
IR56	IR56, IR70	*Pi k*, Pi ta*
IR64	IR46, IR48, IR64, PSBRc28	*Pi ta, Pi 20*
PSBRc10	PSBRc10, PSBRc18, PSBRc20	Other genes[b]

[a]*Pi k** could be one of *Pi k, Pi k-m, Pi k-p, Pi k-h,* and *Pi 1*.
[b]PSBRc10, 18, 20 most probably possess *Pi ta, Pi 20,* and *Pi i* (or *Pi 3*).

Then, we selected three blast isolates "*i.e.*, PO6-6, IK81-25, M36-1-3-10-1," that are avirulent to only one of the three genes. By inoculating with three blast isolates, all IR cultivars were classified into eight possible combinations of the three resistance genes. Some other blast isolates were also chosen to estimate other genes.

4. Inheritance and allelism tests of blast resistant genes in IR varieties

To determine whether the known genes are involved or not, we conducted genetic analysis of blast resistance of IR cultivars by inoculating the seedlings of F_3 or BC_1F_2 lines with several blast isolates. When several resistance genes were expected to exist in one cultivar, BC_1F_2 analysis was conducted to estimate the number of genes (Toriyama *et al.*, 1983). Several avirulent isolates for the expected resistance genes in the cultivar were inoculated to the seedlings of the same BC_1F_2 lines. Monogenically segregating lines were used for further confirmation of the genes in allelism tests and mapping. For each line, 20 plants were tested.

In this report, we only discuss the analysis of the inheritance of blast resistant genes of IR24, but we have done similar analysis for the other 7 varieties. Eight IR cultivars, IR24, IR34, IR36, IR60, IR56, IR64, IR46, and IR74 were tested to evaluate the inheritance of resistance to several isolates (Table 3). Those varieties showed typical reactions to several blast isolates. They are considered to be representative varieties for blast resistant gene components. We conducted inheritance test for the above mentioned eight IR varieties by the inoculation with blast isolates.

In IR24, we have presumed the existence of Pi k-s, Pi b, and an unknown gene according to their reaction pattern to the tested blast isolates. Therefore, 80 BC_1F_2 lines of CO39/IR24//CO39 were inoculated with four blast isolates, V850196 (Av k-s), BN209 (Av b), BN111 (Av $IR24$), and JMB8401102 (Av $IR24$). 77 lines (2 lines with

Table 3. Identified blast resistance genes in IR cultivars

Cultivar	Blast resistance gene						
IR24		Pi 20			Pi b	Pi k-s	Pi a
IR34				Pi z-t	Pi b	Pi k-s	Pi a
IR36	Pi ta			$(Pi$ z-$t)$	Pi b	Pi k-s	
IR60	Pi ta			Pi z-t	Pi b	Pi k-s	
IR56	Pi ta		Pi $k*$	Pi z-t	Pi b		
IR64	Pi ta	Pi 20 $(Pi$ ta-$2)$		Pi z-t	Pi b	Pi k-s	
IR46	Pi ta	Pi 20 $(Pi$ ta-$2)$			Pi b	Pi k-s	
IR74		Pi 20	Pi $k*$		Pi b		Pi a

only a few plants tested and one contaminated line were excluded) segregated in the ratio of 1 segregating line: 1 homozygous susceptible line (including lines with a few resistant plants) for each isolate. These results indicate that one major dominant gene is conferring resistance to each isolate. Reactions of the lines to V850196, BN209, and two other isolates showed independent segregation and reactions to BN111 and JMB8401102 showed cosegregation. Most of the homozygous susceptible lines to V850196 and BN111 had no plants with resistant lesions or no lesions, but around half of the lines susceptible to BN209 and JMB8401102 had more than 10% resistant plants. These results suggest the existence of another recessive gene for BN209 or JMB8401102 besides Pi b or the unknown gene, respectively.

Recombinant inbred lines (RILs) of Asominori/IR24 (F_8, 71 lines, Tsunematsu et al., 1996) were inoculated with V850196, BN111, JMB8401102, and B90190. The RILs showed a segregation ratio of one gene for resistance to V850196, BN111, and B90190, and two genes for JMB8401102. A gene for V850196 was independent of the others and mapped at the end of chromosome 11 where Pi k locus is located. This means that the resistance gene for V850196 in IR24 is Pi k-s. On the other hand, a gene for resistance to BN111, B90190, and one of two genes for resistance to JMB8401102, are linked to each other and mapped on chromosome 12. Another gene for resistance to JMB8401102 might be the recessive gene that was suggested in the BC_1F_2 analysis.

The gene in IR24 that conferred resistance to BN111 was mapped on chromosome 12, where the Pi ta locus is located. Therefore, F_3 lines of IR24/C101PKT

(*Pi ta*) were inoculated with isolates IK81-25 (*Av-ta*), IK81-3 (*Av-ta*), BN111 (*Av-IR24*) and M39-1-2-21-2 (*Av-IR24*). Reactions of the F_3 lines to IK81-25 and IK81-3 were consistent, indicating that resistance to both of them was controlled by the *Pi ta* gene. The new gene also controlled the resistance to both BN111 and M39-1-2-21-2.

All of the 13 F_3 lines resistant to BN111 were susceptible to IK81-25. On the other hand, all of the 22 F_3 lines susceptible to BN111 were resistant to IK81-25. The 47 F_3 lines were segregating for reactions to both of the isolates. Only two lines were resistant to IK81-25, but segregating for resistance to BN111. These results showed that the gene for resistance to BN111 and M39-1-2-21-2 was closely linked to the *Pi ta* locus on chromosome 12. This gene was registered as *Pi 20* (Imbe *et al.*, 1997).

The two lines segregating for resistance to BN111 and homozygous resistant to IK81-25 were inoculated again, and the reactions were confirmed. Resistant individuals are being planted to select lines with both *Pi ta* and *Pi 20*. We presumed that the resistance gene(s) in IR64 was *Pi ta-2* or a combination of *Pi ta* and *Pi 20*. By comparing the reactions of varieties in the IR64 group and those of the lines with both *Pi ta* and *Pi 20,* the presumption will be proved. *Pi 19(t)* is also allelic or closely linked to *Pi ta* in Japanese varieties. We need to conduct an allelism test for *Pi 19(t)* and *Pi 20*.

Allelism tests for *Pi ta, Pi k*, Pi z-t,* and other genes were also conducted. The results were generally consistent with those in Table 2 and 3. Inconsistent results were limited and they were assumed to be mainly due to spore contamination of other blast isolates.

5. Developments of isogenic lines and monogenic lines and their usage for resistant breeding

A simple but efficient differential system to analyze blast resistance genes is necessary for rice breeding programs where many varieties with different resistance genes are used for crossing and many breeding lines are maintained. Genetic analysis for detecting resistance genes usually takes time. The genotypes of major genes for blast resistance can be presumed through their reaction patterns of blast isolates to known resistance genes. By developing these materials, we can select suitable isolates for the differential system, classify rice germplasm, and find out new donors of resistance by using this system.

We are breeding near isogenic lines (NILs) in order to develop international blast differentials for genetic and pathological studies and to breed multiline cultivars for practical use. We have started to incorporate all the known blast resistance genes into the same genetic background by using backcross breeding method. CO 39 (*indica*, broadly susceptible), Lijianxintuanheigu (LTH) (*japonica*, resistance genes are not known), IR24 (a cultivar for irrigated ecosystem), and IR49830-7-1-2-2 (an elite line for rain-fed lowland ecosystem) were used as recurrent parents (RPs).

More than 30 genes for resistance of donor cultivars have been incorporated in the former two RPs and 10 important resistant genes have also been incorporated in the latter two RPs. Resistant BC_4F_1, BC_5F_1, and BC_6F_3 plants have been selected for most of the combinations by inoculating the isolate with avirulence to a target resistance gene and then backcrossing them with RPs. The inoculation methods for the selection of resistant plants were the same as those used for the identification of genotypes.

The BC_6F_2 lines were selected based on their similarity to the respective RPs' in agronomic traits. The homogenity of the blast resistance genes were confirmed at the BC_6F_3 generation. Those NILs can be used for various purposes, such as genetic analysis, gene deployment studies, etc.

We have also developed monogenic lines that have a single resistance gene. Since 1996 second crop season, we have selected lines with a single resistance gene from crosses with LTH and resistant gene donor parents. LTH is expected to have no major genes for blast resistance. So far, all IRRI isolates were compatible to the variety.

Most of the lines probably possess a gene for photoperiod sensitivity from LTH, and therefore, they showed poor growth under natural conditions in the tropics. So, we introduced a lighting system to extend the growing period. Some of the lines can be planted in a lighted screenhouse for seed multiplication. Harvested seeds were distributed as a set of standard differentials to scientists working with blast and collaborating with the International Network for Genetic Evaluation of Rice (INGER).

Pyramiding of major genes, accumulation of minor genes, a combination of major and minor genes, and multilines of major genes are considered to be the methods for developing durable resistance against blast. Knowledge of the genetic constitution of major genes in rice cultivars and the pathogenicity of blast isolates are prerequisite for all of these strategies. We are establishing simple differential systems for the identification of pathogenicity of blast isolates of each country. By using those identified isolates, the resistance genes of different varieties in the world will be identified.

Newly introduced major genes for blast resistances were easily broken down within several years of intensive cultivation in blast prone areas like in Japan (Yamada, 1965) and Korea (Chang, 1994). We consider that this phenomenon is inevitable in the case using the accumulations or pyramidings of major genes. To avoid this risk, it is important to introduce field resistance. In a strict sense, it is measured only under a condition when the effects of the major gene resistances are excluded for the evaluation of the field resistance of the varieties. We can evaluate the field resistance of the varieties by spraying them with a single virulent blast isolate, so that we can compare the degree of development of blast lesions on each variety. For this evaluation of field resistance, knowledge of the presence of the major resistance genes and the pathogenicity of blasts isolates are prerequisite.

6. References

Chang, K.K. (1994) Blast management in high input, high yield potential, temperature rice ecosystem in R.S. Zeigler, S.A. Leong and P.S. Teng (eds.), *"Rice Blast Disease"*, International Rice Research Institute (IRRI), P.O. Box 933, 1099 Manila, Philippines p. 451-463.

Hayashi, N., Ando, I. and Imbe, T. (1998) Identification of a new resistance gene to a Chinese blast fungus isolate in the Japanese rice cultivar Aichi Asahi. *Genetics and Resistance* **88**, 822-827.

Imbe, T., Oba, S., Yanoria, M.J.T. and Tsunematsu, H. (1997) A new gene for blast resistance in rice cultivar, IR24. *Rice Genet. Newsl.* **14**, 60-62.

Inukai, T., Nelson, R.J., Zeigler, R.S., Sarkarung, S., Mackill, D.J., Bonman, J.M., Takamure, I. and Kinoshita, T. (1994) Allelism of blast resistance genes in near-isogenic lines of rice. *Phytopathology* **84**, 1278-1283.

Inukai, T., Zeigler, R.S., Sarkarung, S., Bronson, M., Dung, L.V., Kinoshita, T. and Nelson, R.J. (1996) Development of pre-isogenic lines for rice blast-resistance by marker-aided selection from a recombinant inbred population. *Theor. Appl. Genet.* **93**, 560-567.

Toriyama, K., Ezuka, A., Asaga K. and Yokoo, M. (1983) A method of estimating true resistance genes to blast in rice varieties by testing their backcrossed progenies for rice-specific reactions. *Japan J. Breed.* **33**, 448-456.

Tsunematsu, H., Yoshimura, A., Harushima, Y., Nagamura, Y., Kurata, N., Yano, M., Sasaki, T. and Iwata, N. (1996) RFLP framework map using recombinant inbred lines in rice. *Breeding Science* **46**, 279-284.

Wang, G.L., Mackill, D.J., Bonman, J.M., McCouch, S.R., Champoux, M.C. and Nelson, R.J. (1994) RFLP mapping of genes conferring complete and partial resistance to blast in a durably resistant rice cultivar. *Genetics* **136**, 1421-1434.

Yamada, M. (1965) Breakdown of highly blast resistant foreign varieties. *Plant Protection* **19**, 231-234 [in Japanese].

CLONING OF THE RICE BLAST RESISTANCE GENE *PI-B*

Y. TSUNODA[1,2], N-S. JWA[1], K.AKIYAMA[1], S. NAKAMURA[1], T. MOTOMURA[1],
K. KAMIHARA[1], O. KODAMA[2], S. KAWASAKI[1]*

[1]*National Institute of Agrobiological Resources, Kannondai, Tsukuba, Ibaraki 305*
[2]*School of Agriculture, Ibaraki Univ., Ami, Ibaraki 300-03,*
*corresponding author: kawasa@abr.affrc.go.jp

1. Introduction

As rice is the most important crop in Japan, rice blast has been the most threatening plant disease. Since the first identification of rice blast resistance gene by Sasaki (1922), Japanese breeders have interested in utilization of the resistance genes for plant protection. As the *indica*-derived resistance genes showed wide and strong resistance against Japanese races, they have introgressed some genes from *indica* to Japanese elite cultivars since 1930s, and several NILs have been developed (reviewed by Kiyosawa 1980). Although the practical application of these NILs have experienced severe cases of breakdown in a few years, these provided an excellent environment for tagging the genes with DNA markers, and we have successfully tagged *Pi-b* (Miyamoto et al., 1996) and *Pi-ta²* (Rybka et al., 1997) genes with the cosegregating and flanking markers in rather short time.

Recently, positional cloning has become the most efficient method to clone the genes which have no biochemical clues for conventional cloning, owing to the progress in the methods of genome library construction and providing various markers such as RAPD and AFLP. Although several resistance genes have been cloned by tagging and positional cloning (Baker et al. 1997), no rice blast resistance gene have been cloned yet. Cloning of the true resistance genes is indispensable to understand the mechanism of plant recognition of the infecting pathogens, and induction of various defense genes. The cloned genes will also contribute to construct "multi-lines" of NILs with different resistance genes in short time.

We have established a very versatile framework for positional cloning in plants including rice by constructing;
1) a high quality BAC (Bacterial Artificial Chromosome) library of rice of average

D. Tharreau et al. (eds.), Advances in Rice Blast Research, 9–16.
© 2000 *Kluwer Academic Publishers. Printed in the Netherlands.*

insert size of about 160 kB, and covering 7 genome equivalent (Nakamura et al. 1997),
2) a high capacity binary vector pBIGRZ which can maintain and transfer efficiently
more than 40 kB of insert in T-DNA to rice without rearrangement and with good
fertility (Akiyama et al. in press).

As a model case of application of this system we selected cloning of *Pi-b*.
Utilizing these tools, we could span the flanking markers of *Pi-b*, distant 2.4 cM, with 4
steps of clones, and then *Pi-b* region was narrowed down into a single BAC clone of
180kB. We could also cover the clone with several pBIGRZ clones derived from the
resistant genome library of the cultivar with *Pi-b*, BL-1. Some prospective analogues of
the resistance gene were also found by characterizing the cDNAs expressed in this
region. The complementation test for all the clones is now being done.

2. Construction of high quality BAC library of rice

As almost all of the Japanese cultivars are *japonica*, for the positional cloning
of genes from them the presence of high quality BAC library of *japonica* cv. was
indispensable. Therefore we have constructed a BAC library from the protoplasts of
green leaves of cv. Shimokita with *Pi-ta* gene. Protoplasts were much superior than
nuclei as a material to prepare DNAs of more than megabases. Because there was no
chance of attack by nucleases to the intact nuclear DNAs, and no afraid of
contamination of polysaccharides that may inhibit quantitative digestion by restriction
enzymes. As the BAC vector we used the pBACLAC, having the lacZ of M13mp18
(Asakawa et al. 1997) which is much stronger in enzyme activity than that of Bluescript.

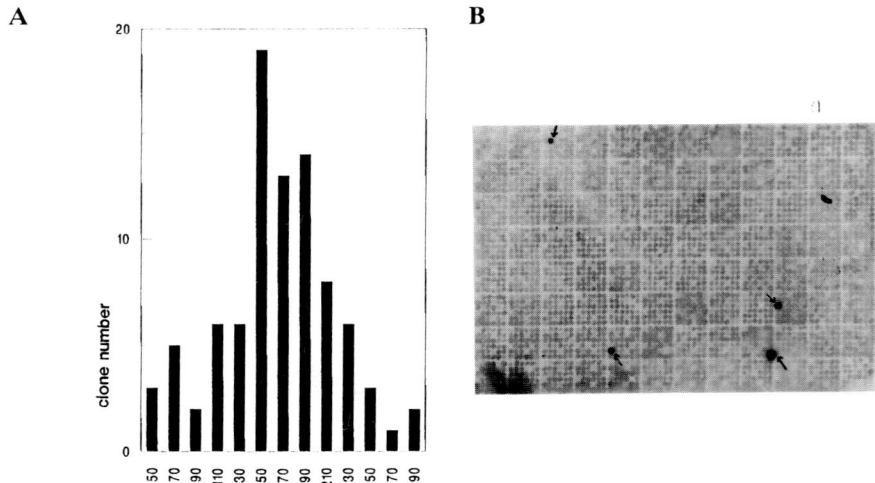

Fig. 1. High quality BAC library of rice. A: Distribution of insert size (kB) of the BAC library of *japonica*
cv. Shimokita. The average insert size is 155 kB, covering 7 genome equivalent, the largest of all the rice
BAC libraries reported. B: High density membrane of the BAC library with 3072 clones, equivalent to 1 rice
genome. The library is open to use; contact to us with an E-mail.

This can differentiate blue/white colonies in one day at conventional concentration of X-Gal even with the single copy of BAC plasmid.

Using these vector and megabase DNAs we could construct a genome library of *japonica* cv. Shimokita with *Pi-ta*. The average insert size was 155 kB, and 21,504 clones, 7 genome equivalent, were arranged on 7 high density membranes of microplate size each with 1 genome equivalent of 3072 clones (Nakamura et al. 1997; Fig. 1A). They were arranged in the 8x12 configuration of 96 holes of microplates (Fig. 1B). One hole area accommodates a 6x6 minimatrix with 4 clones of pFos 1 as position markers no cross-reacting with the BAC vector. ECL detection was convenient because of no need of signal stripping, and repeatability of hybridization more than a few dozens of times.

3. Contig construction spanning the flanking markers of *Pi-b*

The flanking markers of *Pi-b* separate 2.4 cM and the cosegregating markers were used to pick up the corresponding BAC clones, and 5-10 clones were selected for each clones. The arrangement of each clone in a group was estimated from the CHEF (Contour-clumped hexagonal electric field) electrophoresis pattern of *Not* I digest, and confirmed by hybridization of the both end probes of the outermost clone of a single group. The end probes were amplified by vectorrette PCR (Riley et al. 1990).

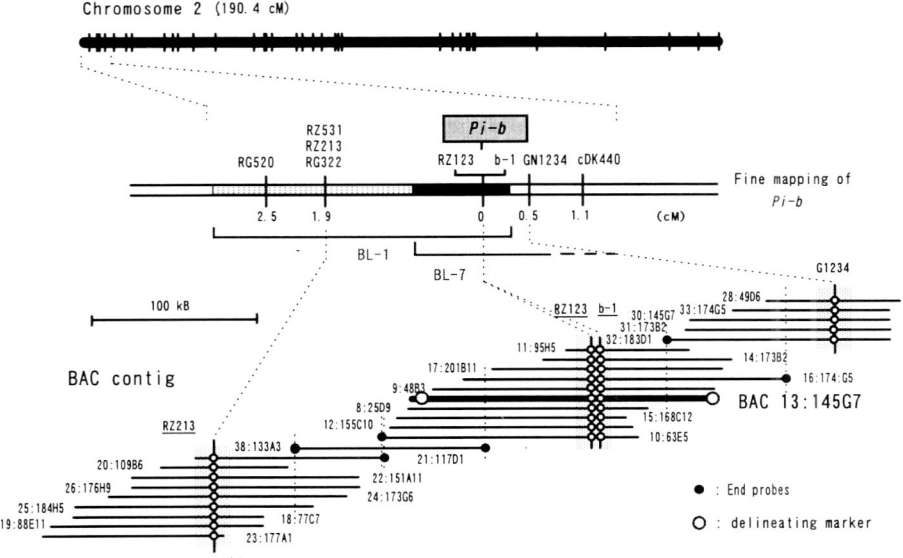

Fig. 2. A contig of BAC clones covering the *Pi-b* region flanked by the both sides markers. The 2.4 cM distant flanking markers were spanned with of 4 steps of contig, and distant 400 kB. The end probes of 145G7, amplified by vectorrette PCR, and some inner fragments of the inserts, were valid as the RFLP markers to narrow down the *Pi-b* region to the single clone.

Only four steps were sufficient to span the flanking markers RZ213 and G1234 (Fig. 2), and their physical distance was 400 kB. The two cosegregating markers b-1, derived from RAPD, and RZ123, a Cornell marker, were both on the same fragment of *Not* I.

The ratio of physical to genetic distances in this region is about 160 kB/cM, about half of the total rice genome, 300 kB/cM. This is in great contrast compared to the case of *Pi-ta^2* located near the centromere of chromosome 12, in which region the ratio of distances was 1000 kB/cM or more than that (Nakamura et al. 1997). Probably the location of *Pi-b* in the near telomeric region of the chromosome may be related to the high recombination rate of the region. The localization of *Pi-b* to the near telomeric end of the long arm of chromosome 2 was also visualized by FISH (Nakamura et al. 1997).

The end probes of the BAC clone 145G7 hit by the cosegregating markers were valid as the RFLP markers, and applied for the analysis of the recombinant F_2s of the flanking markers. As the result, the region of *Pi-b* was narrowed down to this single clone 145G7 of about 180 kB (Fig. 2).

4. Selection of cDNAs expressed in the *Pi-b* region

In most cases of cloned resistance genes, the susceptible cv. retains the homologous counterparts. Therefore, screening of the resistance gene candidate using the BAC clone of the susceptible cv. can be a prospective approach. As the region of *Pi-b* was restricted to a single clone 145G7, selection of the cDNAs expressed in this clone was tried with clone hybridization method developed by Hayashida et al. (1995). The BAC clone was sheared down to about 1 kB by sonication and covalently bound to latex beads through glycidyl methacrylate. On the other hand, the cDNA library was constructed in about 1 kB average size, and inserted in pBR322, which is non homologous to the BACvector. The inserts were PCR-amplified, hybridized to the BAC latex and then reluted after stringent washing. The released inserts were reinserted to pBR322 after reamplification with PCR.

We screened at first 960 clones and then 11,000 clones by subtraction analysis, and 17 species of cDNA were identified by sequencing and mutual hybridization. As the concentration factor was estimated to be about 160 times, from the comparison of the positive clones ratio between the original- and sub-libraries, this screening may be equivalent to that of 2 million clones of the normal library. The 17 clones were all characterized by sequencing and a Ser/Thr-protein kinase homologous to the tomato Pto gene (Martin et al. 1993) was found.

5. High capacity binary vector pBIGRZ and functional genome library

The cDNA approach can not be completely reliable because of the possibility that there is no homologous counterpart in the susceptible clone and its dependence to PCR reactions. Therefore, we have also taken another approach of complementation analysis using a high capacity binary vector to narrow down further the target region. As there was no appropriate high capacity binary vector at that time, we have constructed a binary vector on the basis of pBI series which have *RK2 ori* with low copy number (2-7) and having the plasmid partition apparatus in *E. coli* (Thomas and Smith 1986). As the maintenance of the large insert in *Agrobacterium* was not sufficient with that *Ri ori* was added to supplement the maintenance of it. Hygromycin B resistance (HPT) and intron-GUS (Ohta et al. 1990) were used to select transformants and check the transformation on the course of selection (Fig. 3A). The vector was named as pBIGRZ.

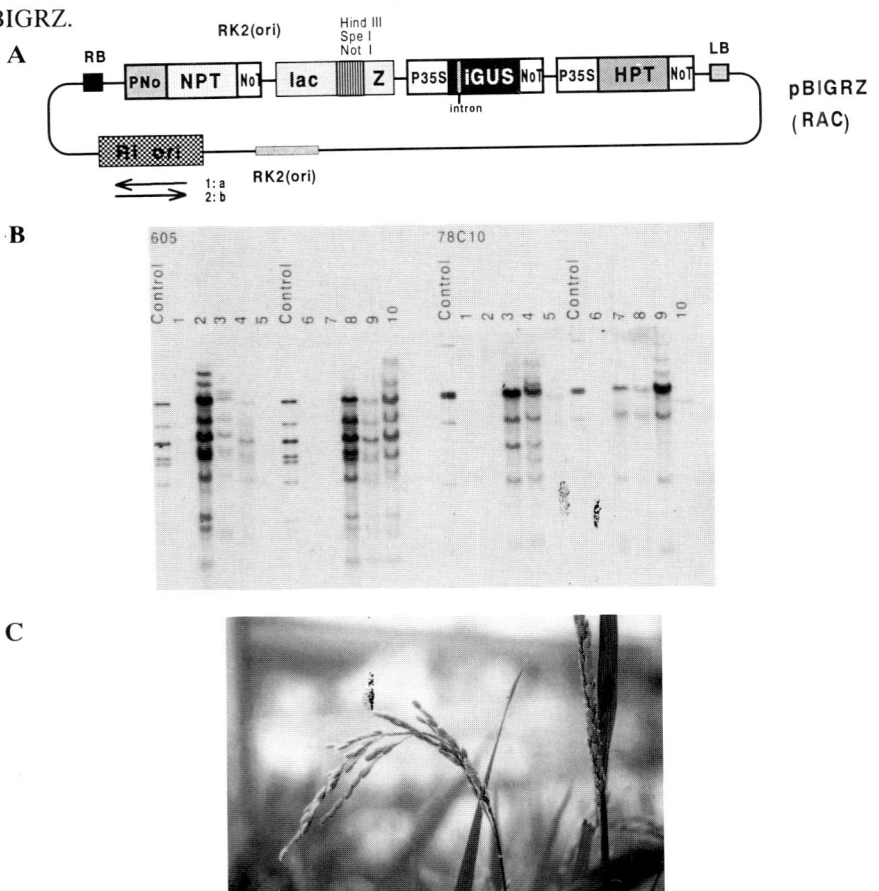

Fig. 3. High capacity binary vector pBIGRZ. A: Construct of the vector with *lacZ* as the insert selection marker, *Hind* III, *Spe* I, and *Not* I as the unique cloning site. Hygromycin resistance (HPT) is for selection of rice transformants, intron GUS is for check of transformation process. B: Southern blotting of the human 40 kB genome fragments transferred to rice with pBIGRZ by *Agrobacterium* coculture. It is noticeable that there are almost no DNA rearrangement during transfer. The control is the single copy equivalent of the original plasmid digestion. C: The R1 siblings retained the original pattern of the transferred fragment. D: The transformants of the human 40 kB genome fragment. The fertility was more than 80% and there was no anomality in the appearance.

The transformation of rice by pBIGRZ with human genome fragments of more than 40 kB was successfully done in high efficiency, 4-5 plants regenerated per 9 cm plate, and the transformed insert showed almost no rearrangements (Fig. 3B). The transformed inserts were stably maintained to the R1 siblings and the fertility of the transformant was more than 80%. These results are far much better than the results of protoplast transformation.

Using this vector, single stroke of electroporation could produce a complementation-ready genome library which covers 1 genome of rice at the average insert size of 50 kB, with the efficiency of 5×10^6 clones/μg insert DNA. This insert size is best suited for the restriction of the region by complementation of the target gene. Although the BiBAC was reported about the same time of our construction (Hamilton et al. 1996), our experience confirmed that *lacZ* is more user friendly as the selection marker than *sucB* and *Hind* III site is more efficient than *Bam*HI.

6. Contig of BL-1 derived clones in pBIGRZ covering the *Pi-b* region

Although the region specific cDNA sublibrary was helpful to pick up the clones from the Pi-b harboring genome library of BL-1, contig formation was not so easy over the *Pi-b* region due to the presence of homologous noises in the genome. Therefore, we first constructed the contig and fine map of the BAC clone 145G7 from its partial digest inserted into the pBIGRZ. A library of the average insert size of about 50 kB was easily constructed from the purified clone, and their *Hind* III map was much easier to construct than the BL-1 library, because of band pattern comparison was

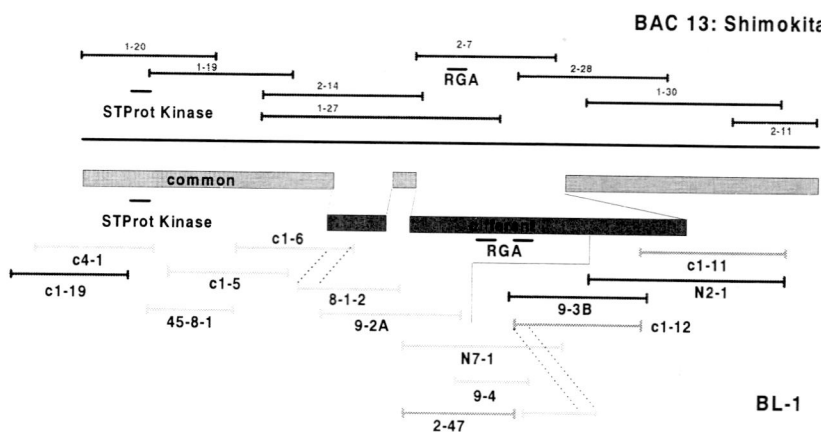

Fig. 4. The contig maps of pBIGRZ sublibraries of the BAC clone 145G7 and BL-1 with *Pi-b*. The clones on the border sides are very common while those of the central regions are very different in their patterns of *Hind* III digestion. . A Ser/Thr-protein kinase was on the border region while some resistance gene analogues (RGAs) were in the central region. All the BL-1 derived clones are being transformed to rice and the complementation test is done by spray inoculating the blast races.

sufficient to verify the overlaps of the clones. After completion of the contig map of 145G7 clone, we could arrange the clones picked up from the BL-1 library more easily by comparison with the guide map. The contiguity between the clones were confirmed by Southern blotting. At the flanking regions, the two maps coincided well, but in the central region, there were great differences between the two maps (Fig. 4). These differences may reflect the genome mobilization leading to the generation of a resistance gene. Analysis of the genome mobilization like this is indispensable to elucidate the evolution of the resistance genes.

There were resistance gene analogues (RGA) with nucleotide binding site at the center of the map, and the Ser-Thr protein-kinase at the rather marginal location. All the clones have been transformed to Nipponbare with a good efficiency, and their harboring the selection genes (HPT) were checked by PCR. The result of the complementation test by infecting some races of rice blasts with and without *avrPi-b* gene and verification of the gene by segregation analysis by R1 generation are now in progress.

7. Conclusion

This system for positional cloning in plants, consisting of 1) BAC library contig formation and 2) complementation with a high capacity binary vector is very versatile for any kind of genes and plants. This will be the easiest method to use for complementation of recessive mutants, because the BAC clone can directly serve as the functional gene source. For expediting the process, we are now constructing a rice genome library of pBIGRZ of medium insert size (about 50 kB) to easily narrow down the target region for open use.

Even for the dominant gene, using BAC clones as references will facilitate the construction of the complementation-ready subcontig greatly. The pBIGRZ plasmid is also appropriate for direct sequencing of the insert.

Acknowledgments

We are grateful to Dr. M. Hara, N. Hayashida, K. Shinozaki of RIKEN, Tsukuba for providing valuable information on the construction of the BAC region specific library. We also appreciate to Dr. S. Asakawa and Prof. N. Shimizu of Keio Univ. for providing human genome fragment clones for transformation, and Professor A. Oka of Kyoto Univ., Prof.. K. Nakamura of Nagoya Univ., and Dr. Hatsuyama of Aomori GreenBio for kind gift of Ri-ori, intron GUS, HPT genes, respectively.

References

Akiyama, K., Tsunoda Y., Kawasaki S. High capacity binary vector pBIGRZ and construction of Functional genome library for complementation analysis. (in press)

Asakawa S., Abe, I., Kudoh, Y., Kishi N., Wang, Y., Kubota, R., Kudoh J., Kawasaki K., Minoshima S., Shimizu N. (1997) Human BAC library: construction and rapid screening. *Gene* **191**, 69-79.

Baker B., Zambrysky P., Staskawicz, B., Dinesh-Kumar S.P. (1997) Signaling in Plant-Microbe interactions. Science **276**, 726-733.

Hamilton CM, Frary A, Lewis C, Tanksley SD (1996) Stable transfer of intact high molecular weight DNA into plant chromosomes. Proc. Natl. Acad. Sci. **93**, 9975-9979

Hayashida N., Sumi Y., Wada T., Handa, H., Shinozaki K. (1995) Construction of a cDNA library for a specific region of a chromosome using a novel cDNA selection method utilizing latex particles. *Gene* **6165**, 155-161.

Kiyosawa, S. 1980 Genetics of blast resistance. In Y. Yamazaki, and T. Kozaka, ed. 1980. Rice blast disease and breeding of resistance. p186-229 Hakuyusha, Tokyo.

Martin GB, Brommonschenkel SH, Chunwongse J, Frary A, Martin W, Ganal RS, Wu T, Earle ED, Tanksley SD (1993) Map-based cloning of a protein kinase gene conferring disease resistance in tomato. Science **262**, 1432-1436

Miyamoto M, Ando I, Rybka K, Kodama O, and Kawasaki S (1996) High resolution mapping of the indica-derived rice blast resistance genes. I. *Pi-b*. *Mol Plant-Microbe Interac*t **9**, 6-13.

Nakamura, S., Asakawa S., Ohmido, N., Fukui,K., Shimizu N., and Kawasaki S. (1997) Construction of an 800-kb contig in the near-centromeric region of the rice blast resistance gene Pi-ta2 using a highly representative rice BAC library, *Mol. Gen. Genet.* **254**, 611-620.

Ohta S, Mita S, Hattori T, Nakamura K (1990) Construction and expression in tobacco of a β-glucuronidase (GUS) reporter gene containing an intron within the coding sequence. Plant Cell physiol. **31**, 805-813

Thomas C.M., Smith C.A. (1986) The *trfB* region of broad host range plasmid RK2: the nucleotide sequence reveals *incC* and key regulatory gene *trfB/korA/korD* as overlapping genes. *Nucleic Acids Res.* **14**, 4453-4469

Riley J, Butler R, Ogilvie D, Finniear R, Jenner D, Powell S, Anand R, Smith JC, Markham AF (1990) A novel, rapid method for the isolation of terminal sequences from yeast artificial chromosome (YAC) clones. Nucl Acids Res **18**, 2887-2890.

Rybka K, Miyamoto M, Ando I, Saito A, Kawasaki S (1997) High resolution mapping of the indica derived rice blast resistance genes. II. *Pi-ta²* and *Pi-ta* and a consideration of their origin. *Mol Plant-Microbe Interact.* **10**, 517-524.

Sasaki, R. 1922. On inheritance of rice blast resistance. *Jpn. J. Genetics* **1**, 81-85 (Jpn)

CHARACTERIZATION OF BLAST RESISTANCE IN THE DURABLY RESISTANT RICE CULTIVAR MOROBEREKAN

CHEN, D. -H[1]., NELSON, R.J.[2], WANG, G. -L[3] , INUKAI, T.[4], MACKILL, D.J.[5], AND RONALD, P.C.[1]

[1] *Department of Plant Pathology, University of California, Davis, 1 Shield Ave, Davis; CA 95616;* [2] *Centro Internacional de la Papa, Lima 12, Peru;* [3] *The Institute of Molecular Agrobiology, The National University of Singapore, 1 Research Link, NUS, Singapore, 117604;* [4] *Faculty of Agriculture, Hokkaido University, Sapporo 060, Japan;* [5] *USDA-ARS, Department of Agronomy and Range Science, University of California, Davis, 1 Shield Ave, CA 95616.*

1. Introduction

Despite tremendous investment on blast research and control, blast remains one of the most widespread and destructive diseases of rice. Growing blast-resistant varieties is the preferred method for protecting rice from blast and for reducing the use of fungicides. Although resistance to blast is often short-lived, especially when resistance is conferred by single major genes, some cultivars possess durable resistance, as defined by Johnson (1981; Parlevliet, 1988). For instance, rice cultivars IR36, IR42, IR64, Milyangv30 and Milyang42 show durable resistance to blast under irrigated or rainfed lowland conditions. Moroberekan, ROK16, LAC23, IRAT13, OS6 and some Brazilian upland rice cultivars show durable resistance to blast in upland conditions (Bonman and Mackill, 1988; Lee *et al*., 1989; Bidaux, 1978; Ahn, 1994; Fomba and Taylor, 1994). Characterization of the genetic basis of the resistance in these cultivars would provide insight into the durable resistance, and would be useful in designing genotypes with novel resistance.

Moroberekan has shown durable blast resistance in West Africa for more than 20 years (Bidaux, 1978; Fomba and Taylor, 1994). Under the auspices of the International Rice Research Institute (Ahn, 1994), sets of rice materials were assayed for both qualitative and quantitative resistance to blast using a 0-9 scoring system in several different countries under high natural blast pressure. Among the 108 trials conducted in the International Rice Blast Nursery, Moroberekan showed qualitative resistance (0-3 scores) in 69% of the trials, which was the highest percentage among the six widely cultivated durably resistant cultivars tested. It further showed the highest level of partial resistance among the six durably resistant cultivars tested (Ahn, 1994). To date, no compatible isolate has been found for Moroberekan, in spite of extensive inoculation analysis conducted in the Philippines and in California, USA (Calvero and Bonman, unpublished; Chen and Ronald, unpublished).

D. Tharreau et al. (eds.), Advances in Rice Blast Research, 17–27.
© 2000 *Kluwer Academic Publishers. Printed in the Netherlands.*

For several years, we have been characterizing the blast resistance in Moroberekan, a *javanica* (upland *japonica*) landrace from Africa. In this paper, we will summarize the results of the following analysis: 1) mapping genes for blast resistance using a recombinant inbred population; 2) analyzing allelism relationships between resistance genes in Moroberekan and known resistance genes, and 3) phenotypic characterization of resistance in Moroberekan.

2. Mapping Genes For Blast Resistance Using A Moroberekan Recombinant Inbred (RI) Population

2.1. IDENTIFYING AND MAPPING RESISTANCE GENES

As part of an effort to improve the understanding of the genetic basis of resistance in tropical rice germplasm, D.J. Mackill and J.M. Bonman initiated a series of crosses aimed at developing mapping populations for quantitative resistance, and for developing near-isogenic lines carrying individual blast-resistance genes. One of the mapping populations was derived from a cross between Moroberekan (of *japonica* subspecies) and CO39 (of *indica* subspecies). By single seed descent, a RI population consisting of 281 F_7 stable lines was developed from this cross, which was used for RFLP mapping of genes resistant to rice blast (Wang *et al.,* 1994).

Each of the RI lines was subjected to RFLP analysis at 127 loci, and was also subjected to phenotypic analysis in greenhouse and in field tests. Five isolates, collected from different regions of the Philippines, were used in greenhouse monocyclic tests for qualitative resistance analysis. Isolate PO6-6 was used to assess the partial resistance (lesion number, lesion size and diseased leaf area) in a polycyclic test of 131 lines that were susceptible to the tested five isolates. To determine the associations between DNA markers and phenotypic effects, regression analysis and MapMaker QTL were used to identify the DNA markers associated with the complete resistance to the five tested isolates, and with partial resistance to the isolate PO6-6.

Two dominant loci conferring qualitative resistance to all five isolates were identified from Moroberekan. One locus, tentatively designated *Pi5(t),* was mapped to chromosome 4, near DNA marker *RG788*. Another locus, tentatively designated *Pi7(t)*, was bracketed by *RG103A* and *RG16* in chromosome 11. Ten putative QTLs from Moroberekan were mapped in eight chromosomes to be associated with the parameters (lesion number, lesion size and diseased leaf area) of partial resistance (Fig. 1).

The five isolates used for identifying *Pi5(t)* and *Pi7(t)* were thought to be genetically diverse because of their different origins and different virulent spectrum. DNA analysis later showed that they represented only three of more than ten genetic lineages identified in the Philippines (Chen, 1993; Chen *et al.*, 1995). Because of the few isolates used for the study, it was expected that more resistance genes could be identified from Moroberekan if diverse isolates are used for phenotypic analysis.

Indeed, three genes conferring complete resistance were identifed when the Moroberekan/CO39 population was inoculated with additional Philippine and Indian isolates (Inukai *et al.*, 1996; Naqvi and Chattoo, 1996). *Pi12(t)* confers complete

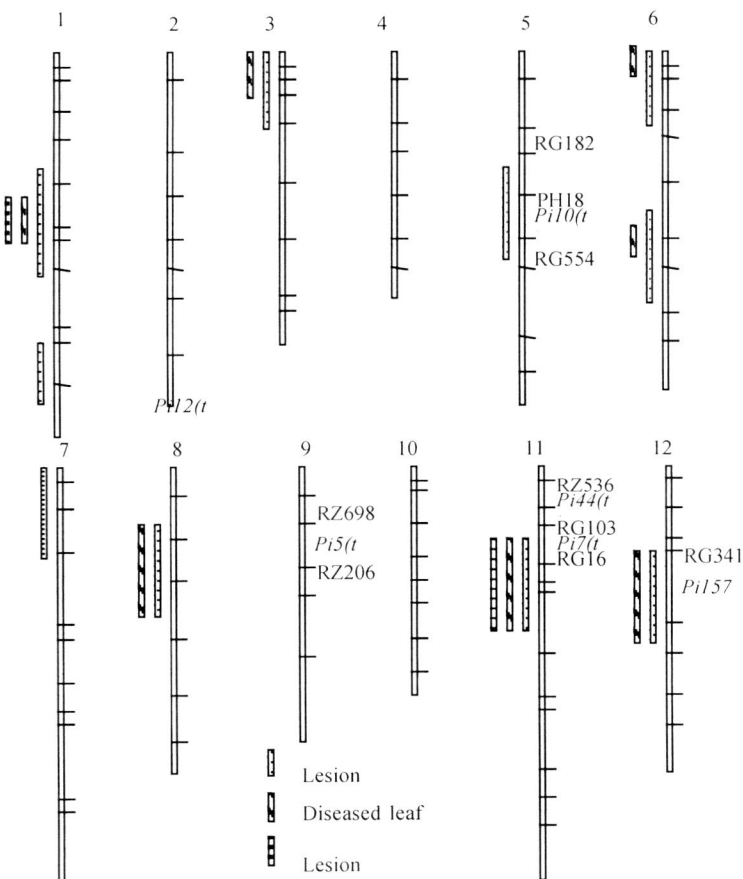

Figure 1. Chromosomal locations of blast resistance genes identified in Moroberekan. The linkage information of *Pi12(t)* on chromosome 2 is not available and listed under the chromosome 2. Bars at the left side of the chromosomes are the region where QTLs conferring partial resistance are located. The map is adapted from Wang *et al.*, 1994; Naqvi and Chattoo, 1996 and Chen *et al.*, 1997.

resistance to isolate Ca65, and was localized on chromosome 2 (Inukai et al., unpublished). *Pi10(t)* confers complete resistance to the Philippine isolates JMB840610. RAPD markers *RRF6* and *RRH18* were found to be tightly linked with the locus *Pi10(t)* (*RRF6*, 2.8 ± 0.9 cM; *RRH18*, 3.9 ± 1.2 cM), and mapped to chromosome 5 (Fig. 1) (Naqvi *et al.*, 1995; Naqvi and Chattoo, 1996). A third gene was detected using the Indian isolate B157, and was found to be linked to RFLP marker *RG341* at 3.0 ± 0.6 cM on chromosome 12 (Fig. 1; Naqvi and Chattoo, 1996).

To allow characterization of individual QTLs, lines carrying QTL(s), but few other alleles from Moroberekan, were selected based on the marker and phenotypic data that were gathered as part of the mapping study of Wang *et al.* (1994). RIL23 and RIL276 were selected as "pre-isogenic lines", each carrying two QTLs for partial blast resistance. One putative QTL was located on chromosome 6, near *RG64*, and another was located on chromosome 1, near CDO920. Because of the segregation distortion observed in the cross, lines could be selected with few introgressions from Moroberekan; RIL23 had 13.9% Moroberekan alleles at the RFLP markers tested, while RIL276 had 9.5% (Wang *et al.*, 1994; Inukai *et al.*, 1996).

While characterizing the effect of QTLs in RIL23 and RIL276 on diverse isolates of *P. grisea*, we found that both lines were qualitatively resistant to a set of isolates of lineage 44 (Chen *et al.*, 1996). This raised the question of whether resistance to lineage 44 in the Moroberekan derivatives was conferred by an additional major gene or by a locus at or near one of the putative QTLs.

To identify and map the gene conditioning resistance to isolates of lineage 44, Chen and Nelson developed an F_2 population consisting of 100 individuals from a cross between RIL276 and CO39. The F_2 plants were then advanced to F_3 families for phenotypic analysis. Genetic analysis of the F_2 and F_3 populations using isolate C9240-1 of lineage 44 showed that resistance in RIL276 to lineage 44 isolates was governed by a single major locus, derived from Moroberekan. This locus was tentatively designated *Pi44(t)* (Chen *et al.*, submitted).

Five RFLP loci (*RG241, RZ192, RG331, RG864* and *RG64*) introgressed from Moroberekan in RIL 276 were used to test whether *Pi44(t)* was carried in one of the known introgressed segments. The RG64 locus, which is linked to a QTL in RIL276, was assayed using two pairs of primers designed based on the sequence of *RG64* (Hittalmani *et al.*, 1995). None of these markers was found to be linked to *Pi44(t)*, so another 63 STS primers (Inoue *et al.*, 1994) were applied, with nine restriction enzymes (*Eco*RV, *Alu* I, *Msp* I, *Hinf* I, *Mva* I, *Hha* I, *Rsa* I, *Nde* II and *Taq* I) used to digest the PCR products to increase the marker polymorphism. Again, no association with *Pi44(t)* was detected.

Continuing the effort to identify DNA markers linked to *Pi44(t)*, 314 AFLP primer combinations (Zabeau and Vos, 1993) were applied to the segregating population, using bulk segregant analysis (Michelmore *et al.*, 1991). Using this strategy, two polymorphic AFLP markers (designated AF_{348} and AF_{349}) associated with resistance were identified, and located at 3.3 ± 1.5 cM and 11 ± 3.5 cM, respectively, from *Pi44(t)*. Two available F_2 mapping populations (Labelle / Black Gora, and IR40931-26-3-3-5 / PI543851), developed respectively by Redoña and Mackill (1996) and Xu and Mackill (1996), were used to locate the linked markers between RFLP markers CDO520 and RZ536 on rice chromosome 11. DNA products at the loci linked to *Pi44(t)* were amplified from RIL276, Labelle and PI543851 using the same primer pairs used to amplify AF_{349} and AF_{348}. Sequence analysis of these bands showed 100% identity between lines.

These results suggest that *Pi44(t)* is distinct from the two QTLs in RIL276, which were located on chromosome 6 (near *RG64*) and on chromosome 1 (near CDO920), respectively. It is also different from *Pi5(t), Pi7(t),Pi10, Pi157* and *Pi12(t),* based on their distinct chromosome locations and resistance spectra (Wang *et al.,* 1994; Naqvi and Chattoo, 1996; Inukai *et al.,* 1996; Chen *et al.,* unpublished).

2.2. DETAILED ANALYSIS OF RESISTANCE LOCI *Pi5(t)* AND *Pi7(t)*

As mentioned above, the major genes *Pi5(t)* and *Pi7(t)* were identified in the F_7 population derived from a cross between Moroberekan (of the *japonica* subspecies) and CO39 (of the *indica* subspecies; Wang *et al.,* 1994). Segregation in the population was heavily skewed, favoring CO39 alleles (overall average of 80% CO39 alleles and 20% Moroberekan alleles; Wang *et al.,* 1994). The distorted segregation population, together with possible epistatic effects of other major and minor genes present in the population, probably affected the accuracy of the chromosomal placement of the genes. Therefore, we further analyzed the locations of *Pi5(t)* and *Pi7(t)* (Chen *et al.,* 1997).

Based on available molecular and phenotypic datasets from the previous study (Wang *et al.,* 1994), three *Pi5(t)* candidate lines (RIL206, RIL249 and RIL260) and two *Pi7(t)* candidate lines (RIL125 and RIL29) were selected. Each of the candidate lines was crossed with CO39. Five F_2 populations, each consisting of about 50 progeny, were obtained from the crosses. The F_2 progeny and their F_3 families were inoculated with the isolate PO6-6 at 21 days after sowing (17 to 20 seedlings per family). Disease was evaluated seven days after inoculation. DNA from 17 to 20 seedlings of each F_3 families was extracted to represent the F_2 individuals. Bulked DNA segregant analysis using AFLP was used to identify DNA markers associated with the resistance to isolate PO6-6.

Genetic analysis showed that all five F_2 populations segregated for resistance to the isolate PO6-6 in a 3:1 ratio, indicating a single locus conferring resistance to isolate PO6-6 each of the selected lines. The F_2 inoculation results were confirmed in two replicated inoculations of F_3 families with the same isolate.

Of the 350 primer pairs surveyed for the RIL206-derived and RIL249-derived F_2 populations, 10 and 11 AFLP markers were identified to be associated with the resistance, respectively. Out of 1024 primers surveyed for the RIL260-derived F_2 population, 12 markers were identified to be associated with resistance. One linked marker showed polymorphism in both the RIL206-derived and RIL249-derived populations, and three markers were associated with resistance in both the RIL249-and RIL260-derived populations. These results suggest that the resistance locus in the three lines was likely to be the same or tightly linked, as anticipated based on the expectation that the three lines all carry *Pi5(t)*. Though six markers co-segregated with the resistance in one of the three small populations, the exact genetic distance should be determined in a larger mapping population.

For the two *Pi7(t)* candidate lines, RIL29 and RIL125, 963 and 579 AFLP primer combinations were screened for markers associated with blast resistance, respectively. Two primer pairs gave polymorphic markers linked with resistance in RIL29 to isolate PO6-6. One primer pair amplified a polymorphic marker (designated S04/G03) associated with resistance in RIL125. Unexpectedly, S04/G03 co-segregated with resistance in RIL260, expected to carry *Pi5(t)* ·(Wang *et al.,* 1994). RIL125 showed moderate resistance to the Californian blast isolate CAL-1, as did RIL260, RIL249 and RIL206. However, RIL29 was completely resistant to CAL-1. This implies that RIL125 might carry a gene that is the same as or tightly linked to the gene in RIL260 that conditions resistance to PO6-6.   That is, RIL125 may carry *Pi5(t)* rather than *Pi7(t)*.

Primer pairs amplifying the linked markers were used to analyze the parents of the mapping population Black Gora/Labell (provided by D.J. Mackill; Redoña and Mackill, 1996; Xu and Mackill, 1996).   Nine primer pairs (B14/M03, S04/G03, S10/F07; S08/G24, S10/M04A, S10/M04B, B11/G19, R01/G23A, R01/G23B) gave rise to polymorphism between Black Gora and Labell at the loci associated with the resistance to PO6-6. The segregation of the 80 F_2 progeny of Black Gora/Labell at these loci fit a 3:1 ratio. Linkage analysis with Mapmaker and Joinmap showed that B14/M03 from the RIL206-derived population mapped to chromosome 11, while other eight markers from the RIL249-derived and RIL260-derived populations were placed on chromosome 9 (Fig. 1). The map location conflict between the marker B14/M03 and other markers could be due to the marker translocation in the mapping population. Considering the similar phenotypic reaction of RIL206 with other *Pi5(t)* candidate lines, and the shared marker with RIL249, we believe that RIL206 also carries the same locus.

Since all markers associated with resistance in the putative *Pi5(t)* candidate lines mapped to chromosome 9 rather than to chromosome 4, we conclude that *Pi5(t)* is located on chromosome 9. The location of *Pi5(t)* on chromosome 9 agreed with the finding of Inukai and Aya (1997), who found that resistance to isolate PO6-6 in the F_2 population of RIL249 / CO39 did not co-segregate with RFLP marker *RG788*, which was shown to be closely linked to *Pi5(t)* on chromosome 4 in the previously study (Wang *et al.*, 1994). The discrepancy between the original report and later findings on the chromosomal location of *Pi5(t)* could be attributable to the skewed mapping population used in previous study, which could lead to the pseudo-linkage between *RG788* and the *Pi5(t)*. Alternatively, Moroberekan could carry additional gene(s) other than *Pi5(t)* conditioning resistance to PO6-6. Of the two markers linked to *Pi7(t)* in RIL29, one was monomorphic in the two mapping populations and another was mapped in chromosome 11, near *RG103A* (Chen and Ronald, unpublished), confirming the previous report of Wang *et al.* (1994).

For fine mapping of the *Pi5(t)* locus, F_4 segregating population derived from RIL260 consisting of over 2,000 individuals was developed.   AFLP markers R01/G23A and S04/G03, which bracketed the *Pi5(t)* locus in a 2.1 *cM* interval in the small F_2 population, were used to identify the rare recombinants in the F_4 population. Fine mapping of the *Pi5(t)* locus is now underway (Chen and Ronald).

3. Allelism Relationships between Resistance Genes in Moroberekan And Other Known Resistance genes

More than 20 major genes for resistance to rice blast have been identified and mapped through classical genetics and/or molecular marker technology (Kiyosawa , 1981; McCouch *et al.,* 1994). Near-isogenic lines have been developed for some of the genes (Mackill and Bonman, 1992, Inukai *et al.,* 1994; Ling *et al.,* 1995), which facilitates allelism analysis. Inukai *et al.* (1996) initiated a study on the allelism relationships among the newly identified resistance genes in Moroberekan and some known genes.

To test the allelic relationship between *Pi5(t)* and *Pi7(t)*, they made crosses between RIL29 and RIL249, RIL29 and C101LAC (carrying *Pi1* on chromosome 11), and RIL249 and C104PKT (carrying *Pi3*). The F_2 progeny were inoculated with isolate PO6-6 for RIL29/RIL249 and RIL29/C101LAC, and with isolate P03-82-51 for RIL249/C104PKT. The results showed that *Pi5(t)* in RIL249 and *Pi7(t)* in RIL29 were non-allelic. *Pi7(t)* was allelic or tightly linked to *Pi1*, and *Pi5(t)* in RIL249 was allelic or closely linked to *Pi3*. The reaction pattern of *Pi12(t)* to Japanese isolates was similar to that of *Pib*. Allelism tests showed that *Pi12(t)* appeared to be allelic or closely linked to *Pib* locus (Inukai et al., unpublished). The reaction pattern *Pi44(t)* in RIL276 was distinct from that of the existing CO39 NILs and the *Pi5(t)* and *Pi7(t)* (Chen and Nelson, unpublished). Since *Pi1*, *Pi7(t)*, *Pi44(t)*, and the multi-allele locus *Pik* were all on the distal segment of chromosome 11, their allelism relationships should be determined.

Clustering of resistance genes has been observed for rice blast [*Pib* and *Pi12(t)* on chromosome 2 and *Pi1*, *Pi7(t)*, *Pi44(t)*, and *Pik* in the same region of chromosome 11], and in other pathosystems (Maisoneuve *et al.*, 1994; Witsenboer *et al.*, 1995). Thus, classical allelism analysis may not be definitive in establishing gene identity. Molecular cloning of resistance loci will eventually elucidate the relationships among resistance loci and alleles.

4. Phenotypic Characteriaztion of Resistance in Moroberekan

Phenotypic characterization of individual major genes and QTLs in different genetic backgrounds using diverse blast isolates can provide insight into the individual contribution of the genes in the durability of resistance in Moroberekan. In the study of Wang *et al.* (1994), RI lines carried *Pi5(t)*, *Pi7(t)* and QTL(s) identified using isolate PO6-6 under greenhouse conditions were tested against field blast populations at two blast screening sites in different countries (Cavinti in the Philippines and Sitiung in the Indonesia). RI lines with *Pi5(t)* and *Pi7(t)* generally had low levels of disease in the two testing sites. Lines carrying more QTLs showed generally less disease than those carrying none or only one QTLs.

Field performance of the lines carrying *Pi5(t)* and *Pi7(t)* suggested that *Pi5(t)* and *Pi7(t)* condition broad resistance spectrum; the pathogen population at the Cavinti screening site was shown to be diverse and broadly virulent (Chen *et al.*, 1995; Zeigler *et al.*, 1995). However, greenhouse tests were necessary to confirm the resistance spectra of the individual genes. We inoculated RI lines carrying *Pi5(t)* and *Pi7(t)* with isolates of different races from different lineages, and found that *Pi5(t)* and *Pi7(t)* were resistant to at least six races belonging to four lineages in the Philippines. *Pi44(t)* only conditioned resistance to a set of isolates of lineage 44 from the Philippines (a rare and peculiar lineage; Chen, 1993) and some Japanese isolates (Inukai, personal communication). Though *Pi44(t)* showed a narrow resistance spectrum, this spectrum complements the resistance of *Pi2* [allele of *Piz* (Inukai et al 1994)] (Chen *et al.*, 1996). The combination of *Pi44(t)*, *Piz* and other genes could give a broad and potentially durable resistance. Little is know about the resistance spectra of *Pi10(t)*, *Pi12(t)* and *Pi157*.

RIL125 and RIL260 showed moderate resistance to 12 California blast isolates, showing type 3 lesions similar to that of the parent CO39. RIL29 was completely

Table 1. Reaction of near-isogenic lines carrying major genes and QTL(s) to compatible isolates of *Pyricularia grisea* in repeated greenhouse tests.

Isolate[1]	Lineage	Number of susceptible lesions relative to CO39					
		C101LAC	C101A51	C104PKT	C101PKT	RIL23	RIL276
		Pil	*Piz*	*Pi3*	*Pita*	QTL	QTL
1	4	0	0	0.07s[2]	0.44s	0.24s	0.31s
		0	0	0.12s	1.14	0.3s	0.67
2	4	0	0	1.68S[3]	0	0.41s	0.88
		0	0	1.02	0	0.42s	0.43s
3	4	0	0	1.02	0.48s	0.42s	nd
		0	0	1.89S	0.40s	0.43s	0.21s
		0	0	1.59S	0.68s	0.34s	0.46s
4	4	0	0	0.94	0.72	0.47s	0.35s
		0	0	1.16	0.41s	0.44s	0.26s
		0	0	1.26	Nd[4]	0.55s	0.62s
5	7	1.74	0	0	1.59S	0.62s	nd
		1.31	0	0	1.42S	0.62s	0.48s
6	14	1.01	0	1.48S	0.86	0.6s	0.72
		1.89	0	1.84S	0.96	0.76	nd
		nd	0	1.32S	0.07s	0.48s	0.55s
7	14	0	0	0	0.68s	0.56s	nd
		0	0	0	0.70s	0.29s	0.56s
8	14	0.45s	0	0	0.68s	0.38s	0.31s
		0.31s	0	0	0.58s	0.62s	nd
9	44	0	2.12S	2.06S	0	0	0
		0	1.61S	1.89S	0	0	0
		0	1.51S	nd	0	0	0
10	44	0	2.01S	1.46S	0	0	0
		0	2.99S	2.18S	0	0	0

[1] = bn107; 2=Ca790; 3=Ca79; 4=Ca89; 5=92325-11; 6=9232-5; 7=9249-3; 8=B90110; 9= C9240-1; 10= C9240-8. [2] s=significantly less disease than CO39; [3] S= significantly more disease than CO39; [4] nd= not determined.

resistant to those isolates. Since reactions of the *Pi5(t)* lines and CO39 were similar in response to the California isolates, it was not clear if the genes conferring partial resistance in CO39 would interfere with the performance of *Pi5(t)* against the California isolates. To understand the effect of genetic background on the gene performance, we crossed RIL125, RIL260 and RIL29 with a blast-susceptible California cultivar M202. F1 plants derived from the RIL125/M202 and RIL260/M202 were inoculated with a California isolate CAL-1, and were susceptible, whereas the F_1 derived from the RIL29/M202 was resistant. These results suggest that the moderate resistance to California isolates in the *Pi5(t)* lines was conferred by the genes from CO39, rather than *Pi5(t)*. Genetic analysis of the *Pi7(t)* resistance in M202 background is in the progress.

For more detailed characterization of the QTL effect on diverse blast isolates, 26 isolates representing six lineages of *P. grisea* were inoculated on RIL23 and RIL276 and a set of CO39 near-isogenic lines. Of these isolates, ten were tested repeatedly. Susceptible lesions were counted seven days after inoculation. Results showed that lesion numbers on the QTL lines (RIL23 and RIL276) were consistently reduced in all interactions, compared with those on susceptible parent CO39, while near-isogenic lines carrying major genes varied in reaction to different isolates (Table 1; Chen, 1993). These two lines also showed low disease levels in the field test (Wang *et al.*, 1994). These results demonstrated that a QTL identified using a single isolate in the greenhouse could reliably predict the performance of lines in the field and greenhouse with diverse isolates.

5. Conclusions and Considerations

Considerable progress towards understanding the genetic basis of durable resistance to rice blast in Moroberekan has been made in a series of studies. The identification of at least six major genes and 10 QTLs in Moroberekan provides strong circumstantial evidence that durability of blast resistance is a function of the combination of major genes for qualitative resistance and QTLs for partial or quantitative resistance. In the past several decades, breeding efforts were mainly devoted to incorporation of single major genes into rice cultivars. However, this strategy often drives the rapid evolution of pathogen populations, resulting in short-lived resistance. Combination of carefully characterized major genes and QTLs would contribute to the consistent development of cultivars with durable resistance. Though pyramiding genes would be difficult through the conventional breeding methodology, it is now possible to combine genes by molecular marker-assisted breeding (McCouch *et al.*, 1996; Huang *et al.*, 1997).

We have illustrated the efficacy of using a recombinant inbred population for identifying and mapping major genes and QTLs in Moroberekan. It is likely that more genes would be identified in Moroberekan by inoculating different isolates. Because of the large number of resistance genes present and the segregation distortion observed in the recombinant inbred population, the exact locations of the genes identified and mapped need to be further refined. Relocation of the *Pi5(t)* locus illustrates the need for continued genetic analysis in later generations.

The epistatic effect of qualitative genes on QTLs is well known. Four additional qualitative genes were identified from the lines that were previously found susceptible to isolate PO6-6 and carrying QTLs for partial resistance. Whether these genes have any effect on the resistance performance and the inferred chromosomal locations of the QTLs needs to be determined. On the other hand, it is not clear how the presence of QTLs affects the identification and mapping the qualitative genes. In the study of Wang *et al.* (1994), a group of RI lines was excluded from the analysis for determining the association between markers and resistance, because of the intermediate reaction of the lines with the test isolate. The intermediate reactions in these lines could result from the presence of multiple QTLs or the epistatic effect of resistance factors. Further molecular analysis of these lines by dissecting the resistance factors using a highly susceptible cultivar might provide needed insight.

Some major genes have much broader resistance spectra than others, and some genes are more difficult for pathogen to overcome. Development of near-isogenic lines with these genes and QTLs are underway (Inukai *et al.*, 1996) and will facilitate the phenotypic analysis of the genes.   Once the genes have been well characterized, strategies for utilizing the genes can be designed.   Molecular markers linked to the genes will facilitate the selection of genotypes carrying desired gene combinations.   As described in this paper, considerable progress has been made in identification and characterization of genes in a rice cultivar that has shown durable resistance. It is hoped that this information will aid in the utilization of these genes.

References

Ahn, S. W. (1994) International collaboration on breeding for resistance to rice blast, in Zeigler, R. S., Leong, S. A. and Teng, P.S. (eds.), *Rice Blast Disease*, CBA and IRRI, Wallingford, Oxon, UK, pp. 137-153.

Bonman, J. M. and Mackill, D.J. (1988) Durable resistance to rice blast disease. *Oryza* 25, 103-110.

Bidaux, J.M.   (1978)   Screening for horizontal resistance to rice blast (*Pyricularia oryzae*), in Buddenhagen, I. W. And Persley, G.J. (eds.), *Rice in Africa*. Academic Press, London, pp. 159-174.

Chen, D. -H. (1993) Population structure of *Pyricularia grisea* at two screening sites and quantitative characterization of major and minor resistance genes. PhD Thesis, University of the Philippines at Los Baños, Laguna, Philippines.

Chen, D., Zeigler, R.S., Leung,H., Nelson, R.J. (1995) Population structure of *Pyricularia grisea* at two screening sites in the Philippines. *Phytopathology* 85, 1011-1-20.

Chen, D. H., Zeigler, R. S., Ahn, S.W., Nelson, R. J. (1996) Phenotypic characterization of the rice blast resistance gene *Pi2(t)*. *Plant Dis.* 80, 52-56.

Chen, D. -H., Wang, G. L., and Ronald. P. C. (1997) Location of the rice blast resistance locus *Pi-5(t)* in Moroberekan by AFLP bulk segregant analysis. *Rice Genetics Newsletter* 14, 95-98.

Chin, D., Meyers, B., Shen K., Lavelle, D., Zhang, Z., Frijters, A., Okubara, P., Arroyo-Garcia, R., Anderson, P., Irwin, S., Ochoa, O., Mazier, M., van Damme, M., and Michelmore, R. (1995) Genetic and Physical Organization of the Major Cluster of Resistance Genes in Lettuce (*Lactuca sativa*) *Plant Genome IV Abstracts*, P12.

Fomba, S. N. and Taylor, D. R. (1994) Rice blast in West Africa: Its nature and control, in Zeigler, R. S., Leong, S. A. and Teng, P.S. (eds.), *Rice Blast Disease*, CBA and IRRI, Wallingford, Oxon, UK, pp. 343-355.

Hittalmani, S., Foolad, M.R., Mew, T., Rodriguez, R.L., Huang, N. (1995) Development of a PCR-based marker to identify rice blast resistance gene, *Pi2(t)*, in a segregating population. *Theor Appl Genet* 91, 9-14.

Huang, N., Angeles, E.R., Domingo, J., Magpantay, G., Singh, S., Zhang, G.,Kumaravadivel, N., Bennett, J., and Khush, G.S. (1997). Pyramiding of bacterial blight resistance genes in rice: marker-assisted selection using RFLP and PCR. *Theor Appl Genet* 95, 313-320.

Inoue, T., Zhong, H.S., Miyao, A., Ashikawa, I., Monna, L., Fukuoka, S., Miyadera, N., Nagamura, Y., Kurata, N., Sasaki, T., Minobe, Y. (1994) Sequence-tagged sites (STSs) as standard landmarkers in the rice genome. *Theor Appl Genet* 89,728-734.

Inukai, T., Nelson, R.J., Zeigler, R.S., Sarkarung, S., Mackill, D.J., Bonman, J.M., Takamure, I., Kinoshita, T. (1994) Allelism of blast resistance genes in near-isogenic lines of rice. *Phytopathology* 84, 1278-1283.

Inukai, T. and Aya, S. (1997) Determination of chromosomal location of blast resistance gene in West African upland cultivar Moroberekan. *Breed. Sci.* 47(Suppl.1), 268

Inukai, T. Nelson, R.J., Zeigler, R.S., Sarkarung, S., Mackill, D.J., Bonman, J.M., Takamure, I., Kinoshita, T. (1996) Development of pre-isogenic lines for rice blast resistance by marker aided selection from a recombinant inbred population. *Theor Appl Genet* 93, 560-567.

Johnson, R. (1981) Durable resistance: definition of, genetic control, and attainment in plant breeding. *Phytopathology* 71, 567-568.

Kiyosawa, S. (1981) Gene analysis for blast resistance. *Oryza* 18, 196-203.

Lee, E.J., Zhang, Q. and Mew, T.W. (1989) Durable resistance to rice disease in irrigated environments, in *Progress in Irrigated Rice Research*, International Rice Research Institute, P.O. Box 933, Manila, Philippines, pp. 93-110.

Ling, Z, Mew, T.V., Wang, J., Lei, C., Huang, N. (1995) Development of near-isogenic lines as international differentials of the blast pathogen. *Int. Rice Res. Newsl.* 20, 13.

Mackill, D.J. and Bonman, J.M. (1992) Inheritance of blast resistance in near-isogenic lines of rice. *Phytopathology* 82, 746-74

Maisoneuve, B., Anderson, P. and Michelmore, R.W. (1994). Rapid mapping of two genes for resistance to downy mildew derived from *Lactuca serriola* to existing clusters of resistance genes. *Theor. Appl. Genet.* 89, 96-104.

McCouch SR, Nelson RJ, Tohme J, Zeigler RS (1994) Mapping of blast resistance genes in rice, in Zeigler, R.S., Leong, S.A., Teng, P.S. (eds.) *Rice blast disease*. CABI and IRRI, Wallingford, Oxon, UK, pp. 167-186.

Michelmore, R.W., Paran, I., Kesseli, R.V. (1991) Identification of markers linked to disease resistance genes by bulked segregant analysis: a rapid method to detect markers in specific genomic regions by using segregating populations. *Proc Natl Acad Sci* 88, 9828-9832.

Naqvi, N.I. and Chattoo, B.B. (1996) Molecular genetic analysis and sequence characterized amplified region-assisted selection of blast resistance in rice, in [IRRI] International Rice Research Institute, Rice Genetics III. *Proceedings of the Third International Rice Genetics Symposium*, 16-20 Oct 1995. Manila (Philippines): IRRI. pp 507-572.

Naqvi, N.I., Bonman, J.M., Mackill, D.J., Nelson, R.J., Chattoo, B.B. (1995) Identification of RAPD markers linked to a major blast resistance gene in rice. *Molecular Breeding* 1, 341-348.

Parlevliet, J.E. (1998) Identification and evaluation of quantitative resistance, in Leonard, K.J., and Fry, W.G. (eds.), *Plant Disease Epidemiology. Genetics, Resistance, and Management*, Vol. 2. McGraw-Hill, New York, pp. 377

Redoña, E.D. and Mackill, D.J. (1996) Mapping quantitative trait loci for seedling vigor in rice using RFLPs. *Theor Appl Genet* 92, 395-402.

Wang, GL, Mackill, D.J., Bonman, J.M., McCouch, S.R., Champoux, M.C., Nelson, R.J. (1994) RFLP mapping of genes conferring complete and partial resistance to blast in a durably resistance rice cultivar. *Genetics* 136, 1421-1434.

Witsenboer, H., Kesseli, R.V., Fortin, M., Stangellini, M., Michelmore, R.W. (1995). Sources and genetic structure of a cluster of genes for resistance to three pathogens in lettuce. *Theor. Appl. Genet.* 91,178-188.

Xu, K. and Mackill, D.J. (1996) A major locus for submergence tolerance mapped on rice chromosome 9. *Mol. Breed.* 2, 219-224.

Zabeau, M. and Vos, P. (1993) Selective restriction fragment amplification : a general method for DNA fingerprinting. European Patent Application No. 0534858 A1. European patent Office, Paris.

Zeigler, R.S., Cuoc, L.X., Scott, R.P., Bernardo, M.A., Chen, D.H., Valent, B., Nelson, R.J. (1995) The relationship between lineage and virulence in *Pyricularia grisea* in the Philippines. *Phytopathology* 85, 443-451.

MAPPING OF LEAF AND NECK BLAST RESISTANCE GENES WITH RFLP, RAPD AND RESISTANCE GENE ANALOGS IN RICE

K.-L. ZHENG[1], R.-Y. CHAI[2], M.-Z. JIN[2], J.-L. WU[1], Y.-Y. FAN[1], H. LEUNG[3] AND J.-Y. ZHUANG[1]

[1] *Biotechnology Department, China National Rice Research Institute, Hangzhou 310006, China;* [2] *Institute of Plant Protection, Zhejiang Academy of Agricultural Sciences, Hangzhou 310021, China;* [3]. *Entomology and Plant Pathology Division, International Rice Research Institute, 1099 Manila,Philippines*

1. Introduction

Rice blast, caused by *Pyricularia grisea* (synonymous *Pyricularia oryzae* Cav.; teleomorph *Magnaporthe grisea*), is considered the most important disease of rice because of its world-wide distribution and the resulting severe yield loss. Growing resistant varieties has been the most effective and economic way to control this disease.

Blast occurs at different growth stages. The fungus produces spots or lesions on leaves, notes and different parts of panicle and grain. The neck blast makes more significant yield and quality losses than leaf blast (Katsube and Koshimizu, 1970). According to a survey of 253 *indica* varieties in a natural condition of Zhejiang Province in China, the correlation between resistance to leaf and neck blast was significantly positive. However, 17 out of 77 varieties resistant to leaf blast were susceptible to neck blast, and 4 out of 164 varieties susceptible to leaf blast were resistant to neck blast (Cai, 1983). In another study, seeds of the same 42 varieties were sowed at different time. The plants at early tillering stage and late booting stage of the same varieties were inoculated with spores from a same race of the fungus at the same time in a same environment. Seven varieties showed different resistance to leaf and neck blast. Five resistant to leaf blast but susceptible to neck blast and 2 susceptible to leaf blast but resistant to neck blast, implying that the mechanisms of the resistance of a same variety might not be the same at different growth stages (Shi et al., 1989).

Recently, a number of blast resistance genes have been located on the molecular genetic map of rice (McCouch et al., 1994), including QTLs (quantitative trait loci) contributing to durable resistance (Wang et al., 1994). Great effort has been made to clone blast resistance genes (Ronald, 1997). However, all the molecular work on rice blast resistance reported so far was based on the resistance to leaf blast.

We have mapped a gene only resistant to leaf blast and a gene resistant to both leaf and neck blast in an recombinant inbred line population with RFLP, RAPD and RGA

D. Tharreau et al. (eds.), Advances in Rice Blast Research, 28–33.

(resistance gene analog).

2. Development of Recombinant Inbred Lines

An F_8 recombinant inbred population was constructed from an *indica* cross Zhong 156/Gumei 2 through single seed descent. The female parent Zhong 156 is a commercial *indica* variety developed recently in China National Rice Research Institute (CNRRI). It has shown to be resistant to a wide spectrum of blast fungus at the vegetative stage, but was found to be susceptible at the reproductive stage (Wu et al., 1992). The male parent Gumei 2 is a semidwarf *indica* rice variety with durable blast resistance, which is widely used as a resistance donor in breeding programs. In a long-term evaluation in natural nurseries of blast "hot-spot", 6 out of the total 38,000 accessions tested showed durable resistance to rice blast, among which Gumei 2 is the only semidwarf variety (Peng et al., 1996). The cross Zhong 156/Gumei 2 was made in 1990, and an F_8 recombinant inbred population of 305 lines was established in 1995. For the first step in gene mapping, 148 lines with similar growth duration were chosen as a subpopulation.

3. Evaluation and Genetic Analysis of Blast Resistance

Resistance to leaf and neck blast of the two parents was screened in 1995. Among 20 races of blast fungus, the isolate   92-183 (race ZC_{15}) was chosen. With this isolate, Zhong 156 was resistant to leaf blast but susceptible to neck blast. While Gumei 2 was resistant to both leaf and neck blast with the same isolate. The resistance to leaf and neck blast in the subpopulation was screened with this isolate in 1996 and 1997. Consistent results of leaf blast resistance and neck blast resistance for the two years were observed for 146 and 127 lines respectively (Table 1), on which further analysis was carried out.

Table 1. Resistance of the RILs to blast race ZC_{15} in two years

Group	Resistance[a]		No. of
	Leaf	Neck	lines
Group 1	SS	SS	19
Group 2	RR	SS	18
Group 3	RR	RR	89
Group 4	-	SS	1
Group 5	RR	-	20
Group 6	-	-	1

[a] RR = resistant in both years
      SS = susceptible in both years

   -   = missing data

The segregation of 127 resistant and 19 susceptible lines to leaf blast fitted the ratio of 7:1 (P=0.800), suggesting that the resistance to leaf blast was controlled by three loci and the presence of resistant alleles at any of the three loci would result in resistance. The susceptibility to leaf blast was due to the complementary effect of susceptible alleles at all the three loci. The segregation of 89 resistant lines and 38 susceptible lines to neck blast agreed with the ratio of 3:1 (P=0.200), suggesting that the neck blast resistance was controlled by two independent dominant genes and presence of resistant alleles at either loci would result in resistance. Since the 89 lines resistant to neck blast were also all resistant to leaf blast, the two genes controlling neck blast resistance should be same with the two out of the three genes controlling leaf blast resistance. In addition, all the 19 lines susceptible to leaf blast were also susceptible to neck blast and 18 lines resistant to leaf blast were susceptible to neck blast, indicating that one of the genes responsible for leaf blast resistance was not effective to neck blast (Zhuang et al., 1997).

4. Mapping of Resistance Genes with RFLP and RAPD

Parents were screened with 190 RFLP probes and 280 RAPD primers. Polymorphism was detected with 35 probes (18%) and 69 RAPD primers (25%). Low level of polymorphism between the parents Zhong 156 and Gumei 2 was not unexpected, because both parents were semi-dwarf *indica* varieties used in Southern China.

All the 35 polymorphic probes and 22 out of the 69 polymorphic RAPD primers were applied to the subpopulation. Two RAPD markers $OPD10_{800}$ and $OPH11_{400}$ having genetic distance of 2.1 cM, co-segregated with the leaf blast resistance in the group lines which were susceptible to leaf and neck blast. The same pattern of co-segregation was not observed among the 18 lines of the group 2 which were resistant to leaf blast and susceptible to neck blast. So this locus is only responsible to leaf blast resistance and the resistance allele is derived from Zhong 156 as expected (Table 2).

Three RAPD markers, $OPK17_{1400}$, $OPA7_{550}$ and $OPB10_{450-750}$ indicated relationship with the resistance phenotype among the 38 lines showing the neck blast susceptibility. It was estimated that the resistance gene was located between $K17_{1400}$ and $A7_{550}$, having genetic distance of 2.4 cM to $K17_{1400}$ and 7.5 cM to $A7_{550}$. This locus is responsible for both leaf and neck blast and as expected, the resistance allele is derived from Gumei 2 (Table 2, Zhuang et al., 1997).

Table 2. Segregation of RAPD markers linked to the resistance genes

| Group | Resistance | | Maker | Marker | | | | |
	Leaf	Neck	type[a]	D10	H11	K17	A7	B10
Group 1	SS	SS	A	0	0	17	16	17
			B	19	19	1	2	1
Group 2	RR	SS	A	13	14	16	15	14
			B	5	4	0	1	2

[a] A = Zhong 156;   B = Gumei 2

Chromosome locations of the genes were determined based on the linkages to RFLP markers in the map of Causse et al. (1994). The gene resistant to leaf blast only was mapped to chromosome 12 and the gene resistant to leaf and neck blast was mapped to chromosome 6 (Fig. 1).

5. Mapping Resistance Genes with RGAs

RGAs (resistance gene analog) are the PCR product amplified with the degenerated primers designed based on the conserved sequences in disease resistance genes against diverse pathogens including viruses, bacteria and fungi. These conserved sequences included leucine rich repeat (LRR), nucleotide binding site (NBS) and kinase. RGAs have been applied to isolate new resistance genes and to develop markers tightly linked to resistance genes in various crop species (Kanazin et al., 1996; Leister et al., 1996; Yu et al., 1996). And the polymorphism level of RGAs could be enhanced when denaturing polyacrylamide gel electrophoresis was used to separate the PCR products (Chen et al., 1998).

Eight pairs of RGA primers were applied for the mapping study (Table 3), all of which were maintained at the Genetics Laboratory of the Entomology and Plant Pathology Division, International Rice Research Institute.

Table 3. A list of RGA primers used in this study

Code	Primer	Sequence(5'-3')
RGA 1	S2	GGIGGIGTIGGIAAIACIAC
	AS3	IAGIGCIAGIGGIAGICC
RGA 2	NLRRinv1	TGCTACGTTCTCCGGG
	NLRRinv2	TCAGGCCGTGAAAAATAT
RGA 3	NLRRfor	TAGGGCCTCTTGCATCGT
	NLRRrev	TATAAAAAGTGCCGGACT
RGA 4	XLRRfor	CCGTTGGACAGGAAGGAG
	XLRRrev	CCCATAGACCGGACTGTT
RGA 5	XLRRinv1	TTGTCAGGCCAGATACCC
	XLRRinv2	GAGGAAGGACAGGTTGCC
RGA 6	Pto kin1	GCATTGGAACAAGGTGAA
	Pto kin2	AGGGGGACCACCACGTAG
RGA 7	Pto kin3	TAGTTCGGACGTTTACAT
	Pto kin4	AGTGTCTTGTAGGGTATC
RGA 8	LMK637	ARIGCTARIGGIARICC
	LMK638	GGIGGIGTIGGIAAIACIAC

All the eight pairs of primers detected polymorphisms between the parents and each generated 3-13 polymorphic bands. They were all applied to the subpopulation. Two RGAs were mapped on chromosome 6 and linked to the resistance gene to leaf and neck blast. RGA7-5was the most tightly linked marker to the gene. Three RGAs were tightly linked to the gene resistant to leaf blast on chromosome 12. The LR. based primer derived marker RGA3-1 was completely co-segregated with the resistance gene (Table 3, Fig. 1).

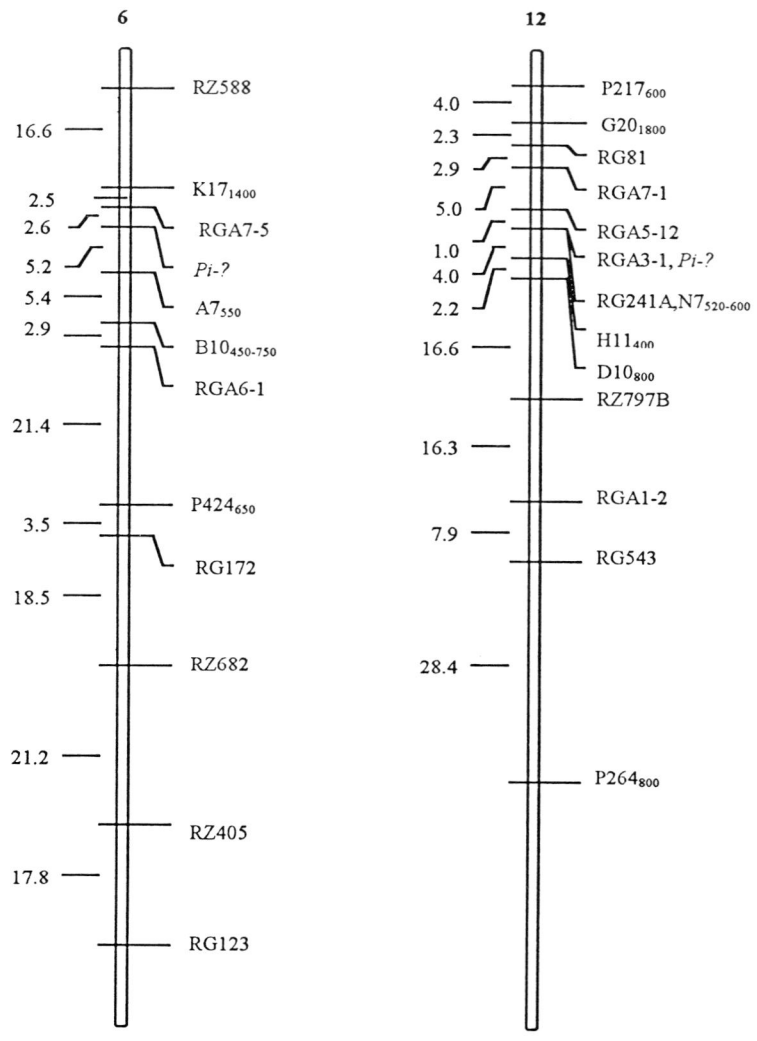

Fig.1 Linkage relationship in the vicinity of the two blast resistance genes
    Left to the chromosome, genetic distance in Kosambi cM
    Right to the chromosome, marker name
    Those with prefix RG or RZ are RFLP markers, with RGA are RGA markers,
    and with A, B, D, G, H, K, P are RAPD markers

References

Cai, G. (1983) Genetic analysis for blast resistance in *indica* rice varieties, Zhejiang Agricultural Sciences (1),
    16-18

Causse, M.A., Fulton, T.M., Cho, Y.G., Ahn, S.N., Chunwongse, J., Wu, K.S., Xiao, J.H., Yu, Z.H., Ronald, P.C., Harrington, S.E., Second, G., McCouch, S.R. and Tanksley, S.D. (1994) Saturated molecular map of the rice genome based on an interspecific backcross population, Genetics 138, 1251-1274

Chen, X.M., Line, R.F. and Leung, H. (in press) Genome scanning for conserved motif of disease resistance gene in rice, barley and wheat by high resolution electrophoresis.

Kanazin, V., Marek, L.F. and Shoemaker, R.C. (1996) Resitance gene analogs are conserved and clustered in soybean, Proc Natl Acad Sci USA 93, 11746-11750

Katsube, T. and Koshimizu, Y. (1970) Influence of blast disease on harvest of rice plants. 1. Effect of panicle infection on yield components and quality, Bulletin of the Tohoku Agricultural Experiment Station 39, 55-96

Leister, D., Ballvora, A., Salamini, F. and Gebhardt, C. (1996) A PCR-based approach for isolating pathogen resistance genes from potato with potential for wide application in plants, Nature Genet 14, 421-429

McCouch, S.R., Nelson, R.J., Tohme, J. and Zeigler, R.S. (1994) Mapping of blast resistance genes in rice, in R.S. Zeigler, S.A. Leong and P.S. Teng (eds.), Rice Blast Disease, CAB International in association with International Rice Research Institute, pp. 167-186

Peng, S.Q., Huang, F.Y., Sun G.C., Liu, E.M., Sun, Y.J., Ai, R.X., Bai, S.Z. and Xiao, F.H. (1996) Studies on the durable resistance to blast disease in different latitudes for rice, Scientia Agricultura Sinica 29, 52-58

Ronald, P.C. (1997) The molecular basis of disease resistance in rice, Plant Mol Biol 35,176-186

Shi, D., Sun, S.Y. and Shen, Z.T. (1989) Preliminary studies on the resistance reaction and resistance genetics of leaf blast and neck blast for various rice varieties, Acta Agriculturae Zhejiangensis 1, 94-96

Wang, G.L., Mackill, D.J., Bonman, J.M., McCouch, S.R., Champoux, M.C. and Nelson, R.J. (1994) RFLP mapping of genes conferring complete and partial resistance to blast in a durably resistant rice cultivar, Genetics 136, 1421-1434

Wu, M.L., Zhao, M.L., Li, X.M., Zhuang, J.Y., Ma, L.Y., Peng, Y.C. and Xia, X.J. (1992) Zhong 156, a high-yieldign early-season indica rice variety, CNRRI Annual Report 1991, China Agriculture Press, Beijing, pp.19-20

Yu, Y.G., Buss, G.R. and Saghai-Maroof, M.A. (1996) Isolation of a superfamily of candidate disease-resistance genes in soybean on a conserved nucleotide- binding site, Proc Natl Acad Sci USA 93, 11751-11756

Zhuang, J.Y., Chai, R.Y., Ma, W.B., Lu, J., Jin, M.Z. and Zheng, K.L. (1997) Genetic analysis of the blast resistance at vegetative and reproductive stages in rice, Rice Genetics Newsletter 14, 62-64

IDENTIFICATION OF DNA MARKERS LINKED TO PARTIAL RESISTANCE FOR BLAST DISEASE IN RICE ACROSS FOUR LOCATIONS

PRASHANTH G. BAGALI, SHAILAJA HITTALMANI*,
SRINIVASACHARY Y,. SHASHIDHAR AND H.E.SHASHIDHAR
Department of Genetics and Plant Breeding,
University of Agricultural Sciences,
GKVK Bangalore-560 065, India

Abstract

One hundred and fourteen doubled haploid plants of IR64/Azucena population were used to identify QTL controlling partial resistance to rice blast disease. The plants were scored for leaf and neck blast reaction in three locations in South India and at IRRI (Philippines). Seedlings were evaluated in nursery for leaf blast and in field for neck blast reaction. The RFLP marker data consisting of 175 DNA markers was used for mapping of QTL (MAPMAKER/QTL, LOD>2.50). Among the several QTL were identified, the major QTL flanked by RG958-RZ816 on chromosome 12 was common in two locations, IRRI, Philippines and Ponnampet, India with a range LOD score of 2.44-8.87, and 10.8% - 27.4% variation. Another QTL on chromosome 5 linked to RZ649-RZ225 markers was identified for leaf blast resistance in Ponnampet location (LOD=2.67, 16.5% variation) and for neck blast resistance in Mudigere location in India (LOD=2.42, 11.1% variation). A major QTL on chromosome 9 linked to Amy3ABC-RG667 markers was identified for neck blast resistance in Bangalore (LOD=2.03, 8.5% variation). The occurence of common QTL could be due to pleiotropic effect of QTL. The tightly linked markers will be fine mapped and used in MAS and in developing isogenic lines.

1. Introduction

Blast disease caused by *Pyricularia grisea* Sacc is the most serious fungal disease of rice (*Oryza sativa* L.). Host resistance is controlled by both major genes and quantitative trait loci (QTL) (Wand *et al.,* 1994). The rapid changes in virulence characterisctics of pathogen population is a continuous threat to effectiveness of existing blast resistance varieties. However, breeding for blast resistance is still considered the most economical

D. Tharreau et al. (eds.), Advances in Rice Blast Research, 34–42.

and effective strategy for dealing with the pathogen, since resistant cultivars add no cost to the farmer. Therefore identifying QTL and breeding for durable resistance is the priority for plant breeders and pathologists alike.

The mechanism of durable resistance is not yet well established. But partial resistance is attributed to minor genes or QTL. The first study on restriction fragment length polymorphism (RFLP) mapping of genes conferring partial resistance to blast was reported by Wang et al.,(1994). Two major genes and ten QTL for blast resistance were identified in *japonica* cultivar Moroberekan. The relative scarcity of conventional markers along the chromosome made it difficult to locate QTL (Paterson *et al.,* 1991). However, the construction of high density saturated RFLP linkage map helped overcome this problem, thus allowing a more thorough coverage of the genome. Such a framework will be useful in fine mapping and locating major and minor genes closely associated with for the use in marker aided selection.

In this study QTL for leaf and neck blast were identified and located on the molecular map of rice. The differences among QTL in their relative effects on components of partial resistance across four environments and the commonly occuring QTL were identified.

2. Materials and methods

Plant material
One hundred and fourteen lines of doubled haploid (DH) rice population developed from a cross between IR64, an *indica* variety well adapted to irrigated conditions and Azucena, an upland *japonica* variety (Guiderdoni *et al.,* 1992) were used for mapping. The RFLP map for this population was established by Huang *et al.,* (1994) using 175 DNA markers. Rice varieties IR50, HR12 and CO39 were used as susceptible check varieties.

Evaluation the genotypes for leaf and neck blast disease
The plants were raised in two locations. Seedling leaf blast was scored at four locations in India (Bangalore, Mudigere and Ponnampet, in South India in 1995 and 1996) and at IRRI (1997). In both the locations the plants were grown in Blast nursery where high natural inoculum was present. The DH population, along with the parents IR64 and Azucena were sown (ten grams of seeds of each genotype) in rows of 30 cm length with 5 cm spacing between rows. All around these entries, two spreader rows of IR50, HR12 and CO39 were sown to trap fungal spores and enhance the natural inoculum.

Method of sampling and recording of observations
The entire row of each genotype was considered for visual scoring. Leaf blast was scored using 0-5 scale (Mackill and Bonman, 1992). The plants were scored at regular intervals from seventh day after inoculation to transplanting date when the spreader rows

were completely infected with blast disease. The three parameters i.e., percentage diseased leaf area (DLA%), number of susceptible lesions (LSNN) and susceptible lesion size (LSSI) were scored in all four locations. Percent neck blast infection was scored prior to harvest (Anonymous, 1988) in the main field.

Statistical analyses and Interval mapping
MAPMAKER/EXP and MAPMAKER/QTL V1.1b (Lincoln *et al.*, 1992) were used to establish RFLP map. This map contains a total of 175 polymorphic DNA markers (146 RFLP's, 3 isozymes, 14 RAPD's and 12 cloned genes) Huang *et al.* 1994). The markers associated with three parameters of leaf blast resistance and neck blast were identified by interval mapping using threshold LOD$\geq$2.50.

3. Results and discussion

The phenotypic evaluation for blast disease indicated IR64 to possess high level of leaf blast resistance across four locations. The other parent Azucena was moderately resistant in all locations studied. Sanger *et al.* (1997) observed resistant lesions on IR64 in highly favorable blast disease environment on the Northern Hill zone of Chhatisgarh, Madhya Pradesh, India. Ghesquiere *et al.* (1996) studied IR64 x Azucena DH population and identified few resistant lines in France using six diverse strains of *Magnaporthe grisea*. IR64 is known to possess partial resistance to leaf blast as reported by Ghesquiere *et al.*, (1996) and Bagali (1997). In the nursery evaluation at IRRI, Philippines, showed thirty two partially resistant DH lines (data not shown) whereas nineteen exhibited resistance in single spore inoculation using isolate PO6-06. Nine DH lines and IR64 parent showed resistant lesions in the presence of high level of inoculum in the nursery at Ponnampet, India. These lines could possess major genes as well as QTL for blast resistance as they exhibited resistance at the hotspot where several pathogenic races are present.
Interval mapping with threshold LOD>2.50 identified a total sixteen QTL across three locations for leaf blast (Table 1.) and two QTL (LOD>2.00) for neck blast resistance (Table 2). Among sixteen QTL for leaf blast resistance, twelve QTL were detected at blast nursery, IRRI and four at Ponnampet in India (Table 1.). Thirteen QTL were identified for DLA and three QTL for LSNN. The putative QTL conferring partial resistance to leaf blast were mainly located on chromosomes 1, 3, 4, 5, 6, 9 and 12 with the phenotypic variation ranging from 8.80% to 27.40%. Eight QTL had major effect (LOD>3.00) for leaf blast resistance and were localized on RFLP map of rice (Fig. 1). The QTL flanked by RG 958-RZ816 was identified in both IRRI and Ponnampet with a range of LOD score of 2.44-8.87 and 10.80%-27.40% variation. This loci is likely to host both major and minor genes. The presence of major QTL in both the diverse locations indicates the stability of this QTL for blast with little environmental effect. Thus this QTL need to be fine mapped to identify closely linked markers for the use in MAS for QTL.

Table 1: QTL identified for leaf blast resistance in IR64/Azucena DH population across locations (Threshold LOD>2.50)

Sl. No.	Locations	QTL*	Chro.#	Flanking Markers	Peak LOD	Variation(%)
1	**ARS, Ponnampet**	qDLA'96-4	4	RG218 - RG908	2.63	10.90
	(1995 & 1996)	qDLA'95-5	5	RZ70 - RZ225	2.67	16.50
	(Nursery Evaluation)	qDLA'95-9	9	RZ228 - RG451	2.84	24.10
		qLSNN'96-9	9	RG451 - RZ404	2.96	24.80
2	**IRRI, Philippines**	qDLA-3-1	3	RG179 - RZ394	2.58	8.80
	(1997)	qDLA-3-2	3	CDO337 - RZ394	2.78	9.60
	(Nursery Evaluation)	qDLA-3-3	3	CDO337 - RZ394	3.01	10.50
		qDLA-12-1	12	Sdh1 - RZ816	3.37	11.40
		qDLA-12-2	12	RG901 - RG574	8.87	27.40
		qDLA-12-3	12	RG463 - RZ816	5.34	17.50
		qDLA-12-4	12	Sdh1 - RZ816	3.78	12.70
3	**IRRI, Philippines**	qDLA-1-1	1	RG381 - RG331	5.34	22.10
	(1997)	qDLA-1-2	1	RZ19 - RG331	5.26	21.00
	(Single Spore	qLSNN-1	1	RZ730 - RG331	2.78	12.70
	Inoculation)	qLSNN-6	6	RG653 - Cat1	3.02	10.90
		qDLA-12-5	12	Sdh1 - RZ816	2.87	10.80

*= QTL nomenclature is according to McCouch et al. (1997)

Table 2: QTL for neck blast resistance in IR64/Azucena DH population across locations (Threshold LOD>2.00)

Locations	QTL*	Chro.#	Flanking Markers	Peak LOD	Variation(%)
1. Agricultural Research Station (1996) Mudigere	qNBL-5	5	RZ649 - RZ225	2.42	11.10
2. Main Research Station (1996) Bangalore	qNBL-9	9	Amy3ABC - RG667	2.03	8.50

*= QTL nomenclature is according to McCouch et al. (1997)

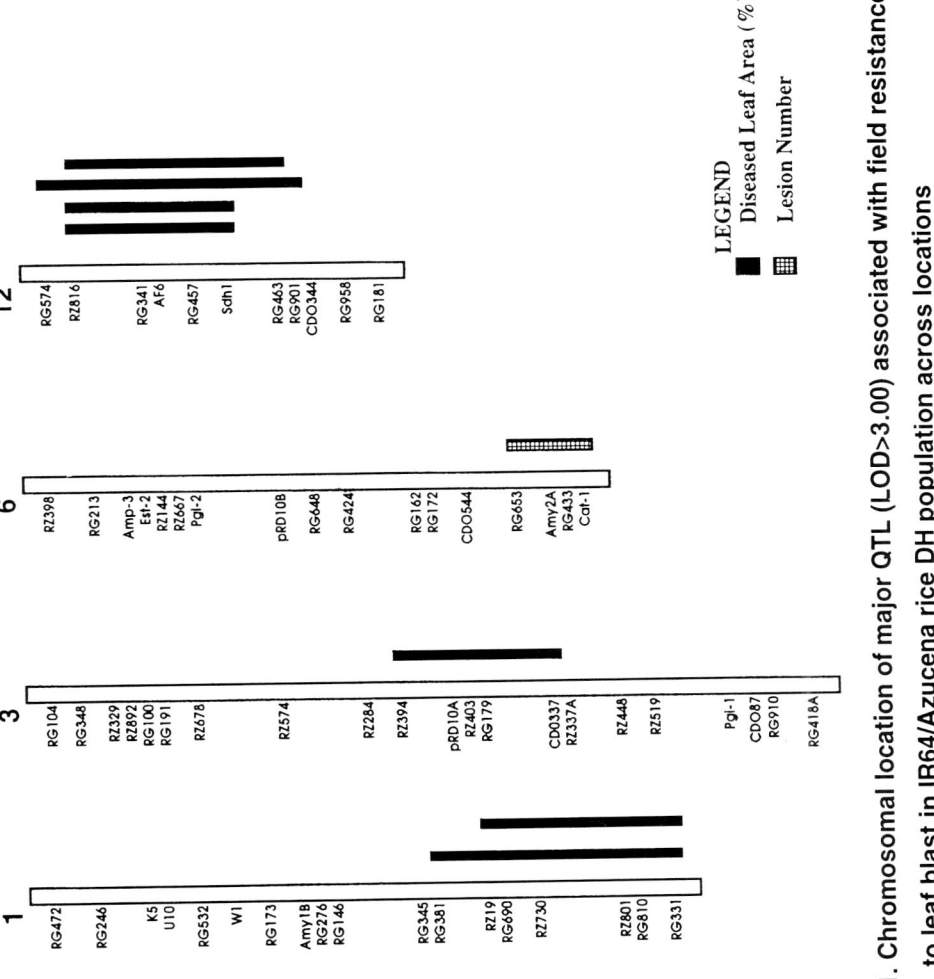

Figure 1. Chromosomal location of major QTL (LOD>3.00) associated with field resistance to leaf blast in IR64/Azucena rice DH population across locations

Table 3: Multiple Effect of QTL identified across locations (Threshold LOD>2.00)

Sl.No	Locations	QTL	Chro.#	Flanking Markers	Peak LOD	Variation(%)
1	Ponnampet	qDLA'95-5	5	RZ649 - RZ225	2.67	16.50
	Mudigere	qNBL-5	5	RZ649 - RZ225	2.42	11.10
2	*IRRI (Single Spore Inoculation)*	qLSNN-6	6	Amp2A-Cat1	2.02	8.02
	Ponnampet	qDLA-6	6	Amp2A-Cat1	3.02	10.90
3	IRRI *(Nursery Evaluation)*	qDLA-11	11	RG103-RZ536	2.28	7.90
	Ponnampet	qLSSI-11	11	RG103-Npb186	2.02	9.50

*= QTL nomenclature is according to McCouch et al. (1997)

Multiple effect of QTL

Several QTL were identified to control more than one parameter for partial resistance to blast disease across locations (Table 3). Based on the QTL locations, multiple effect of certain chromosomal segments/loci conferring blast resistance is hypothesized. The results insinuate that pleiotropism rather than close linkage of different QTL and could be the main reason why QTL for different blast related parameters were frequently detected in the same marker intervals across locations. QTL controlling neck blast resistance (qNBL) identified in Bangalore location overlapped with leaf blast resistance QTL flanked by RG 358-RZ12 (LOD=2.76, 6.00% variation) identified in another recombinant inbred population RIL population, CO39/Moroberekan screened in Ponnampet (unpublished data). This indicated tne commonality of the QTL not only across location but also in different populations. A minor QTL for lesion size (qLSSI) bracketed by RG103-*Npb*186 flanking markers (LOD=2.02, 9.50% variation) on chromosome 11 partially overlapped the major gene *Pi-7* (t) (Wang *et al.,* 1994) linked to RG103-RG16 markers identified in Moroberekan, a resitant *japonica* variety.

The presence of QTL in one location and its absence in another location could be because of the different races of the blast pathogens operating in the regions. While, the QTL with major effect identified across wide geographic locations, and at similar chromosomal loci indicate the stability of the QTL for blast resistance. Such markers associated with QTL that are common across locations could be used for selecting blast resistant genotypes in field conditions. The DH lines, P013, P022, P047, P048 and P055 were found to be partial or completely resistant to both leaf and neck blast across four locations during the years 1995, 1996 and 1997. Such genotypes that are stable in their performance with respect to blast have pertinent utility in breeding program. This study provides an insight into the differential response of genotypes derived from an *indica* x *japonica* cross and will give a reliability of the RFLP markers to be used in Marker-assisted selection, towards selection of rice cultivars that have durable, broad spectrum resistance to the blast disease.

Acknowledgments

The authors are thankful to. Dr. Ning Huang, IRRI, Philippines for the seeds and molecular map of IR64/Azucena DH population. We acknowledge the help of colleagues of MAS, Lab., Department of Genetics and Plant Breeding, UAS, Bangalore. This research work was supported by the Rockfeller Foundation, USA (RF 95001 #321) to Shailaja Hittalmani.

4. References

Anonymous, (1988) Standard evaluation system for rice. International Rice Testing Programme (3rd edition). international Rice Research Institute, Los Banos, Philippines. pp 52.

Bagali, P.G. (1977) RFLP mapping of Quantitative Trait Loci controlling yield related traits and resistance to Leaf Blast disease in Rice (*Oryza sativa* L.). M. SC. (Agri.) Thesis submitted to the University of Agricultural Sciences, GKVK, Bangalore 560 065, India.

Causse M., Fulton T.M., Cho Y.G., Ahn S.N. Chunwongse J., Wu K., Xiao J., Yu Z., Ronald P.C., Harrington S.B., Second G., McCouch S.R., Tanksley S.D. (1994) Saturated molecular map of the rice genome based on an inter specific backcross population. *Genetics* **138**: 1251-1274

Ghesquiere A., Lorieux M., Roumen E., Albar L., Huang N. and Notteghem J.L. (1996) *Indica / japonica* doubled haploid population as a model for mapping rice yellow mottle virus and blast resistance genes. *IRRN* **21**(2-3), 47-49.

Guiderdoni A., Gallinato E., Luistro J., Vergara G. (1992) Another culture of tropical *japonica / indica* hybrids of rice (*Oryza sativa* L.) *Euphytica* **62**:219-224

Huang N., McCouch S., Mew T., Parco A., Guiderdoni E. (1994) Development of an RFLP map from a doubled haploid population in rice. *Rice Genetics News Lett.* **11**:134-137.

Lincoln S., Daly M., Lander E. (1992) Mapping genes controlling quantitative traits with MAPMAKER/QTL 1.1., Whitehead Institute Technical Report (manual) 2nd edition. Mackill D.J., Bonnman J.M., 1992. inheritance of blast resistance in near-isogenic lines of rice. *Phytopathology* **82**:746-749.

Mackill D.J., Bonnman J.M. (1992) Inheritance of blast resistance in near-isogenic lines of rice. *Phytopathology* **82**: 746-749.

McCouch S.R., Cho Y.G., Yano M., Paul E., Blinstrub M., Morishima H., Kinoshita T. (1997) Report on QTL nomenclature. *Rice Genetics Newsletter* **14**:11-13

Paterson A.H., Damon S., Hewitt J.D., Zamir D., Rabmowitch H.D., Lincoln S.E., Lander E.S., Tanksley S.D., (1991) Mendelian factors underlying quantitative traits in tomato: comparison across species, generations and environments. *Genetics* **127**:181-197.

Rossman A.Y., Howard R.J., Valent B. (1990) *Pyricularia grisea,* the correct name for the rice blast disease fungus. *Mycologia* **82**: 509-512.

Sanger R.B.S., Agarwal K.C., Srivastva M.N. and Sarangi A.K. (1997) Reaction of promising rice genotypes to blast (short communication). *Oryza* **34**, 83-84.

Wang G.L., Mackill D.J., Bonnman M., McCouch S.R., Champoux M.C., Nelson R.J. (1994). RFLP mapping of genes conferring complete and partial resistance to blast in a durably resistant rice cultivar. *Genetics* **136**: 1421-1434.

MARKER ASSISTED BACKCROSS GENE INTROGRESSION OF MAJOR GENES FOR BLAST RESISTANCE IN RICE

K. GIRISH KUMAR, SHAILAJA HITTALMANI*,
SRINIVASACHARY AND SHASHIDHARHE
*Department of Genetics and Plant Breeding,
University of Agricultural Sciences,
GKVK, Bangalore-560 065, India*

Abstract

Isogenic lines carrying the two major genes for blast resistance *Pi-1* and *Pi-2* were crossed to popular rice varieties IR36, IR64 and IR72 to transfer the resistance. The closely linked markers linked to the two genes c481 for *Pi-1* and RG64 for *Pi-2* were used for identifying the genes in the parents, F_1s and in the F_2 segregating generations. The resistant plants identified and selected by PCR based molecular markers were further screened using twenty RAPD primers for identification of plants nearer to reccurent parent for backcrossing. Among the ten plants identified to contain the two major genes in each cross the comparative efficiency of identifying plants more nearer to reccurent parent for backcrossing was estimated by txenty RAPD primers. An average of 16-18% higher efficiency was observed by using RAPD than by conventional estimates. It is expected that in three backcrosses entire recurrent parent genome can be retrieved along with resistance genes using markers as the selection criteria.

1. Introduction

Rice blast disease caused by fungal pathogen *Pyricularia grisea* is a serious biotic constraint in rice production throughout the rice growing regions of the world. Efforts are being made in this direction to transfer the resistance genes into agronomically superior blast susceptible varieties. Conventionally transfer of the genes into such varietis is made through hybridization and backcross breeding methods followed by selection of resistant plants. This is not always easy as every time the lants have to be exposed to the disease either naturally or by artificial inoculation using specific isolates. Hybridization followed by identification of true F1s and identification of resistant plants especially in the backcross progeny and other later segregating generations using

D. Tharreau et al. (eds.), Advances in Rice Blast Research, 43–53.
© *2000 Kluwer Academic Publishers. Printed in the Netherlands.*

specific isolate becomes laborious anad also need 5-6 generations of backcrossing. When more than one gene has to be transferred into a variety several different compatible isolates have to be tested with the same genotype which is very cumbersome and not possible many a times. Presence of more genes in a single genotype my make the interpretation of the disease reaction on the phenotype difficult and misleading. In such situations DNA markers tightly linked to the genes tagged by mapping would be of great help in identifying and selecting the genotypes carrying the resistance genes. In this study the two major blast resistance genes *Pi-1* and *Pi-2* were transferred by hybridization to IR36, IR64 and IR72 rice varieties and selected by tightly linked markers in F_1 and F_2 generation for the presence of genes. RAPD based markers identified genotypes simular to reccurent parent that up to 18 percentage more efficiently than conventional prediction methods.

2. Material and methods

The tree popularly cultivated varieties IR36, IR64 and IR72 were crossed with blast isogenic lines C101LAC, C101A51 carrying i-1 and *Pi-2* genes respectively and its combinations *Pi-1* + *Pi-2* to transfer resistance to selected varieties. Initially the parents and later the F_1 individuals were screened for *Pi-1* and *PI-2* genes using PCR based DNA markers closely linked to the genes. The F_1 individuals were grown for DNA isoltion and later the plants were selfed to produce F_2 individuals. The F_2 individuals were selfed as well as crossed to their respective recurrent parent to generate F_2BC1 plants. The DNA from F_2 and F_2BC1 individuals was extracted to diagnoze for the presence of *Pi-1* and *Pi-2* genes in these individual crosses.

The resistant F_2 individuals identified as shown by linked markers were subjected to RAPD analysis for identification of plants that were genetically more nearer to reccurent parent for the use in backcrossing

Plant DNA isolation:

Total genomic DNA from the parental varieties and blast isogenic lines was extracted according to the method of Dellaporta *et al.,* (1983) using fresh frozen leaf tissues. The quality and quantity of DNA was determined spectrophotometrically. The DNA of F_1, F_2 and F_2BC_1 individuals were extracted using 21 day-old seeding as per Zheng *et al.,* (1995).

Scoring of RAPD generated bands

The bands generated by random primers were scored by binary code using '0' for absence of band '1' for presence of band for each genotype.

Statistical analysis

Chi-square tests were performed to examine the godness of fit between expected Mendelian ratio an the STS based *Pi-1* gene marker and SAP based for *Pi-2* gene (data not shown), STATISTICA analysis package was used for cluster analysis (tree clustering) with row input data each population. The main parameter guiding the joining tree clustering process linkage rule was unweighted pair group average and the distance was computed using Euclidean distance (Fig. 7).

For calculation of comparataive efficiency of RAPD and conventional methods to identify genotypes more similar to reccurent parent genome for back crossing the following estimations were made :

a. To calculate the similarity of F_2 plants with Recurrent parent (RP)

$$\text{Similarity of plant with RP (\%)} = \frac{\text{Bands / no band marker levels produce by line}}{\text{Bands / no band marker levels produce by RP}} \times 100$$

b. To calculate recovery of RP in BC_1 generation

$$\text{Recovery of RP in } BC_1 \text{ (\%)} = \frac{\text{Similarity with RP (\%)}}{2} + 50$$

c. To calculate efficiency percent of RAPD marker over conventional prediction method

$$1. \text{ Efficiency (\%) over conventional method} = \frac{(\text{Percent recovery of RAPD in } BC_1 - 75)}{75} \times 100$$

$$2. \text{ Average efficiency percent for presence and absence of bands (\%)} = \frac{\text{Efficiency (presence of band) + efficiency (absence of band)}}{75}$$

PCR primers used for genetic analysis:

The sequences used, chromosome number on which the gene and marker are present and the source of these primers are presented in below.

Gene	Marker	Chro. #	Sequence of the primers	Reference
Pi-1	C481F	11	5'CTCCTTCTCCGACCGTGCTC3'	———
	C481R		5'CAATGATGTGCTTCTATGCT3'	
Pi-2	RG-64			
	431F	6	5'GTTGTTTCAGCTCTCCAATGCC TGTTC3'	Hittalmani *et al.* 1995a
	432R		3'GGACCGGCATGTAACGTGACG TC5'	

The marker linked to *Pi-1* gene is a sequence tagged site (STS) while for *Pl-2* it is a sequence amplicon polymorphism (SAP). The printers were synthesized by (OPERON

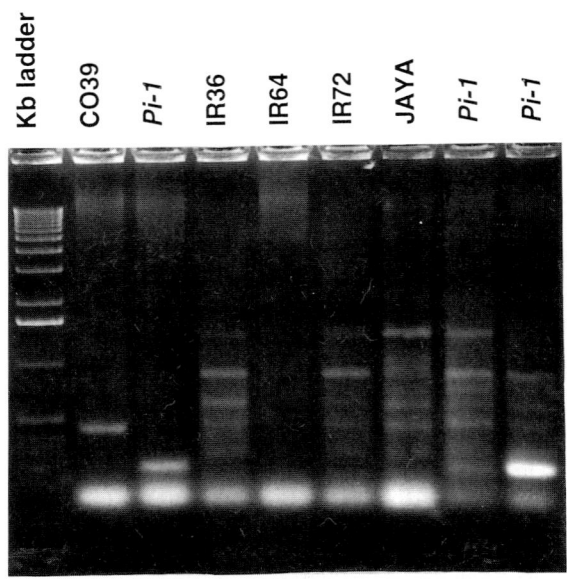

Figure 1. Parental survey for the presence of *Pi-1* gene using c481 markers in four rice varieties.

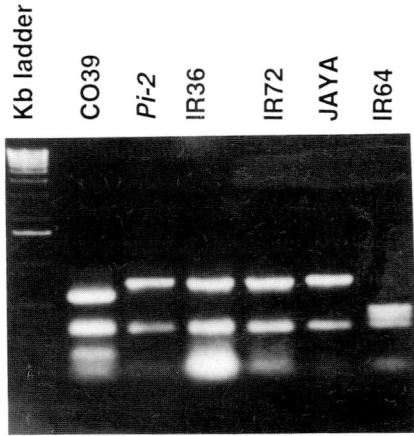

Figure 2. Parental survey of RG64 marker linked to *Pi-2* in four rice varieties.

Screening for presence of Pi-1 and Pi-2 genes in F₁ and F₂ generation

Screening for *Pi-1* gene using RG64 (SAP) IR64/*Pi-1+Pi-2+Pi-4* cross (Fig. 3) showed that plant 2 contained the resistant band similar to C101A51 parent while plant no. 1 did not have the resistant band. The screening for *Pi-1* gene by c481 STS primers in F_1

plants was done and an example is shown in Fig. 4. The segregation pattern of *Pi-1* and *Pi-2* gene using marker c481 and RG64 in all the F_2 populations was analyzed and is shown in Fig. 5 and Fig. 6. Chi-square test for co-dominant RG64 marker segregation in IR64/*Pi-1+Pi-2+Pi-4* and in F_2 population agreed with 1:2:1 ratio (data not shown).

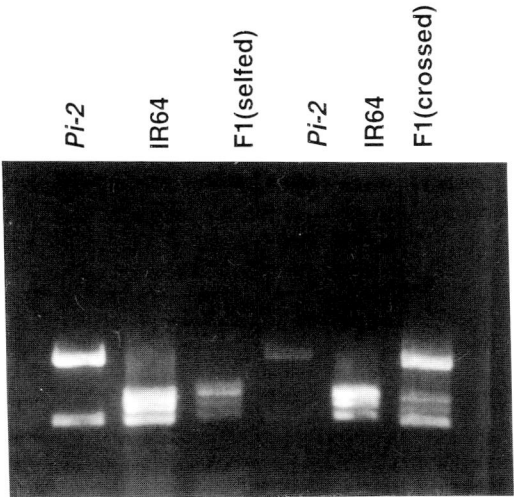

Figure 3. Detection of *Pi-2* gene F_1 of the cross IR64/ *Pi-1+Pi-2+Pi-4* using RG64 SAP marker.

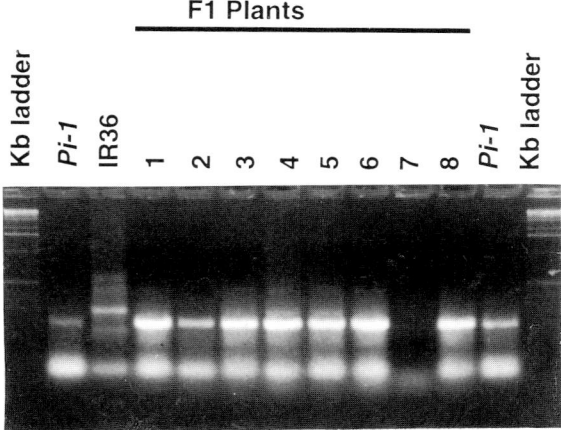

Figure 4. PCR banding pattern depicted by c481 marker for *Pi-1* gene in parents IR36/*Pi-1+Pi-2* and F_1s of the IR36/*Pi-1+Pi-2* cross

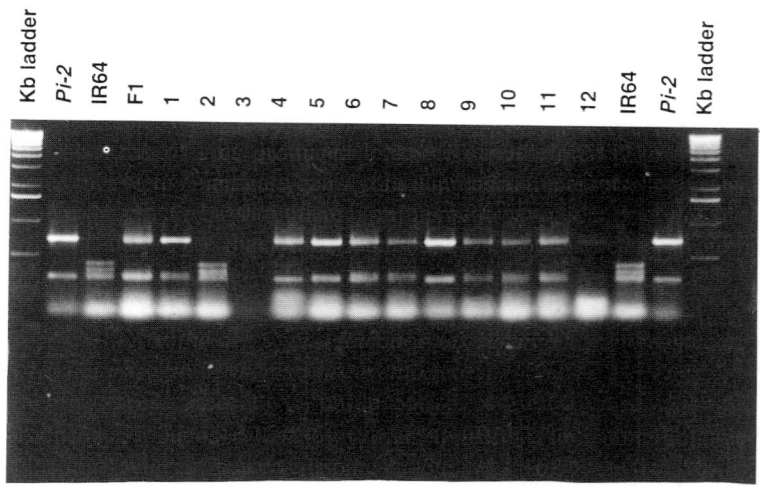

Figure 4. Segregation of RG64 marker in F_2 individuals of the cross between IR64/*Pi-1*+*Pi-2*+*Pi-4*.

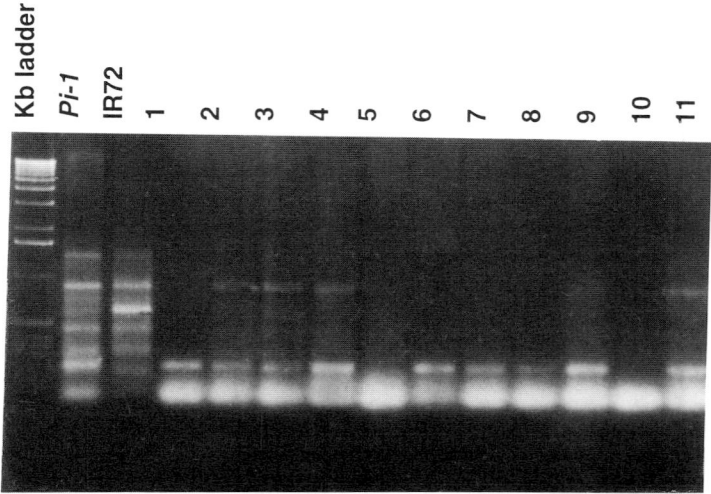

Figure 6. Segregation of c481 marker associated with *Pi-1* gene
in F_2 progenies of the cross IR72/*Pi-1*+*Pi-2*.

Identification of the resistant plants similar to that of the recurrent parent using RAPD markers

While the PCR marker closely linked to the resistant gene were used to identify the resistance gene *Pi-1* and *Pi-2* genes, the RAPD markers helped in identifying the plants that were genetically more closer to the recurrent parents, IR36, IR64 and IR72. Cluster analysis of the bands produced by the RAPD primers also revealed the grouping of plants that were genetically similar to the recurrent parents. The resistant plants in all the three crosss that nearer to the recurrent parents were identified for backcrossing. In IR36 cross all the F_2 resistant individuals clustered into one group while the male parent grouped into other group (Fig. 7).

Comparison of the selection efficiency for recurrent parent genome recovery between RAPD marker and conventional methods

The recurrent parent IR36 produced 76 bands (100%), while the F_2 plant no. 10 produced 69 bands (90.98%) similar to IR36. The maximum average efficiency of selecting recurrent parent genome in BC_1 was 16.5% in line 10. In the F_2 generation involving IR64 recurrent parent plant no. 1 produced 85% bands of 119 marker levels produced by RAPD primers using IR72 as recurrent parent, the maximum efficiency was 17.57% more than in conventional method for plant no. 3 (Table 1).

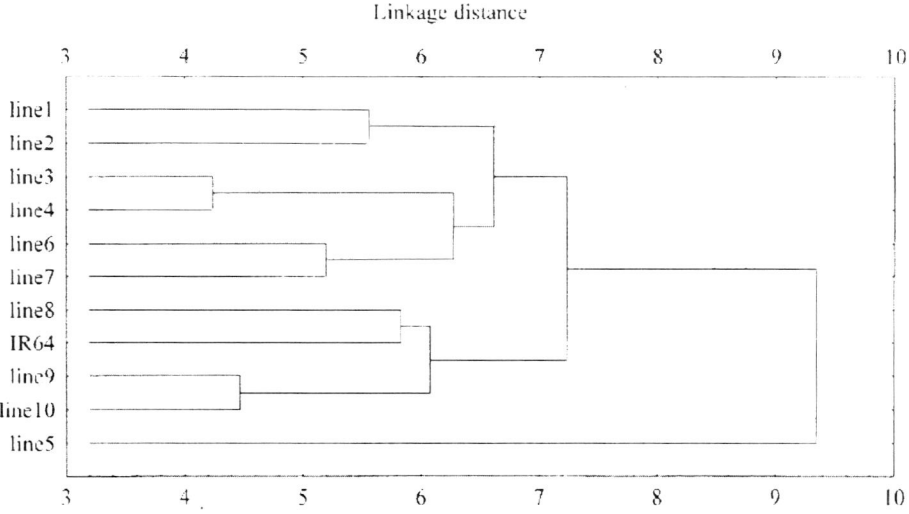

Figure 7. Tree diagram of 10 resistant F_2 lines with recurrent parent IR64
in IR64/*Pi-1*+*Pi-2*+*Pi-4* population

TABLE 1. Comparison of efficiency of selection of RP using RAPD and conventional methods in identification of Recurrent Parent genome in three crosses involving IR36, IR64 and IR72 recurrent parents.

Plant No.	Total no. of bands produced	Similarity with R.P.	Recovery of R.P. in BC1 (%)	Efficiency over conventional method (%)	Absence of band marker levels	Similarity with R.P. (%)	Recovery of R.P. in BC1 (%)	Efficiency over R.P in conventional method (%)	Average efficiency over conventional method
IR36	76	100.00	–	–	41	100.00	–	–	–
#1	57	75.00	87.50	16.66	21	51.21	75.60	0.81	8.73
#2	58	76.30	88.15	17.53	26	63.41	81.70	8.94	13.23
#3	65	85.52	92.76	23.68	20	48.70	74.35	-0.86	11.41
#4	58	76.30	88.15	17.53	27	65.85	82.92	10.56	14.04
#5	59	77.63	88.81	18.42	24	58.53	79.26	5.68	12.05
#6	58	76.30	88.15	17.53	22	53.65	76.82	2.43	9.98
#7	55	72.36	86.18	14.90	25	60.97	80.48	7.31	11.1
#8	61	80.26	90.13	20.17	26	63.41	81.70	8.94	14.55
#9	65	85.52	92.76	23.68	23	56.09	78.04	4.06	13.87
#10	69	90.98	95.49	27.32	24	58.53	72.26	5.69	16.5
IR64	140	100.00	–	–	55	100.00	–	–	–
#1	119	85.00	92.50	23.33	37	67.27	83.63	11.51	17.42
#2	112	80.00	90.00	20.00	35	63.63	81.81	9.09	14.54
#3	96	68.57	84.28	12.38	38	69.09	84.54	12.72	12.55
#4	100	71.42	85.71	14.28	40	72.72	86.36	15.14	14.71
#5	79	56.42	78.21	4.28	24	43.63	71.81	-4.24	0.02
#6	107	76.42	88.21	17.61	38	69.09	84.54	12.72	15.16
#7	113	80.71	90.35	20.47	41	74.54	87.27	16.36	18.415
#8	116	82.28	91.41	21.52	45	81.81	90.90	21.20	21.36
#9	118	84.28	92.14	22.85	43	78.18	89.09	18.78	20.81
#10	112	80.00	90.00	20.00	43	78.18	89.09	18.78	19.39
IR72	119	100.00	–	–	41	100.00	–	–	–
#1	106	89.07	94.50	26.00	13	31.70	65.85	-12.20	6.9
#2	109	91.59	95.79	27.72	17	41.46	70.73	-5.69	11.015
#3	115	96.63	98.32	31.08	23	56.09	78.05	4.06	17.57
#4	112	94.11	97.05	29.40	21	51.21	75.60	0.81	15.1
#5	110	92.24	96.12	28.16	18	43.90	71.95	-4.06	12.05
#6	109	91.59	95.79	27.72	20	48.78	74.39	-0.81	13.45
#7	111	93.27	96.63	28.84	16	39.02	69.51	-7.32	10.76
#8	110	92.43	96.21	28.28	19	46.34	73.17	-2.44	12.92
#9	108	90.75	95.37	27.16	21	51.21	75.60	0.80	13.98
#10	109	91.59	95.79	27.72	20	48.78	74.39	-0.81	13.45

R.P. = Recurrent Parent

4. Discussion

Resistance to rice blast is considered durable when it remains in cultivars despite wide sprad cultivation in an environment favoring the disease. In various pathosystems durable is generally controlled by single genes, multiple genes with cumulative effect, polygenes and also resistance is generally controlled by single genes, multiple genes with cumulative effect, polygenes and also resistance may be incomplete (Johnson, 1983, Parlerliet, 1988). *Pi-2* gene has broad spectrum of resistance and in combination with *Pi-1* gene has complementary resistance spectrum to most lineages (Chen *et al.,* 1995, 1996). The parental survey of the *Pi-1* gene using c481 marker revealed absence of the gene in the three varieties observed while the *Pi-2* gene was present in IR36 and IR72 varieties. This study also revealed that markers can effectively identify *Pi-1* and *Pi-2* genes separetely and also the F_1 individuals carrying the genes.

The three class segregation of 1:2:1 ratio and two class segregation of 3:1 for *Pi-2* and *Pi-1* gene marker respectively agreed with the Chi-suare test indicating the genes segregated in dominant fashion as already reported in similar studies for blast (Hittalmani *et al.,*) and for bacterial leaf blight (Abenes *et al.,* 1993 and Ready *et al.,* 1997). In this study it is evident that molecular markers helped in detection of one gene idependent of the other even though they are present in the same plant which is normally difficult by conventional methods. This has helped in a great way for pyramiding of the genes without pathotyping (Hittalmani *et al.,* manuscript in process) in which three major genes *Pi-1*, *Pi-2* and *Pi-4* were pyramided in different combinations. Of which *Pi-1/Pi-2* was very effective against Philippine blast races (Chen *et al.,* 1996) and Indian races (Srinivasachary and Shailaja Hittalmani, personnal communication). Similarly Yoshimura *et al.,* (1995) and Huang *et al.,* 1997 pyramided bacterial blight resistance genes and found enhanced resistance and can delay the breakdown (Chen *et al.,* 1995 and Zeigler *et al.,* 1994). The theoretical average of the recurrent parent genome in the F_2 population is 50% but variation exists from the means in individual plants towards either of the parents. Hospital *et al.,* (1992) opined that molecular markers and later among the resistant plants, the plants that were more closer to the recurrent parents as revealed by the similarity of the bands for the selection in backcrossing to the tune of 16.5% more than the average compared to the conventional methods. Stuber and Edwards (1986) in the study with maize using isozymes identified plants that produced 20% more grain than in unselected F_2 population. This would help in recovery of all the recurrent parental alleles by three backcrosses using markers.

This study reveals the power of the molecular markers in identifying the resistance genes in a new gentic background, identify more than one gene simultaneously as well as identify the plants taht are more similar to recurrent parent. The study demonstrates the efficient use of PCR based markers in identifying desired genotypes in segregating generations to be effective and quicker by reducing the number of breeding cycles.

5. Acknowledgment

The study was conducted as part of the project funded by The Rockefeller Foundation USA (RF95001 #321). We are thankful for the Rice Genome Project, Japan for supplying the sequence of the clone c481.

6. References

Abens M.L.P., Angeles E.P., Khush G.S., Huang W. (1993) Selection of bacterial blight resistant plants in F_2 generation via their linkage to molecular markers. *Rice Genet. Newslett.* **10**, 120-123

Chen D.H., Zeigler R.S., Ahn S.W., Nelson R.J. (1996) Phenotypic charachterization of the rice blast resistance gene *Pi-2(t)* . *Plant disease* **80** (1), 52-56

Chen D.H., Zeigler R.S., Leung H., Nelson R.J. (1995) Population structure of *Pyricularia grisea* at two screening sites in the Philippines. *Phytopathology* **85**, 1011-1020

Dellaporta S.L., Woods J., Hicks J.B. (1983) A plant DNA mini preparation *Plant Mol. Biol. Rep.* **1**, 19-21

Hittalmani S., Foolad M., Mew T., Rodriguez R.L., Huang N. (1995a) Identification of blast resistance gene *Pi-2(t)* in rice plants by flanking DNA markers. *Rice Gent. Newslett.* **11**, 144-146

Hittalmani S., Foolad M.R., Mew T., Rodriguez R., Huang N. (1994) Development of PCR based on markers to identify blast resistance gene, *Pi-2(t)* in a segregation population. *Theor. Appl. Genet.* **91**, 9-14.

Hittalmani S., Mew T.V., Khush G.S., Huang N. (1995b) DNA marker assisted breeding for disease resistance in rice. In: *Fragile lives in fragile ecosystems.* Proceedings of the international Rice Research Conference, 1317 Feb. 1995, Los Banos, Philippines. Manila (Philippines), International Rice Research Institute pp 949-961.

Hospital F., Cheralet C., Muolsant P. (1992) Using markers in gene introgression breeding programs *Genetics* **132**, 1199-1210.

Huang N., Angeles S.E.R., Domingo J., Maspantay G., Singh S., Bennett J., Khush G.S. (1997) Pyramiding of bacterial blight resistance genes via DNA marker-aided selection in rice. *Theor. Appl. Genet.* **95**, 313-315.

Johnson R., (1983) Genetic background of durable resistance pp 5-26 in Durable Resistance in crops, Edited by F. Lamberti, J.M. Waller and N.A. Vander Graff Plenum Press, New York.

Parlevliet J.E. (1988) Identification and evaluation of quantitative resistance by Leonard K.J. and Fry W.G. (eds) *Plant disease epidemiology, Genetics, Resistance and Management* **Vol. 2** McGraw-Hill, New York pp 377.

Reddy J.N., Baraoidan M.R., Bernardo M.A., George M.L.C; Sridhar R. (1997) Application of marker assisted selection in rice for bacterial blight resistance gene, *Xa-21 Curr. Sci* **73 (10)**, 873-875

Stuber C.W., Edwards M. (1986) Genotypic selection for improvement of quantitative traits in corn using molecular marker loci. *Proc. Annu. Seed Trade Assoc.* 41, 70-83.

Yoshimura S., Yoshimura A., Iwata N., McCouch S., Abenes M.L., Baraoidan M.R., Mew T., Nelson R.J. (1995) Tagging and combining bacterial blight resistance genes in rice using RAPD and RFLP markers. *Mol. Breed.* 1, 375-387

Zeigler R.S., Tohme J. Nelson J. Levy M., Correa F. (1994) Linking blast population analysis to resistance breeding : A proposal strategy for durable resistance pp 267-292 in: Rice blast Disease. R.S. Ziegler, S. Leong and P.S. Teng, eds CAB International, Wallinford, UK

ZhengK., Huang N., Bennett J., Khush G.S. (1995) Rapid micro-level extraction of DNA in rice, *IRRI discussion paper series* No 12 p. 24.

NEW TOOLS FOR RESISTANCE GENE CHARACTERISATION IN RICE.

D. THARREAU[1], E. ROUMEN[1], M. LORIEUX[2], A. PRICE[3], W. DIOH[4], A. GHESQUIERE[2], M.H. LEBRUN[5], J.L. NOTTEGHEM[6]
[1] CIRAD-CA, BP 5035, 34032 Montpellier Cedex 1, France
[2] IRD–LRGAPT, BP 5045, 34032 Montpellier Cedex 1, France
[3] University of Aberdeen, Aberdeen, UK
[4] Université Paris-Sud, 91405 Orsay, France
[5] CNRS-RPA UMR41, Physiologie végétale, 14/20 rue P. Baizet, 69263 Lyon Cedex 9, France
[6] ENSA.M, 2 place Viala, 34060 Montpellier Cedex 1

Abstract

Rice progeny used for the construction of genetic maps could be used to identify and map the resistance genes present in their parental cultivars. Using a large number of isolates avirulent on at least one of the two parents, such crosses facilitate the mapping of new blast resistance genes. We inoculated rice progeny from the cross IR64 x Azucena using different *Magnaporthe grisea* isolates with differential responses on the two parental cultivars. QTL mapping allowed the identification of 7 independent loci conferring resistance to these isolates that could correspond to specific 7 resistance genes. Some of these QTL/gene mapped at chromosomal locations where no resistance gene were previously located. We used a similar approach to map the resistance gene corresponding to *avr*1-Irat7, an avirulence gene recently cloned (MH Lebrun personal communication). The two strains, 2/0/3 virulent on Bala and Azucena, and PH14 virulent on IR64 and Azucena, were transformed with the cloned avirulence gene *avr*1-Irat7. For the cross Bala x Azucena, we inoculated on all progeny both 2/0/3 and its *avr*1-Irat7 transformant being avirulent Bala. For the cross IR64 x Azucena, we inoculated on all progeny both the isolate PH14 and its *avr*1-Irat7 transformant being avirulent on IR 64. The analysis of cosegregation between the resistance gene specific of *avr*1-Irat7 transformants and rice molecular markers indicates that it is located on chromosome 8 in both crosses.

1. Introduction.

Rice resistance genes to *Magnaporthe grisea*, the causal agent of the rice blast disease were first described in Japan by Sasaki (1922). Kiyosawa *et al.* (1984) described reference differential cultivars with one or two resistance genes named Pi- followed by a different letter for each gene. Some genes have several alleles, such as loci Pi-z and Pi-ta with two alleles and Pi-k with five alleles. When testing Japanese differentials

D. Tharreau et al. (eds.), Advances in Rice Blast Research, 54–62.
© 2000 *Kluwer Academic Publishers. Printed in the Netherlands.*

cultivars with exotic isolates having new avirulence genes, Imbe and Matsumoto (1985) discovered a new resistance gene, named Pi-sh. This suggest that uncharacterised resistance genes could be easily detected when using a large set of isolates from diverse geographic or genetic origins. Many comparative inoculations of reference differentials and resistant cultivars were carried out, demonstrating that resistance genes are present in almost all rice cultivars. In most cases, resistance patterns of tested varieties differed from those of reference resistant cultivars such as Japanese differentials. Therefore, new resistant genes or combinations of genes might be present in these cultivars. The characterisation of these new genes was frequently not completed since allelism tests were in general not carried out. Some resistance genes from tropical cultivars were introduced into isogenic lines using the susceptible cultivar Co39 as recurrent parent (Imbe et al., 1999). Allelism test showed that most of these introduced genes were allelic or identical to five known Pi genes. For each new rice cultivar, the identification of its resistance genes to blast is a long task. It requires the inoculation of a large set of differential isolates to compare its resistant pattern to those of reference cultivars with known resistance genes. Allelism tests must be also performed using progeny from crosses with reference cultivars having similar resistance patterns. Such a work is tedious when the number of resistance genes to be tested is increasing. As new strategies for resistance genes deployment are being tested (lineage exclusion), it is of importance to characterise the resistance gene diversity in rice. First, we examined how many different resistance genes were present in parents from the genetic maps crosses. Second we set up an efficient way to identify a resistance gene when the corresponding M. grisea avirulence gene was already cloned.

2. Material and Methods

2.1. RICE CULTIVARS AND PROGENY

We used the progeny from the two crosses IR64 x Azucena and Bala X Azucena. The cross IR64 x Azucena is a reference cross between an improved semi dwarf indica (IR64) obtained by IRRI and an upland japonica from The Philippines (Azucena). IR64 has a very complex genealogy including the following parents: Tetep, Tadukan, Peta, Taichung Native 1, CP-SLO, Chow Sung, BPI 76, NM S4, PTB 18, TKM6, GP 15, MUDGO, 17-1LT, W1263, Dee-geo-woo-gen, and O. nivara.. A population of doubled haploid lines was obtained by Guiderdoni et al. (1992) and used to map 200 molecular markers (Causse et al., 1994). The other cross Bala X Azucena was realised by A. Price who obtained and mapped 120 SSD (single seed descent) with the original objective of studying resistance to drought (A. Price personal communication).

2.2. RICE CULTIVATION

Rice plants were grown in a greenhouse in trays of 40 x 29 x 7 cm filled with compost (Motex compost n 7, Inter humus S.A., Lunel, France). Ten to 15 seeds of each DH line was sown in rows in trays containing 14 lines. Soil was kept moist with water and, once a week, with nutritive solution. Nitrogen fertilisation with 8.6 g of nitrogen equivalent was done at 10, 3, and 1 days before inoculation to increase susceptibility to blast. .

2.3. INOCULATION METHOD

Inoculations were performed by two methods: injection or spray of conidia suspensions. For the spray method, 30 ml of a 50,000 conidia.ml^{-1} suspension with 0.5% gelatin were sprayed on each tray. Then rice plants were stored for one night in a controlled climatic chamber at 24 C and more than 95% relative humidity. Then they were transferred back to the greenhouse. Each inoculation was performed on two series of DH lines sown at different times. For the injection method, plants were inoculated by injecting 0.1 ml of a 25,000 conidia.ml^{-1} suspension with a syringe into leaf sheaths. Two repetitions of parents and DH lines were grown for inoculation at different times. After seven days, lesions types on rice leaves were observed and scored 1 (resistant) to 6 (susceptible) according to a reference scale (Silué *et al.*, 1992). Progeny with scores between 1 and 3 were considered as resistant and progeny with scores from 4 to 6 were considered as susceptible. In order to use QTL mapping, the exact score of each progeny (1 to 6) was used as a quantitative variable.

2.4. ISOLATES

Isolates from the CIRAD plant pathology collection (1800 field isolates from 55 countries) were chosen to maximise their diversity according to their country of origin, year of collect, rice cultivar sampling. To identify isolates able to detect resistance gene in progeny, parents of the cross IR 64 X Azucena and 13 arbitrarily chosen DH lines were inoculated with 29 isolates from 20 countries representative of Latin America, Asia, and Africa. Isolates showing clear differential reactions on DH lines were then inoculated to the whole set of 105 available DH lines. To map the resistance gene corresponding to the avirulence gene *avr*1-Irat7, we used the following pair of isolates. The isolate 2/0/3 is a laboratory isolate obtained by crossing the two field isolates Guy11 and Ml25 (Silué *et al.*, 1992). 2/0/3 is virulent on cultivars Bala and Azucena. 2/0/3 was transformed with a 40 kb cosmid which confers avirulence on the cultivar IRAT7 and Bala (Dioh *et al.*, 1997). Isolate PH14 is a field isolate from IRRI that is virulent on IR64, Bala and Azucena. PH14 was transformed with a 40 kb cosmid which confers avirulence on the cultivar IR64 and Bala (Dioh *et al.*, 1999).

2.5. MOLECULAR MARKERS

Plant DNAs were isolated from lyophilised leaves using the CTAB method (Murray & Thompson 1980). In this study, a core map developed at IRRI (Huang *et al.*, 1994) made with RFLP, RAPD and isozyme markers was used. Regions of specific interest were saturated with the following protocols. For RFLP markers, we used probes from the saturated interspecific rice map (Causse *et al.*, 1994). Probes were kindly provided by Dr. S. McCouch (Cornell University, USA). Southern transfers, hybridisations and non-radioactive DNA labelling used for revelation of hybridisations, were done according to IRRI or CIMMYT protocols (Hoisington *et al.*, 1994). Six restriction enzymes were used for the core map: DraI, EcoRI, EcoRV, HindIII, ScaI, XbaI. Fourteen additional enzymes were used to test for polymorphism with probes showing a monomorphic pattern. For RAPD markers, PCR amplifications were carried out in 25μl, with the

Table 1. Loci identified for resistance to six blast isolates by a QTL approach in IR 64 x Azucena cross.
In regular font: *LOD score*, in italic: % of explained variance by the locus

chr.	markers	Map position (cM)	Blast isolates						Possible correspondance with known loci
			br26	ph68	cd69	ch66	ch72	cl6	
1	rg173	147.4						2.9	
	rg532	177.8						24.1 (a)	
2	rg520	0	5.8 20.5 (i)			3.0 11.6 (i)	2.3 9.5 (i)		
2	rz123	16.6	5.7						
	rg654	27.4	18.8 (i)						
5	rz70	132.8		3.1					
	rz225	152.5		15.2 (a)					
6	rg213	20.7				3.0 13.8 (i)			Pi-2(t) or Pi-9(t)
8	g104	34.6						6.1	Pi-11(t)
	rz617	41.6						22.2 (i)	
11	r g 1094	100.2				4.8	4.2		Pi-a
	rg118	111.8				21.6 (i)	20.2 (i)		
12	o10	73.5		10.8					Pi 4(t)
	rg341	76.4		37.0 (i)					Pi–ta
12	rg869	76.4	6.1		25.0				Pi 20(t)
	rz816	103.2	23.2 (i)		90.0 (i)				

chr. : chromosome number
(i): resistance locus brought by IR64
(a): resistance locus brought by Azucena

following reagents concentrations: primer 0.4 µM, Taq polymerase (Appligene) 0.02 U/µl, buffer mix (Appligene) 1X, dNTP 150 µM. Conditions were 95 °C -5 min; 45 x (95 °C -1 min; 35 °C -1 min; 72 °C -2 min); 72 °C -6 min. Migrations of PCR products were carried out on 1.5 % agarose gels and stained with ethidium bromide. STSs published by Inoue *et al.* (1994) were also used. These primers correspond to probes of the saturated map of the Rice Genome Research Program – Japan (Kurata *et al.*, 1994). PCR conditions were the same as in Inoue *et al.* (1994). When polymorphism between parents was not detected directly, digestion of amplification products was done with 4bp restriction enzymes. Migrations of PCR/digestion products were carried out on 2-3% agarose or 8% polyacrylamide gels and stained with ethidium bromide.

2.6. MAP CONSTRUCTION

The map was constructed using the multipoint functions performed by MapMaker 3.0 (Lander *et al.*, 1987). The two-point LOD score threshold was equal to 5, and r_{max} to 0.3. Ordering of markers was done using the 'order' and 'try' commands, which calculate likelihood ratios for the different possible multipoint orders. Conversion of recombination fractions into centimorgans (cM) was obtained with Kosambi's mapping function (Kosambi, 1944).

2.7. COSEGREGATION ANALYSIS BETWEEN MARKERS AND RESISTANCE TRAIT

A QTL detection approach was used in order to localise loci involved in the resistance to different *M. grisea* isolates. The interval mapping method (Lander & Botstein, 1989) was used, and results were confirmed by the distribution-free Kruskall & Wallis test. All calculations were done using MapQTL for Unix v.2.4 (Van Ooijen, 1992).

3. Results

3.1. DIVERSITY OF ISOLATES FOR THEIR AVIRULENCE

Inoculation of the 29 *M. grisea* isolates on parents and on a subset of 13 DH lines from the cross IR64 x Azucena revealed an important variability in their resistance patterns. We did not detect isolates virulent on both parents. Seventeen isolates were avirulent on both parents. Among them, seven were avirulent on all 13 DH lines and 10 were virulent on some of the DH lines. The twelve remaining isolates were avirulent on one of the two parents and virulent on the other. These 12 isolates were virulent to some of the DH lines. All isolates differ for their virulence to the subset of isogenic lines tested. These results strongly suggest the existence of a combination of avirulence genes in *M. grisea* isolates and of the corresponding resistance genes in the DH lines. Among the 29 isolates tested, six were kept for further studies as they cause clear susceptible reactions on at least 25% of DH lines. Such proportion of susceptible DH lines could correspond to the segregation of at most two resistance genes. These six isolates were inoculated on the 105 available DH lines.

3.2. RESISTANCE GENES/QTL MAPPED IN THE CROSS IR64 X AZUCENA

We mapped the resistance genes revealed by the 6 tested *M. grisea* isolates, using the scoring of each progeny for its resistance on a 1 to 6 scale, as a quantitative variable for QTL mapping. We considered that resistance is dominant and epistasic on suceptibility and that resistance phenotypes provided by any resistance gene give the same score. We also assumed that resistance genes mapping at different chromosomal locations were different and that resistance genes mapping on the same chromosomal locations could be allelic. If we consider resistance QTLs with a LOD score higher than 2,5, we detected at least 7 isolate-specific resistance QTL (table 1). We may consider such a QTL as a **putative Resistance Gene** (pRG). Each isolate reveal 1 (CD69) to 3 pRGs. Each pRG is efficient against 1 (5 pRG) to 3 (1 pRG) isolates. Five pRG are present in the indica cultivar IR64 and 2 pRG are present in the japonica cultivar Azucena. The 2 pRG from Azucena mapped on chromosome 1 and 5 in chromosome regions where there is no known resistance gene. The 5 pRGs from IR 64 mapped in chromosomal locations where known resistance genes were previously mapped. The simplest situation is the resistance to isolate CD 69 that is controlled by only one pRg located on chromosome 12 near Pi-ta or Pi 20. In this case, the LOD score is very high (25) and this pRg explains 90% of the resistance. Therefore, it is likely that the resistance of IR64 to *M. grisea* isolate CD69 is controlled by one dominant gene which could be Pi-ta or Pi 20.

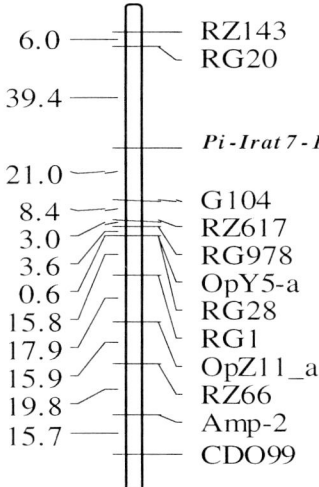

Fig 1: Localization of Pi-Irat7-1 on chromosome 8 using progeny of cross IR64 x Azucena

3.3. MAPPING THE RESISTANCE GENE CORRESPONDING TO THE AVIRULENCE GENE *avr1-Irat7*

We inoculated IR64, Azucena, Bala, progeny of crosses IR64 x Azucena and Bala x Azucena with 2 pairs of *M. grisea* isolates. Isolate 2/0/3 which is avirulent on IR 64 but virulent on Bala and Azucena and its transformant 2/0/3-T carrying *avr1-Irat7* (Lebrun M-H, personal communication) which confers avirulence to Bala were both inoculated on progeny from cross Bala x Azucena. Isolate PH14 which is virulent on IR64 and Azucena and its transformant PH14-T carrying *avr1-Irat7* (Lebrun M-H, personal communication) which confers avirulence to IR64 were both inoculated on progeny from cross IR64 x Azucena. In both crosses, we scored only progeny for which we detected changes from suceptibility to resistance when inoculated with the *avr1-Irat7* transformant compared to wild-type. We observed a segregation 40 R:40 S in the cross IR64 x Azucena and 37 R:47 S in the cross Bala X Azucena.

For these two crosses, the segregating resistance gene was located at the same position on Chromosome 8 at a locus where the resistance gene Pi-11(t) was previously mapped by Zhu *et al.* (1993). We can conclude that the gene present in IR64, Bala and IRAT 7 and conferring resistance to isolates 2/0/3-T or PH14-T is Pi-11(t) or another resistance gene closely linked to Pi-11(t).

4. Discussion and conclusion.

DH lines progeny are a very useful tool for the characterisation of resistance genes from the 2 parents of the cross. As more DH lines are being produced and more markers being mapped using the cross IR64 x Azucena, a more accurate characterisation of the existing pRGs will be possible. Allelism test between reference cultivars with known resistance genes and some DH lines having only one pRG will be necessary to demonstrate that pRGs are new genes or not. When we inoculated a limited set of 13 DH lines from this cross with 29 *M. grisea* isolates, each isolate gave a different resistance pattern. Such an apparent complex situation may only reflect the different combinations of at least 7 segregating resistance genes in the progeny. The presence of five pRGs in IR 64 is unexpected, since most rice cultivars seemed to have one to two known resistance gene. Such a unplanned pyramiding of resistance gene was only found in rice cultivar Hama Asahi (Kiyosawa *et al.*, 1991). We may have not identified all pRGs present in the 2 parents, using six isolates and additional isolates could probably reveal others pRGs. The origin of the 5 pRGS from IR64 is not known since the genealogy of this cultivar involves different resistant parental cultivars. Progeny of such crosses were tested in IRBN nurseries (with susceptible spreaders and a natural local population of *M. grisea*). This assay of resistance may have led to the selection of progeny that had accumulated different resistance genes. IR 64 is resistant in many countries in irrigated conditions and in Asia it has a good level of resistance even when virulent isolates are present. The 5 pRGs and probably other resistance genes still unidentified present in IR64 could have an important role in the resistance of this cultivar in the field. Finally, mapping resistance genes using a set of isolates as diverse as possible for their avirulence or isolates modified by transformation with known avirulence genes are powerful tools for the characterisation of resistance genes..

Acknowledgements

We thank Dr. N. Huang (IRRI), for having kindly providing us with the core map data.

Abbreviations
CAPS: Cleaved Amplified Polymorphic Sequence
QTL: Quantitative Trait Locus
RAPD: Random Amplified Polymorphic DNA
RFLP: Restriction Fragment Length Polymorphism
STS: Sequence-Tagged Site

References

Causse, M.A., T.M. Fulton, Y.G. Cho, S.N. Ahn, J. Chunwongse, K. Wu, J. Xiao, Z. Yu, P.C. Ronald, S.E. Harrington, G. Second, S.R. McCouch & S.D. Tanksley, 1994. Saturated molecular map of the rice genome based on an interspecific backcross population. Genetics 138: 1251-1274.

Dioh W. , Tharreau D., Nottéghem J.L., Orbach M., and Lebrun MHL 1999. Mapping of avirulence genes inthe rice blast fungus, *Magnaporthe grisea* using RFLP and RAPD markers. Accepted for publication in Molecular Plant-Microbe Interactions.

Guiderdoni, E., E. Galinato, J. Luistro & G. Vergara, 1992. Anther culture of tropical japonica x indica hybrids of rice (*Oryza sativa* L.). Euphytica 62: 219-224.

Hoisington, D., M. Khairallah & D. González-de-León, 1994. Laboratory Protocols: CIMMYT Applied Molecular Genetics Laboratory. Second Edition. Mexico, D.F.: CIMMYT.

Huang, N., S.R. McCouch, T. Mew, A. Parco & E. Guiderdoni, 1994. Development of an RFLP map from a doubled haploid population in rice. Rice Genet. News. 11: 134-137.

Imbe T., Matsumoto S., 1985, Inheritance of resistance of rice varieties to the blast fungus strains virulent to the variety « Reheio», Jap. J. Breed. 35:332-339.

Imbe T., Tsunematsu H., KatoH ., Kush G., 1999, Genetic analysis of blast resistance in IR varieties and resistance breeding strategy (this volume).

Inoue, T., H.S. Zhong, A. Miyao, I. Ashikawa, L. Monna, S. Fukuoka, N. Miyadera, Y. Nagamura, N. Kurata, T. Sasaki & Y. Minobe, 1994. Sequence-tagged sites (STSs) as standard landmarkers in the rice genome. Theor. Appl. Genet. 89: 728-734.

Kosambi, D.D., 1944. The estimation of map distance from recombination values. Ann. Eug. 12: 172-175.

Kiyosawa S.1984. Establishment of differential rice varieties for pathogenicity test of trice blast fungus. Rice Genetics Newsletter 1,53-67

Kiyosawa S., Ando I, Ishigura K, Nemoto F., Hashimoto A., 1991 Three exemples of yearly changes of unnnecessary virulende genes under field conditions where resistance genes for rice blast are nor present. Bull. Fukushima Prefect. Exp. Stn. 30:23-36

Kurata, N., Y. Nagamura, K. Yamamoto, Y. Harushima, N. Sue, J. Wu, B.A. Antonio, A. Shomura, T. Shimizu, S.-Y. Lin, T. Inoue, A. Fukuda, T. Shimano, Y. Kuboki, T. Toyama, Y. Miyamoto, T. Kirihara, K.

Hayasaka, A. Miyao, L. Monna, H.S. Zhong, Y. Tamura, Z.-X. Wang, T. Momma, Y. Umehara, M. Yano, T. Sasaki & Y. Minobe, 1994. A 300 kilobase interval genetic map of rice including 883 expressed sequences. Nature Genetics 8: 365-372.

Lander, E.S. and D. Botstein, 1989. Mapping mendelian factors underlying quantitative traits using RFLP linkage maps. Genetics 121:185-199.

Lander, E.S., P. Green, J. Abrahamson, A. Barlow, M.J. Daly, S.E. Lincoln and L. Newburg, 1987. Mapmaker: an interactive computer package for constructing primary genetic linkage maps of experimental and natural populations. Genomics 1: 174-181.

Murray, M.G. & W.F. Thompson, 1980. Rapid isolation of high molecular weight plant DNA. Nucleic Acids Res. 8: 4321-4325.

Sasaki 1922. Inheritance of resistance to *Pyricularia oryzae* in different varieties of rice. Japanese genetics, Japan, 1,81-85 (in japanese)

Silue D., Notteghem J.L. and Tharreau D. 1992. Evidence of a gene-for-gene relationships in the pathosystem *Oryza sativa-Magnaporthe grisea*. Phytopathology 82: 577-580

Van Ooijen, J.W., 1992. Accuracy of mapping quantitative trait loci in autogamous species. Theor. Appl. Genet. 84: 803-811.

Zhu L.H., Chen Y., Xu Y.B., Xu J.C., Cai H.W., Ling Z.Z., 1993. Construction of a molecular map using a double haploid population of a cross between indica and japonica varieties. Rice Genetics Newsletter: 10:132-135 .

GENETIC DISSECTION OF DISEASE RESISTANCE PATHWAY(S) IN RICE

WANG G.L.[1], HE C.Z.[1], WU C.J.[2], YIN Z.C.[1], BARAUIDAN M.[2], RONALD P.C.[3], KHUSH G.S[2]. AND LEUNG H[2].

[1]*The Institute of Molecular Agrobiology, The National University of Singapore, 1 Research Link, 117604 Singapore*
[2] *The International Rice Research Institute, PO Box 933, 1099, Manila Philippines.*
[3] *Department of Plant Pathology, University of California, Davis, 1 Shield Ave, CA 95616 USA.*

1. INTRODUCTION

Most plants have developed sophisticated mechanism to protect themselves from microbial attack in their natural environment. One of the effective way is through the gene-for-gene interactions in which, a specific resistance gene in plants recognizes a corresponding avirulence (*avr*) gene of the pathogen. Subsequently, this recognition process activates a cascade of defense genes that inhibit the pathogen's ingress (Keen, 1990; Staskawicz *et al.*, 1995). During the last 5 years, over 20 disease resistance genes from different plants have been cloned (Staskawicz *et al* 1995; Baker *et al.* 1997; Ellis and Jones, 1998). This has dramatically advanced our understanding of the molecular basis of disease resistance in plants. Sequence analysis of the predicted proteins reveals that resistance genes of diverse origin and pathogen specificity share similar structural motifs such leucine-rich repeats (LRR), kinase and nucleotide binding site (NBS) (reviewed by Baker et al., 1997). The structural similarity of different resistance genes leads us to speculating the existence of a common or few resistance pathways in plants. If this is true, one might ask how many genes are required in the signal transduction pathway before defense genes are activated. The second question is what the function of each gene is in switching on the signaling pathway.

Genetic and molecular approaches to identify genes involved in the resistance gene signal transduction was first initiated in *Arabidopsis*, tomato and barley. Three strategies have been used: screening for mutants affecting resistance phenotype, screening for mutants affecting specific defense response and yeast two-hybrid screening for proteins interacting with the cloned genes (Innes, 1998). In the last few years, all these three approaches have made significant progress, providing invaluable information towards the understanding of the resistance signal transduction pathways. For example, a gene required for the *Pto*-mediated

D. Tharreau et al. (eds.), Advances in Rice Blast Research, 63–72.

resistance, *Prf*, was identified in a mutagenesis screen (Salmeron *et al.*, 1994). Interestingly, *Prf* encodes an NBS/LRR type protein, a characteristic feature of several disease resistance genes (Salmeron *et al.*, 1996). Using *Pto* as a bait in the yeast two-hybrid screen of a tomato cDNA library, a protein kinase (*Pti1*) and three putative transcription factors (*Pti4, 5* and *6*) have been found to interact specifically with *Pto* (Zhou *et al.*, 1995; 1997). Recently, two novel genes, *NDR1* and *EDS1*, have been cloned in *Arabidopsis* which are signal transduction components shared by multiple resistance genes and encode possible membrane protein and lipase, respectively (Century et al., 1997; Innes, 1998). Due to the difference in their gene structure, it appears that these two genes may play different roles in the resistance signal transduction pathways (Innes, 1998).

To investigate the role of hypersensitive response (HR) in the activation of disease resistance, lesion mimic mutants have been characterized and the genes responsible for the mutations have been cloned (Dietrich *et al.*, 1994; Greenberg *et al.*, 1994; Dangl *et al.*, 1996; Dietrich *et al.*, 1997; Gray *et al.*, 1997). For instance, the *Arabidopsis* lesion mimic mutant, *lsd1*, was isolated by a map-based cloning strategy (Dietrich *et al.*, 1997). The predicted *LSD1* protein contains three zinc finger domains. It is possible that *LSD1* regulates transcription via either repression of a pro-death pathway or activation of an "anti-death" pathway, in response to signals emanating from cells undergoing pathogen-induced hypersensitive cell death. The non-race specific resistance gene *mlo*, in barley has been also cloned which exhibits a spontaneous cell death phenotype under pathogen-free conditions (Buschges *et al.*, 1997). The gene encodes a 60 kDa protein which is predicted to be membrane-anchored by at least six membrane-spanning helices. Isolation of these genes has provided us with important information on the induction and biochemistry of the HR.

The information generated from *Arabidopsis* and other crops can certainly provide insight on the mechanism of disease resistance signaling, but its direct application in rice disease control may not be possible. Since 1996, we have initiated two projects aiming to isolate and characterize mutants in rice that are required for the resistance pathway(s) specific to bacterial blight and blast or common to both the diseases. One project deals with the identification of mutants required for the *Xa21* resistance using the mutagenesis approach. The second project involves the characterization of 8 lesion mimic mutants in order to investigate whether some of them have enhanced resistance to blast and bacterial blight diseases due to the mutation. Genetic analysis and cloning of genes involved in the *Xa21* resistance pathway and genes responsible for the lesion mimic mutations will not only lead to identification of genetic components controlling disease resistance and cell death, but also allows us to manipulate these non-pathogen specific genes to achieve broad spectrum resistance in rice.

2. IDENTIFICATION OF MUTANTS REQUIRED FOR THE *XA21* DISEASE RESISTANCE

2.1. GENERATION AND SCREENING FOR MUTANTS ALTERED IN THE RESISTANCE TO *XANTHOMONAS ORYZAE* PV. *ORYZAE (XOO)*

The cloned gene *Xa21* is a member of a small multigene family with at least 7 members. Most of these family members are linked suggesting that *Xa21* is part of a complex locus (Ronald *et al.* 1992; Song *et al.* 1995). When compared with other cloned plant resistance genes, the structure of *Xa21* represent a previously uncharacterized class. The deduced amino acid sequence of *Xa21* encodes a receptor kinase like protein carrying leucine rich repeats (LRRs) in the putative extracellular domain, a single pass transmembrane domain and a serine threonine kinase intracellular domain. To determine if the multi-isolate resistance observed in line IRBB21 is due to a single gene or multiple genes at the *Xa21* locus, transgenic plants expressing the cloned *Xa21* gene were inoculated with 29 diverse isolates from 8 countries. It was found that the plants were resistant to 29 isolates indicating that the single cloned gene was sufficient to confer multi-isolate resistance.

The *Xa21* gene has been used to engineer several susceptible cultivars into highly resistant ones in several countries (Tu et al., 1998, L. Zhu, personal communication). This makes us speculate that the genes required for the *Xa21* resistance are conserved in all rice cultivars. Isolation of these genes should provide us information on how the *Xa21* gene interacts with other genes and these interactions activate the defense pathway in rice. Manipulation of these genes in transgenic rice may produce novel and broad-spectrum resistance against pathogens. In this paper, we report the preliminary results on screening for mutants in the *Xa21* resistance pathway using a mutagenesis approach.

Diepoxybutane (DEB) has been reported to cause deletion mutations at high frequency in *Drosophila* (Reardon *et al.*, 1987) and has been used to generate mutants in several plant species (Salmeron *et al.*, 1994). DEB primarily causes deletions of less than 250 base pairs although deletions of up to 8 kb have also been observed (Reardon *et al.*, 1987). Fast neutrons (FN) induce both base pair changes and deletions. Salmeron *et al.* (1994; 1996) isolated several *Pto* mutants using DEB and FN as mutagens and cloned the *Prf* gene which is a common component for the transduction of signals for *Pto*-mediated resistance to *Pseudomonas syringae* pv. *tomato.* as well as for sensitivity to the insecticide Fenthion. To identify the genes required for the function of *Xa21* resistance to *Xoo*, Pam Ronald, Susan McCouch and Ning Huang (1994-95) have generated about 4500 M2 families from IRBB21 (*Xa21* donor) with DEB (3500 families) and FN (1000 families).

In November 1996, we planted all 4500 M2 families at IRRI, Philippines. About 10 seeds of each family were sown. Philippine *Xoo* race 6 (PXO99) was used to inoculate the two-month old plants. Plants were scored two weeks after inoculation. Thirty-two susceptible plants (lesion length more than 8 cm) from 29 families were identified and transplanted into large pots to produce M3 seeds. M3 plants were grown at both IRRI and IMA for re-inoculation with race 6 to confirm its susceptibility. We confirmed that 7 lines were highly susceptible (>20 cm), 17 lines were partially susceptible (6-20 cm) and 5 lines were as resistant as the *Xa21* donor line IRBB21 (<5.0 cm).

2.2. MOLECULAR ANALYSIS OF THE MUTANTS IN THE *XA21* LOCUS

Both PCR and Southern hybridization methods were used to detect if there are any deletions or re-arrangements in the *Xa21* coding region. Two pairs of specific primers in the LRR region and two pairs of specific primers in the kinase region of the *Xa21* gene were designed. It was found that some mutants such as 3453, 3469, 3954 and N20-247 lost both LRR and kinase DNA fragments. Some lines such as N18-116 and N18-238 have re-arrangements in the kinase region.

For Southern analysis, DNA was digested with the restriction enzyme HindIII. Amplified fragments from the *Xa21* LRR and kinase regions were used as probes. As shown in Figure 1, six FN-induced mutants had similar hybridization pattern with three missing LRR-hybridizing bands. Three DEB-induced lines had deletions in the *Xa21* gene when the LRR fragment was used as a probe. Line 3453 had lost three bands and the size of the remaining bands were different from that of the *Xa21* donor line IRBB21. Lines 3469 and 3954 lost 7 bands and had the same hybridization pattern as the recurrent parent IR24. When the kinase fragment was used in Southern hybridization, all 9 mutants with deletion in the LRR region showed only one hybridizing band. Surprisingly, all these 9 lines which had deletions in the *Xa21* gene cluster were highly susceptible to *Xoo*. Mutants with partial resistance to race 6 had no apparent deletion in the *Xa21* locus which was detected using both PCR and Southern analysis methods. These results suggested that genes required for the *Xa21* resistance might be redundant or branched, since all the mutants without deletion in the *Xa21* locus were partially resistant to Xoo and this is consistent with the findings in other plants (Innes, 1998).

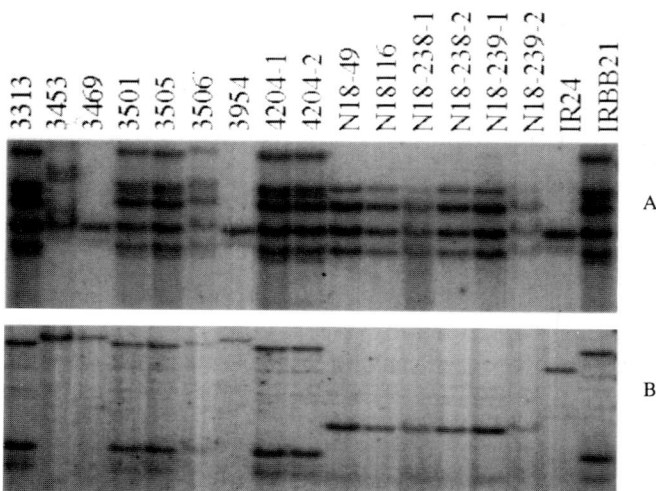

Figure 1. Southern blot analysis of representative mutants required by the Xa21 mediated resistance. Total genomic DNA extracted from rice leave was digested with restriction enzyme HindIII and hybridized with the amplified fragments of the Xa21 LRR (A) and kinase (B) domains.

2.3. MUTANTS ALTERED RESISTANCE SPECIFICITY TO BOTH BACTERIAL BLIGHT AND BLAST FUNGUS

To investigate if the resistance pathway mediated by the *Xa21* gene is converged or shared with those required by the resistance genes to blast, all mutants were inoculated with 4 Philippine blast isolates belonging to 4 different genetic lineages. Most of lines showed no difference in their resistance specificity to all the isolates. Only two lines (3453 and 3469) were found to be susceptible to isolate CBN9219-25 which was incompatible to IRBB21. As discussed above, these two lines have deletions in the *Xa21* locus. Whether the *Xa21* gene confers dual resistance function (to bacterial blight and blast) or whether a blast resistance gene is located within the cluster is under investigation.

3. CHARACTERIZATION OF EIGHT LESION MIMIC MUTANTS IN RICE
3.1. LESION MIMIC MUTANTS IN RICE

Eight spotted leaf (*spl*) mutants (*spl1, spl2, spl3, spl4, spl5, spl6, spl9* and *spl11*) were chosen to study the relationship between cell-death like lesion mimics and disease resistance to pathogens. With the exception of *spl11*, which is an EMS mutant from IR68 , the other seven *spl* mutants are near-isogenic lines in IR36 background (BC3F3) (Khush *et al.* unpublished; Singh *et al.*, 1995). These mutants show different types of mimic blast lesions on leaves at different growth stages. The gene responsible for these mutations have been mapped on different rice chromosomes (Table 1). Interestingly, three *spl* genes are located on chromosome 7 where *spl5* and *spl11* are closely linked (Singh *et al.*, 1995). These mutants have different lesion phenotypes. For instance, *spl1* mutant has large lesions on leaves which are similar to the type 4-5 susceptible blast lesions. As reported by Marchetti *et al.* (1983), we observed that the lesions (or called Sekiguchi lesion) in the *spl11* plants were first visible two weeks after sowing as 1-2 mm diameter spots. Several brown spots were visible in a fully-expanded leaf which enlarged rapidly until the entire leaf collapsed. Except *spl2* mutants which have big stripe and while-color lesions, the other 6 mutants showed small/medium and brown-color lesions similar to the resistant lesion types (Table 1). As compared with the recurrent parent IR36, some *spl* mutations affect plant growth and flowering. For example, the *spl1, spl2* and *spl5* plants are short and flower at least two weeks earlier than IR36. *spl4* plants are almost sterile and have an abnormal seed shape. *spl11* plants flower about one week later than the wildtype plants IR68.

3.2. REACTIONS OF LESION MIMIC MUTANTS TO BLAST INOCULATION

A common feature of plant defense responses is the rapid formation of brown lesions at the infection site or hypersensitive reaction, which is assumed to confine pathogen growth within the collapsed tissue. Since most of these *spl* mutants show the characteristics of resistant reaction after blast infection, we reason that the mutations may occur in the pathway leading to hypersensitive reaction and/or resistance response. To assess whether the eight rice lesion mimic mutants are similar to the lesion mimic mutants observed in *Arabidopsis* and barley

(Buschges *et al., 1997*; Dietrich *et al., 1997*), we inoculated them with the blast isolate PO6-6 at the seedling stage (21 days

TABLE 1. Chromosomal location and lesion characteristics of the 8 lesion mimic mutants

Mutant	Lesion type	Chromosome	Reference
spl1	big, few, brown	12	Khush *et al.*, 1984
spl2	big, few, white	2	Kinoshita, 1995
spl3	small, many, brown	3	Kinoshita, 1995
spl4	small, many, brown	6	Kinoshita, 1995
spl5	small, many, brown	7	Kinoshita, 1995
spl6	small, few, brown	1	Kinoshita, 1995
spl7	small, few, brown	5	Kinoshita, 1995
spl9	small, few, brown	7	Sanchez and Khush, 1994
spl11	medium, many, brown	7	Singh *et al.*, 1995

after sowing). We found that three mutants in the IR36 background (*spl1*, *spl5*, *spl7* and *spl9*) showed smaller or no lesions seven days after inoculation as compared with the susceptible recurrent parent IR36. This is similar to the cloned *LSD1* gene in *Arabidopsis*. Since lesion mimics are seen in the first or second leaf of mutant plants before inoculation, it is possible that an early initiation of lesion development may have triggered the defense response in the plants, resulting in enhanced resistance against blast. Except lesion mimics, no disease lesions were detected in *spl11* plants. Since the wild type plants (IR68) are resistant to most of the isolates, a genetic approach has been designed to eliminate the endogenous resistance gene in the *spl11* plants (see next section).

It has been previously shown that mutants or recombinant alleles at the maize *Rp1* locus exhibit a lesion mimic phenotype and lesion formation is triggered by rust or other biotic stimuli (Hu *et al.*, 1996). In our experiments with isolate PO6-6 inoculation, we found that the lesion mimics increased dramatically in plants of *spl1*, *spl5*, *spl7*, *spl9* and *spl11* compared with that in control plants (spraying water only). Similar observation was made when *spl11* plants were inoculated with another 9 isolates, indicating that lesion mimic induction by blast inoculation was non-isolate specific. Whether the molecules are secreted by the fungus or whether appressorium penetration triggers the lesion formation requires further investigation.

3.3. MUTANT *spl11* IS RESISTANT TO BOTH BACTERIAL BLIGHT AND RICE BLAST FUNGUS

To eliminate endogenous resistance genes in the *spl11* mutant, a cross was made between the mutant and CO39, a parental line highly susceptible to bacterial blight and rice blast. In the F2 population, non-lesion mimic plants susceptible to both blast and bacterial blight were selected and selfed to produce F3 families. Plants in two F3 families which segregated for the lesion mimics were advanced into F4 families for phenotype analysis. Ten F4 families homozygous for the lesion mimics or homozygous for the wild type were used for blast and bacterial inoculation experiments. Ten plants of each family were used for inoculation with the isolate PO6-6. No disease lesions were seen in the lesion mimics plants, whereas the plants without lesion mimics were highly susceptible similar to the susceptible control CO39. Two month-old plants were inoculated with the bacterial blight strain race 6 and the lesion length was scored two weeks after inoculation. The average lesion length in lesion mimic plants was about 8-10 cm which was much shorter than that in plants without lesion mimics (20-25 cm). These results indicate that the *spl11* mutants confer enhanced resistance to both blast and bacterial blight.

3.4. Three defense related genes are constitutively expressed in lesion mimic mutants

Several classes of defense related genes are induced during the resistance response to pathogens which are believed to contribute to limiting pathogen growth (Lamb *et al.,* 1992). Dietrich *et al.* (1994) reported that six *Arabidopsis* lesion mimic mutants constitutively express biochemical markers of plant defense responses, which is correlated with enhanced resistance to normally virulent plant pathogens. We have used three defense-related genes of rice, i.e., *POX22.3* (a peroxidase gene, Chittor *et al.*, 1997), *PBZ1* and *PR1* (gift of K. Shimamoto) in Northern analysis of the 8 mutants. *POX22.3* was reported to be induced by bacterial blight only in the resistant plants (Chittor *et al.*, 1997). *PBZ1* was found to be induced by probenazole which induces disease resistance in rice to the blast fungus (Midoh and Iwata, 1996). PR1 gene has been shown to be associated with systemic acquired resistance in several plants species (Uknes *et al.*, 1992). Total RNA was isolated from leaves showing obvious lesion mimics from two month-old mutant, IR36 and IR68 plants. *PBZ1* was the only gene constitutively expressing in all the 8 mutants. Compared to other mutants, *spl4* and *spl11* had the highest expression level. The expression of *POX22.3* in the *spl3* and *spl11* plants was at least 5-10 times higher than that in control plants. Except in *spl7*, the *PR1* gene had elevated expression in all mutants. *spl1*, *spl5* and *spl11* had relatively higher expression compared to other mutants. These results suggest that the mutation in these 8 mutants had activated at least one of the defense-related genes. Different expression patterns and expression levels among these mutants might indicate that the mutations may represent different genes in the disease resistance signal transduction pathways.

4. CONCLUSIONS

In the last few years, a systematic search of genes involved in the signal transduction pathways have been performed in *Arabidopsis,* tomato and barley. Only a few loci was identified and some screens even failed to yield a single mutation in genes required in resistance. This indicates that the number of genetically identified components in the pathway is low. In addition, the majority of identified mutants did not lose of resistance completely (Innes, 1998). In our screen for genes required for the *Xa21* resistance, we employed two different types of mutagens (DEB and fast neutrons) and produced large populations (total 4500 M2 families) to maximize the chance of finding mutants. Twenty nine putative mutants were identified when inoculated with *Xoo* race 6 and an additional mutant was identified when the whole population was inoculated again with two avirulent blast isolates. Among them, 9 mutants were found to have mutation at the *Xa21* locus and all of them were highly susceptible to *Xoo*. The remaining mutants possibly carry mutation in the *Xa21* mediated resistance pathway. This result is consistent with findings in other plant species where most of the identified mutants do not lose resistance completely. As pointed out by Innes (1998), the failure to identify complete susceptible mutants could attribute to three reasons. Firstly, the signal transduction components may be encoded by redundant genes. Secondly, deletion of some genes in the resistance pathway may affect plant viability. The third explanation is that the resistance gene transduction pathways are highly branched and the deletion of one branch produce only the immediate phenotypes.

It has been shown in *Arabidopsis* that some genes in the signal transduction pathways such as *NDR1* and *EDS1* are required for multiple disease resistance genes (Innes, 1998). In our study, three mutants susceptible to *Xoo* were also found to be susceptible to two of the blast isolates. In addition, these three mutants have deletion in the *Xa21* locus, detected by PCR and Southern analysis. However, it is unclear whether the *Xa21* confers dual resistance function or whether a blast resistance gene is located very closely with the *Xa21* gene. Detailed genetic analysis of these mutants is in progress to distinguish these possibilities.

Lesion mimic mutants have been identified and characterized in several plant species. Whether the genes responsible for spontaneous cell death lesions are related to the disease resistance pathways is currently debatable (Dangl *et al.*, 1996). Dietrich *et al.* (1994) identified 7 *Arabidopsis* lesion mimic mutants of which only two showed enhanced resistance to pathogens. Eight rice lesion mimic mutants were characterized by analyzing the expression of three defense related genes in the mutant plants and challenging them with rice pathogens. Interestingly, the probenazole-induced gene *PBZ1* showed an elevated expression level in all the mutants. The other two genes were expressed only in few mutants. *spl11* mutant is the only one constitutively expressing all three genes and showed enhanced resistance to both blast and bacterial blight pathogens. Hence, we speculate that the lesion mimic formation in *spl11* plants may have triggered pathogen non-specific resistance resembling SAR. The gene responsible for the *spl11* mutation might be involved in the resistance pathway required against for both the pathogens. Since the gene is similar to the barley recessive gene *mlo*, isolation of the *spl11* gene would be invaluable in understanding the relationship between cell death and disease resistance at molecular level and also for designing new strategies for disease control in crop plants.

References

Baker, B., Zambryski, P., Stakawicz, B. and Dinesh-Kumar, S.P. (1997). Signaling in plant-microbe interactions. *Science* **276**, 726-732.

Buschges, R., Hollricher, K., Panstruga, R., Simons, G., Wolter, M., Frijters, A., van Daelen, R., van der Lee, T., Diergaarde, P., Groenendijk, J., Topsch, S., Vos, P., Salamini, F. and Schulze-lefert, P. (1997) The barley *mlo* gene: a novel control element of plant pathogen resistance. *Cell* **88**, 695-705.

Chittoor, J.M., Leach, J.E. and White, F.F. (1997) Differential induction of a peroxidase gene family during infection of rice by *Xanthomonas oryzae* pv. *oryzae*. *Mol Plant-Microbe Interact* **10**, 861-871.

Century, K.S., Shapiro, A.D., Repetti, P.P., Dahlbeck, D., Holub, E.and Staskawicz, B.J. .(1997) NDR1, a pathogen-induced component required for *Arabidopsis* disease resistance. *Science* **278**, 1963-1965.

Dangl, J.L., Dietrich, R.A. and Richberg, M.H. (1996) Death don't have no mercy: cell death programs in plant-microbe interactions. *Plant Cell* **8**, 1793-1807.

Dietrich, R., Delaney, T.P., Uknes, S.J., Ward, E.J., Ryals, J.A. and Dangl, J. (1994) *Arabidopsis* mutants simulating disease resistance response. *Cell* **77**, 565-578.

Dietrich, R., Richberg, M.H., Schmid, R., Dean, C., Dangl, J. (1997) A novel zinc finger protein is encoded by the *Arabidopsis LSD1* gene and functions as a negative regulator of plant cell death. *Cell* **88**, 685-694.

Ellis, J. and Jones, D. (1998) Structure and function of proteins controlling strain-specific pathogen resistance in plants. *Current Opinion in Plant Biology* **1**:288-293.

Gray, J., Close, P.S., Briggs, S.P. and Johal, G.S. (1997) A novel suppressor of cell death in plants encoded by the Lls1 gene of maize. *Cell* **89**, 25-31.

Greenberg, J.T., Guo, A., Lessig, D.F. and Ausubel, F.M. (1994). Programmed cell death in plants: a pathogen-triggered response activated coordinately with multiple defense functions. *Cell* **77**, 551-564.

Hu, G., Richter, T., Hulbert, S.T. and Pryor, T. (1996) Disease lesion mimicry caused by mutations in the rust resistance gene rp1. *Plant Cell* **8**:1367-1376.

Innes RW. (1998) Genetic dissection of R gene signal transduction pathways. *Current Opinion in Plant Biology* **1**:229-304.

Keen, N.T. (1990) Gene-for-gene complementarity in plant-pathogen interaction. *Annu Rev Genet* **24**, 447-463.

Khush, G.S., Singh, R.J., Sur, S.C. and Librojo, A.L. (1984) Primary trisomics of rice. Origin, morphology, cytology, and use in linkage mapping. *Genetics* **107**, 141-163.

Kinoshita, T. (1995) Report of committee on gene symbolization, nomenclature and linkage groups. *Rice Genet Newsl* **12**, 9-115.

Salmeron, J.M., Oldroyd, G.E.D., Rommens, C.M.T., Scofield, S.R., Kim, H.S., Lavelle, D.T., Dahlbeck, D. and Staskawicz, B.J. (1996) Tomato *Prf* is a member of the leucine-rich repeat class of disease resistance genes and lies embedded within the *Pto* kinase gene cluster. *Cell* **86**, 123-133.

Sanchez, A.C., and Khush G.S. (1994) Chromosomal location of some marker genes in rice using the primary trisomics. *J. Hered.* **85**, 297-300.

Singh, K., Multani, D,S, and Khush, G.S. (1995) A new spotted leaf mutant in rice. *Rice Genetics Newsletter* **12**, 192-193.

Song, W.-Y., Wang, G.-L., Chen, L., Kim, H.-S., Pi, L.-Y., Gardner, J., Wang, B., Holsten, T., Zhai,W.-X., Zhu, L.-H., Fauquet, C., and Ronald, P.C. (1995). A receptor kinase-like protein encoded by the rice disease resistance gene *Xa21. Science* **270**, 1804-1806.

Staskawicz, B.J., Ausubel, F.M., Baker, B., Ellis, J.G., and Jones, J.D.G. (1995). Molecular genetics of plant disease resistance. *Science* **268**, 661-667.

Tu, J., Ona, I., Zhang, Q., Mew, T.W., Khush, G.S. and Datta, S.K. (1998) Transgenic rice variety "IR72" with Xa21 is ressitant to bacterial blight. *Theor. Appl. Genet.* **97**, 31-36.

Uknes, S., Mauch-Mani, B., Moyer, M., Potter, S., Williams, S., Dincher, S., Chandler, D., Slusarenko, A., Ward, E. and Ryals, J. (1992) Acquired resistance in *Arabidopsis. Plant Cell* **4**, 645-56.

Zhou, J., Loh, Y., Bressan, R.A., and Martin, G.B. (1995) The tomato gene Pti1 encodes a serine/threonine kinase that is phosphorylated by *Pto* and is involved in the hypersensitive response. *Cell* **83**, 925-935.

Zhou, J., Tang, X., Martin, G.B. (1997) The *Pto* kinase conferring resistance to tomato bacterial speck disease interacts with proteins that bind a cis-element of pathogenisis-related genes. *The EMBO Journal* **16**, 3207-3218.

INDUCTION OF ACQUIRED RESISTANCE IN RICE TO RICE BLAST BY SYRINGOLIN, AN ELICITOR FROM *PSEUDOMONAS SYRINGAE* PV. *SYRINGAE*

U. WÄSPI[1], T. WINKLER[2], AND R. DUDLER[1]
[1]Institute of Plant Biology, University of Zurich
Zollikerstrasse 107, CH-8008 Zurich, Switzerland
[2]Novartis Crop Protection AG, 4002 Basel, Switzerland

Recognition by rice plants (*Oryza sativa*) of the non-host pathogen *Pseudomonas syringae* pv. *syringae* leads to local acquired resistance to the rice blast fungus *Pyricularia oryzae* accompanied by the accumulation of a set of transcripts. We have identified and isolated a compound secreted by the bacteria that, when applied to rice leaves, elicits accumulation of defense gene transcripts and increases resistance to *P. oryzae*, but is not toxic for the fungus. This compound, which we name syringolin, is a novel peptide in which the two non-proteinogenic amino acids 5-methyl-4-amino-2-hexenoic acid and 3,4-dehydrolysine form a 12-membered ring that is attached by peptide bond to a valine that in turn is linked to a second valine via a urea moiety. Thus, syringolin appears to be one of the determinants of the non-host pathogen *P. syringae* pv. *syringae* that can be perceived by rice plants.

1. Introduction

Inoculation of rice leaves with the non-host pathogen *Pseudomonas syringae* pv. *syringae*) causes necrotic lesions typical for a hypersensitive response and induces local acquired resistance to a following infection by the causing agent of the rice blast disease (*Pyricularia oryzae*) (Smith and Métraux, 1991). Resistance induction is accompanied by the accumulation of a set of gene transcripts whose products may play an active role in the defense against pathogens. Several cDNAs corresponding to such transcripts have been cloned (Reimmann and Dudler, 1993; Reimmann et al., 1995; Reimmann et al., 1992; Wäspi et al., 1998). Whereas a number of *P. syringae* pv. *syringae* strains caused the hypersensitive response and resistance induction, the set of accumulating rice transcripts was not identical for all strains. In particular, *Pir7b* gene transcripts, which encode an esterase (Wäspi et al., 1998), did not accumulate upon inoculation of rice leaves by a strain dubbed SM, in contrast to all other defense-related transcripts tested (Reimmann et al., 1995). Experiments with *P. syringae* pv. *syringae* strains carrying defined mutations showed that the ability of the bacteria to induce *Pir7b* transcript accumulation in rice leaves depended on the *lemA* two-component regulatory system (Reimmann et al., 1995) which also controlled the secretion of the phytotoxin syringomycin and pathogenicity in some systems (Hrabak and Willis, 1992; Hrabak and Willis, 1993). This lead us to speculate that this two-component system controlled the

D. Tharreau et al. (eds.), Advances in Rice Blast Research, 73–78.

production of a secreted elicitor of *Pir7b* transcript accumulation in rice leaves. Thus, using *Pir7b* transcript accumulation as an assay, the postulated elicitor, which we named syringolin, was isolated and characterized, and its structure was solved (Wäspi et al., 1998).

2. Results and Discussion

2.1. ISOLATION AND STRUCTURE OF SYRINGOLIN

For the characterization and isolation of the *Pir7b* eliciting activity, *P. syringae* pv. *syringae* strain B 301D-R (Xu and Gross, 1988) was chosen, which was previously shown to induce *Pir7b* transcript accumulation in rice (Reimmann et al., 1995). Experiments summarized in Table 1 showed that the bacteria indeed secreted *Pir7b* inducing activity into the medium under appropriate culture conditions. Elicitor activity was only detected in the medium if bacteria were grown in still cultures with SRM_{AF} medium (1% D-glucose, 0.1 % fructose, 100 µM arbutin, 0.4% L-histidine, 0.8 mM $MgSO_4$, 10 µM $FeCl_3$, 0.8 mM potassium phosphate, pH 7), but not if grown in aerated rich (Luria-Bertani) medium. SRM_{AF} medium has been defined for efficient *in vitro* production of the phytotoxin syringomycin (Gross, 1985; Mo and Gross, 1991). The *Pir7b* eliciting activity was heat and protease resistant and appeared to have a molecular mass between 500 and 3000 Da. This elicitor was termed syringolin.

TABLE 1. Characterization of *Pir7b*-eliciting activity in conditioned culture media
[1] (Gross, 1985; Mo and Gross, 1991).

Infiltration of rice leaves with	*Pir7b* elicitor activity
B 301 D-R (wild type) suspension in water	yes
B 301 D-R killed by heat or formaldehyde	no
culture filtrate (o/n shaken in Luria-Bertani medium)	no
culture filtrate (5d still culture in SRM_{AF} medium[1])	yes
5d SRM_{AF} conditioned filtrate boiled (5 min.)	yes
5d SRM_{AF} conditioned filtrate proteinase K-treated	yes
3000 Da size exclusion 5d SRM_{AF} conditioned filtrate	yes
500 Da size exclusion 5d SRM_{AF} conditioned filtrate	no

Syringolin was isolated to homogeneity from *P. syringae* pv. *syringae* strain B 301 D-R still cultures in SRM_{AF} medium after 5 to 8 days by a combination of ultrafiltration, gel permeation, and HPLC chromatography as described (Wäspi et al., 1998). The molecular mass of syringolin as determined by mass spectrometry was 493 amu. The structure of syringolin, which was solved by NMR experiments described elsewhere

(Wäspi et al., 1998), is shown in Figure 1. Syringolin contains a cyclic dipeptide consisting of the two non-proteinogenic amino acids 5-methyl-4-amino-2-hexenoic acid and 3,4-dehydrolysine, which both have to our knowledge not been described to occur in nature. The α-amino group of 3,4-dehydrolysine is linked to the carboxyl group of a valine that in turn is linked to a second valine by a urea group. This urea group is unusual and has only rarely been described in natural compounds (e.. g. (Schmidt et al., 1997). Syringolin very likely is produced by a peptide synthetase, as are many other non-ribosomally synthesized bioactive peptides.

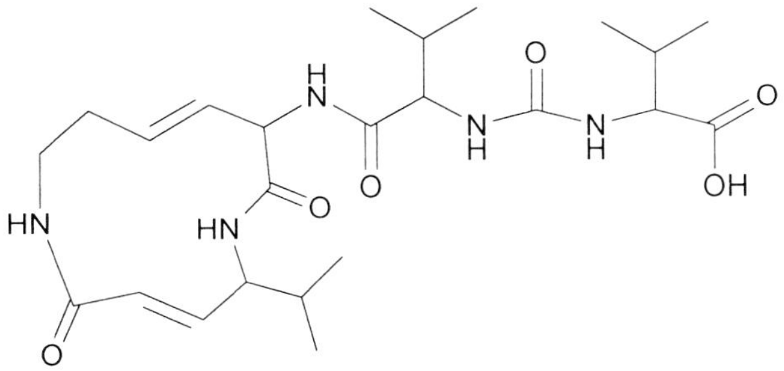

Figure 1. Structure of syringolin

2.2. RESPONSES OF RICE LEAVES TO SYRINGOLIN

Application of syringolin solutions to rice leaves by spotting eight 5-µl droplets on the surface of a 4-cm-long leaf segment did not induce visible symptoms, as e. g. necrosis, at all syringolin concentrations tested (up to 200 µM). However, accumulation of the Pir7b protein as detected by a specific antiserum raised against recombinant protein was observed with concentrations as low as 6.25 nM and appeared to be essentially saturated at 5 µM, as application of a ten times higher concentration did not lead to much larger amounts of Pir7b esterase (Figure 2A). Interestingly, syringolin not only induced accumulation of *Pir7b* mRNA, but, as shown in Figure 2B, also of transcripts corresponding to other defense-related gene that were probed for (*Rir1a*, encoding a cell wall protein (Mauch et al., 1998); *Pir2* and *Pir3*, encoding a thaumatin-like protein and a peroxidase, respectively (Reimmann and Dudler, 1993; Reimmann et al., 1992)).

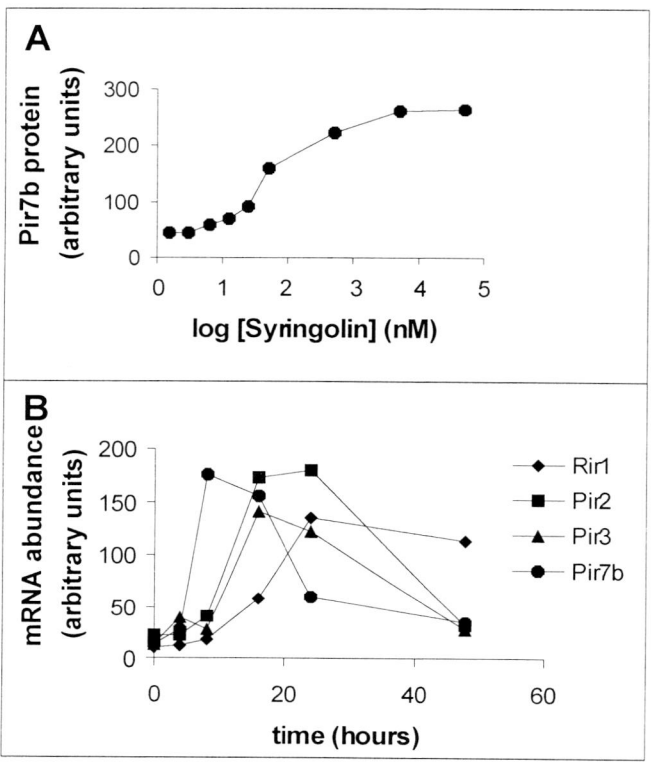

Figure 2. A. Dose response of Pir7b protein accumulation in rice leaves after treatment with syringolin. Eight droplets of syringolin solutions of the indicated concentrations were applied to the surface of 4-cm-long leaf segments that were incubated on agar plates.. Proteins were extracted 48 hours later and Pir7b was detected with a specific antibody by protein gel blot analysis using a chemiluminescent immunodetection system. Pir7b was densitometrically quantified using a graphical software package (National Institutes of Health Image, Macintosh version 1.60). B. Accumulation of putative defense gene transcripts in rice leaves after spraying plants with a 40 μM syringolin solution. The data shown were obtained by RNA gel blot analysis using the indicated cDNA probes. Autoradiograms were densitometrically processed and bands were quantified as described in A.

The ability of syringolin to increase resistance of rice plants to rice blast was tested by spraying 3-week-old rice plants with a 40 μM solution of syringolin followed by a spray inoculation with conidiospores of *P. oryzae* 24 h later. Scoring of the forth leaf 5 days after the inoculation showed that the number of blasts on syringolin treated plants was significantly reduced to about 25% of that on untreated control plants (1.17 ± 0.47 versus 4.40 ± 0.79 lesions/leaf on syringolin treated and control plants, respectively; Mean number ± standard deviation is given from two independent experiments with 6 pots containing approximately 15 plants per pot).

Syringolin did not appear to be toxic for the fungus. No growth or germination inhibition was observed around filter disks soaked with a 250 μM solution of syringolin or a control solution on agar plates that were inoculated with conidiospores. In addition, the germination rate of conidiospores on plates sprayed with a 250 μM syringolin solution was indistinguishable from the on observed on plates sprayed with the control solution. Interestingly, the compounds that are structurally most similar to syringolin are the glidobactins and cepafungins, a group of antifungal compounds isolated from the Gram-negative bacteria *Polyangium brachysporum* and *Pseudomonas* species, respectively (Oka et al., 1988; Shoji et al., 1990). These compounds contain also a cyclic dipeptide that consists, however, of 4-amino-2-pentenoic acid and 4-hydroxylysine linked to an acylated threonine residue. The fatty acid moiety was shown to be essential for the antifungal activity, because its elimination caused complete loss of toxicity (Oka et al., 1988; Terui et al., 1990). These results are in agreement with the observation that syringolin, which is not acylated, appears not to be toxic for *P. oryzae*.

2.3. CONCLUSIONS

Based on the observations that application of syringolin leads to the accumulation of putative defense gene transcripts, and that the compound appears not to be directly toxic for the fungus, we conclude that syringolin induces resistance of rice plants against *P. oryzae*. Thus, syringolin is a determinant of the non-host pathogen *P. syringae* pv. *syringae* that is perceived by rice plants, which, as a result, launch defense responses that have a certain efficacy towards the rice blast fungus. With respect to perception of *P. syringae* pv. *syringae* by rice plants, syringolin is redundant, because strains that do not produce it are still recognized and induce resistance (Reimmann et al., 1995; Wäspi et al., 1998). Thus, this non-host pathogen must possess more than one determinant that can be perceived by the rice plants, presumably by means of specific receptors. Interestingly, the defense responses that are triggered upon recognition of different determinants are apparently not identical, as is evident from the fact that the *Pir7b* gene is activated only by syringolin, but not by a syringolin-negative strain, in contrast to other defense-related genes. Presently one can only speculate about the nature of the putative receptors of non-host determinants and their relation to receptors involved in race-specific recognition events. Receptors of non-host determinants may not be readily amenable to genetic analysis, if their redundancy with respect to recognition of the non-host pathogen by the plant is the rule. Therefore, we are currently trying to use syringolin as a tool in experiments aimed at the identification of its putative receptor.

3. Acknowledgments

We are indebted to Drs. J. Smith, D. M. Gross, and D. K. Willis for the gift of *Pseudomonas* strains. Financial support by KTI grant 3157.1 and Novartis Crop Protection AG is acknowledged.

4. References

Gross, D. C. (1985) Regulation of syringomycin synthesis in *Pseudomonas syringae* pv. *syringae* and defined conditions for its production, *J. Appl. Bacteriol.* **58**, 167-174.

Hrabak, E. M. & Willis, D. K. (1992) The *lemA* gene required for pathogenicity of *Pseudomonas syringae* pv. *syringae* on bean is a member of a family of two-component regulators, *J. Bacteriol.* **174**, 3011-3020.

Hrabak, E. M. & Willis, D. K. (1993) Involvement of the *lemA* gene in production of syringomycin and protease by *Pseudomonas syringae* pv. *syringae*, *Molecular Plant Microbe Interactions* **6**, 368-375.

Mauch, F., Reimmann, C., Freydl, E., Schaffrath, U. & Dudler, R. (1998) Characterization of the rice pathogen-related protein Rir1a and regulation of the corresponding gene, *Plant Mol. Biol.* **38**, 577-586.

Mo, Y.-Y. & Gross, D. C. (1991) Plant signal molecules activate the *syrB* gene, which is required for syringomycin production by *Pseudomonas syringae* pv. *syringae*, *J. Bacteriol.* **173**, 5784-5792.

Oka, M., Yaginuma, K., Numata, K., Konishi, M., Oki, T. & Kawaguchi, H. (1988) Glidobactins A, B and C, new antitumor antibiotics. II. Structure elucidation, *J. Antibiot.* **41**, 1338-1350.

Reimmann, C. & Dudler, R. (1993) Complementary DNA cloning and sequence analysis of a pathogen-induced thaumatin-like protein from rice, *Plant Physiol.* **101**, 1113-1114.

Reimmann, C., Hofmann, C., Mauch, F. & Dudler, R. (1995) Characterization of a rice gene induced by *Pseudomonas syringae* pv. *syringae*: Requirement for the bacterial *lemA* gene function, *Physiol. Molec. Plant Pathol.* **46**, 71-81.

Reimmann, C., Ringli, C. & Dudler, R. (1992) Complementary DNA cloning and sequencing of a pathogen-induced putative peroxidase from rice, *Plant Physiol.* **100**, 1611-1612.

Schmidt, E. W., Harper, M. K. & Faulkner, D. J. (1997) Mozamides A and B, cyclic peptides from a theonellid sponge from Mozambique., *J. Nat. Prod.* **60**, 779-782.

Shoji, J., Hinoo, H., Kato, T., Hattori, T., Hirooka, K., Tawara, K., Shiratori, O. & Terui, Y. (1990) Isolation of cepafungins I, II and III from *Pseudomonas* species, *J. Antibiot.* **43**, 783-787.

Smith, J. A. & Métraux, J. P. (1991) *Pseudomonas syringae* pathovar *syringae* induces systemic resistance to *Pyricularia oryzae* in rice, *Physiol. Molec. Plant Pathol.* **39**, 451-461.

Terui, Y., Nishikawa, J., Hinoo, H., Kato, T. & Shoji, J. (1990) Structures of cepafungins I, II and III, *J. Antibiot.* **43**, 788-795.

Wäspi, U., Blanc, D., Winkler, T., Ruedi, P. & Dudler, R. (1998) Syringolin, a novel peptide elicitor from *Pseudomonas syringae* pv. *syringae* that induces resistance to *Pyricularia oryzae* in rice, *Mol. Plant-Microbe Interact.* **11**, 727-733.

Wäspi, U., Misteli, B., Hasslacher, M., Jandrositz, A., Kohlwein, S. D., Schwab, H. & Dudler, R. (1998) The defense-related rice gene *Pir7b* encodes an "alpha/beta hydrolase fold" protein exhibiting esterase activity towards naphthol AS-esters, *Eur. J. Biochem.* **254**, 32-34.

Xu, G. W. & Gross, D. C. (1988) Physical and Functional Analyses Of the Syr-a and Syr-B Genes Involved In Syringomycin Production By Pseudomonas-Syringae Pathovar Syringae, *J. Bacteriol.* **170**, 5680-5688.

A BLAST LESION MIMIC MUTANT OF RICE

S.G. PARK, S.O. KIM, H.J. KOH[1], and Y.H. LEE
*Dept. of Agricultural Biology and Agronomy[1], and RCNBMA,
Seoul National University, Suwon 441-744, Korea*

1. Introduction

Plant reacts to pathogen attack by activating multiple defense mechanisms as a result from the interaction of resistance gene in the plant and a corresponding avirulence gene in the pathogen. Interaction of the resistance and avirulence genes would stimulate a signalling cascade leading to the activation of defense mechanisms; hypersensitive response (HR), production of reactive oxygen species, cell wall fortification, accumulation of salicylic acid and phytoalexin, and induction of PR genes (reviewed in Bent, 1996; Hammond-Kosack and Jones, 1996).

In certain instances, however, some of these defense mechanisms could be activated, and resulted in lesion appearance in the absence of a pathogen (reviewed in Dangl *et al.*, 1996). Lesions are similar in appearance to necrotic lesion or HR lesion incited by a pathogen attack. These mutants are termed "disease lesion mimics" and have been reported from a variety of plants such as maize, barley, tomato, tobacco and *Arabidopsis* (Gray *et al.*, 1997; Dietrich *et al.*, 1997; Buschges *et al.*, 1997). Mutation of *Lls1* in maize and *Lsd1* in *Arabidopsis* results in spontaneous activation of cell death pathway in the absence of pathogen (Gray *et al.*, 1997). On the other hand, the mlo mutation of barley results in enhanced resistance to all common races of *Erysiphe graminis* f. sp. *hordei* and thus confer a broad spectrum resistance that differs from race-specific incompatibility to single pathogen strains (Freialdenhoven *et al.*, 1996). Lesion mimic phenotypes have also been observed in some transgenic plants (Becker *et al.*, 1993; Mittler *et al.*, 1995; Willekens *et al.*, 1997).

It has been suggested that lesion mimic mutants may affect mechanisms which control initiation or propagation of multiple biochemical events in pathogen-triggered R-gene specific resistance (Dangl *et al.*, 1996). Elucidating how these lesion mimic mutants lead to spontaneous cell death will be a help to reveal the regulation of defense mechanisms against pathogens.

We report here a disease lesion mimic mutant of rice termed as 'blast lesion mimic (*blm*)'. This mutant developed spontaneous necrotic lesions similar to blast lesions

D. Tharreau et al. (eds.), Advances in Rice Blast Research, 79–85.

without any pathogen attack, and exhibited race non-specific response to blast pathogen, *Magnaporthe grisea*. This mutant also displayed altered lesion type to infection of brown spot pathogen, *Bipolaris oryzae*.

2. Materials and methods

A blast lesion mimic mutant (*blm*) of rice was acquired from rice cv. Hwacheong by treatment of N-methyl-N-nitrosourea (MNU), as previously described (Kim *et al.*, 1991). Mutant rice plants were grown in the greenhouse as well as in the field for the observation of spontaneous lesion formation. For inoculation tests by several rice and non-rice pathogens, the 3 - 4 leaf stage of greenhouse-grown rice plants were used. Inoculation of rice pathogens [*Magnaporthe grisea* (race KI197, KJ201, KJ401, and 70-15), *Bipolaris oryzae*, and *Rhizoctonia solani*] and non-rice pathogens [*Colletotrichum graminicola* and *M. grisea* 2539] was conducted as standard procedures.

For observation of callose and phenolic compounds, inoculated leaves were decolorized by boiling in alcoholic lactophenol, stained by aniline blue for at least 48h, and cleared in the distilled water. Autofluorescence was observed in the decolorized leaves without any staining by fluorescent microscope.

For enzyme assay, rice leaves were macerated in liquid nitrogen, and suspended in extraction buffer [10 mM tris-HCl, pH 7.5, 250 mM sucrose, 1 mM Na_2EDTA and 1 mM PMSF]. The suspension was filtrated through the 3-layers of cheese cloth and centrifuged at 10,000 g for 20 min at 4 °C. Peroxidase and catalase activities were measured by spectrophotometer following the method described by Rao *et al.* (1997). Phenylalanine ammonia lyase (PAL) activity was determined by measuring the production of trans-cinnamic acid from L-phenylalanine (Nagarathna *et al*, 1993).

3. Results and discussion

3.1. THE *blm* DEVELOPED SPONTANEOUSLY NECROTIC LESIONS SIMILAR TO BLAST LESION

One of rice breeding lines from cv. Hwacheong, mutagenized by MNU, was severely damaged with many blast-like lesions in the research farm at Seoul National University, Korea, while the mother line did not exhibit any symptom. It was thought that this line became highly susceptible against blast fungus infection. However, the same peculiar lesions were developed in the greenhouse. A few lesions were first observed when the seedlings were 2 - 3 weeks old, and many lesions appeared mainly near the tip of the oldest leaf on 4 weeks after planting. The lesion was initiated as appearance of small chlorotic spot, similar to the susceptible lesion of the blast in earlier symptom development. And the lesions expanded to some degree, and thereafter were restricted (Fig. 1).

All the effort to isolate any pathogen from spontaneously developed lesions was failed. We named this mutant as blast lesion mimic (*blm*). This lesion was predorminantly observed during long-day season, but not during the winter in the greenhouse. The *blm* displayed many *blm*-specific lesions at 48h after deposition under continuous light, while Hwacheong showed minute dark brown spots. However, the number of *blm*-specific lesions was strikingly reduced when the *blm* plants had dark condition at least 2h. The effect of light on spontaneous lesion formation has also been reported in other lesion mimic mutants (Dietrich *et al.*, 1994; Dangl *et al.*, 1996). Lesion formation in *lls1* in maize requires red light, whereas long-day condition of continous light for *lsd1* of *Arabidopsis*.

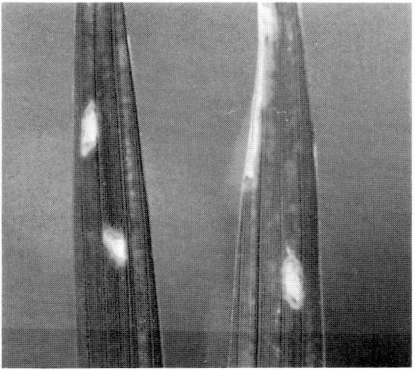

Figure 1. Spontaneously occurred lesions in the *blm* without any pathogen in the greenhouse.

3.2. RACE NON-SPECIFIC RESPONSE OF *blm* TO *Magnaporthe grisea*

Inoculation by four different races of *M. grisea* (KJ201, KJ401, KI197, and 70-15) developed the same disease symptom on the leaves of the *blm*. The lesion was initiated as chlorotic spots, developed as distinct and uniform necrotic spots, similar to those developed spontaneously, within 5 days. No further expanding of the lesions was observed until 2 weeks after inoculation. However, Hwacheong exhibited clear race-specific response; compatible with KI197 and KJ201, and incompatible with KJ401 and 70-15 (Fig. 2). Although the *blm* exhibited enhanced resistance to virulent races of blast pathogens, no brown spots, as a result of HR response, was induced in when inoculated with even avirulent races of the pathogen.

How race non-specific response is accomplished without HR cell death is not well understood. Enhanced resistance in mlo mutant against *E. graminis* f. sp. *hordei* correlates with enhanced accumulation of PR genes upon pathogen challenge (Peterhansel *et al.*, 1997) and with the formation of cell wall apposition that may prevent fungal penetration. (Wolter et al, 1993). Accumulation of callose and phenolic compounds was not constitutively observed on the leaves of the *blm*. However, these compunds were abundantly accumulated in fully developed lesions incited by blast fungus or spontaneously occurred. The *blm* mutant appears to acquire the late-acting but strongly inducing defense mechanisms mediating the accumulation of callose or phenolic compounds.

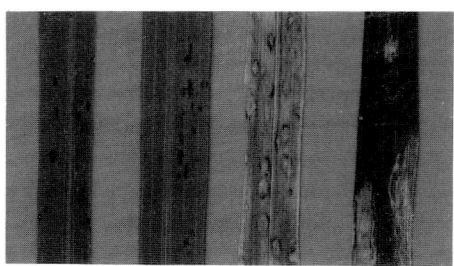 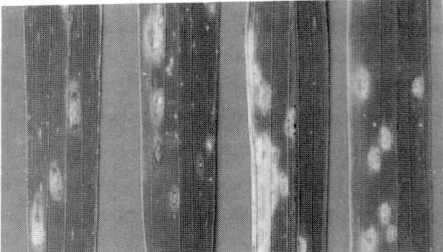

Figure 2. Response to different races of *Magnaporthe grisea*. Left: Response of Hwacheong against four races (from the left, KJ401, 70-15, KI197, and KJ201). Incompatible (left two) and compatible (right two) interactions were clearly observed 7 days after inoculation. Right: Response of the *blm*. Four races incited the similar symptom and they are indistinguishable.

3.3. PEROXIDASE ACTIVITY IN THE *blm*

Peroxidase activity was increased in Hwacheong when inoculated with an incompatible *M. grisea* KJ401 at 24h after inoculation, but not with a compatible *M. grisea* KJ201. It suggests that peroxidase activity may function on HR-like cell death. However, peroxidase activity abruptly increased and keeping pace during lesion development in *blm* when inoculated with both races of blast pathogens (Fig. 3). High level of peroxidase activity was also observed in the leaves of the *blm* when lesions developed spontaneously. Extracellular peroxidase was postulated to play an integral role in the polymerization of cell wall components including lignin and suberin with consuming of H_2O_2 (Lewis and Yamamoto, 1990). Highly activated peroxidase in *blm* may function in similar reaction, and resulted in suppression of pathogen growth. However, further research is required to elucidate the clear function of peroxidase in the *blm*. No significant difference of the activities of the catalase and PAL was detected between the *blm* and Hwacheong.

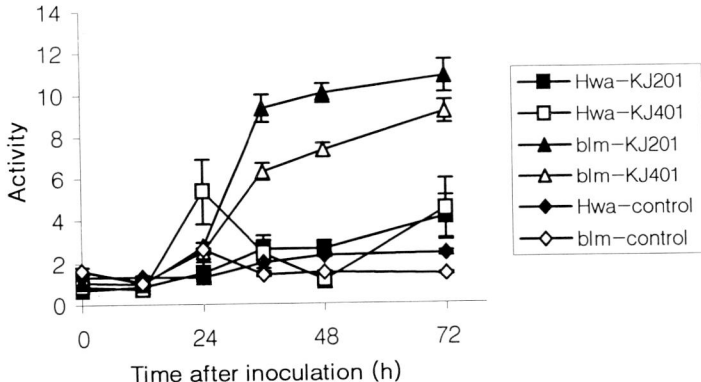

Figure 3. Changes of peroxidase actitivity in the *blm* and Hwacheong after inoculation with a compatible and incompatible races of *M. grisea*.

3.4. THE *blm* BECAME SUPER-SUSCEPTIBLE TO *Bipolaris oryzae*

The *blm* displayed enhanced resistance to virulent races of the blast pathogen, but *blm* was severely diseased in the field with many lesions similar to blast lesions. Although no pathogen was isolated from spontaneously developed lesions in the greenhouse, *Bipolaris oryzae* was isolated from the lesions fully developed in the field. To test whether *blm* became supersuceptible to *B. oryzae* or it is merely the result of secondary infection on the spontaneously occurred necrotic lesions, the seedlings of the *blm* were inoculated by *B. oryzae*. No typical brown spot lesion was developed in the *blm*, but similar lesions incited by *M. grisea* were developed. The lesions were continuously expanded and developed into large lesion with gray center, and the leaves of the *blm* were all died. However, many typical brown spots were developed in Hwacheong. No difference was observed when inoculated by *R. solani* between the *blm* and Hwacheong. Non-rice pathogens such as *C. graminicola* amd *M. grisea* 2539 did not cause any symptom both in *blm* and in Hwacheong. These data indicate that the *blm* mutation disrupted race specificity against rice blast pathogen and symptom development against brown spot pathogen, but not host specificity. Preliminary genetic data indicate that the mutation of *blm* was controlled by a single locus as recessive manner. Understanding of the mechanisms involved in lesion mimic phenomenon would be a help to elucidate R-gene specific as well as race non-specific resistance mechanisms in plant-pathogen interactions.

4. Acknowledgment

This work was supported in part by the Korea Science and Engineering Foundation through the Research Center for New Bio-materials in Agriculture at Seoul National University.

5. References

Bent, A.F. (1996). Plant disease resistance genes : Function meets structure. Plant Cell 8, 1757-1771.

Becker, F., Buschfeld, E., Schell, J., and Bachmair, A. (1993). Altered response to viral infection by tobacco plants perturbed in ubiquitin system. Plant J. 3, 875-881.

Buschges, R., Hollricher, K., Panstruga, R., Simon, G., Wolter, M., Frijters, A., Daelen, R., Lee, T., Diergaade, P., Groenendijk, J., Topsch, S., Vos, P., Salamini, F., and Schulze-Lefert, P. (1997). The barley *Mlo* gene: A novel control element of plant pathgen resistance. Cell 88, 695-705.

Dangl, J.L., Dietrich, R.A., and Richberg, M.H. (1996). Death don't have no mercy: cell death programs in plant-microbe interactions. Plant Cell 8, 1793-1807.

Dietrich, R.A., Delaney, T.P., Uknes, S.J., Ward,E.J., Ryals, J.A., and Dangl, J.L. (1994). Arabidopsis mutants simulating disease resistance response. Cell 77, 565-578.

Dietrich, R.A., Richberg, M.H., Schmidt, R., Dean, C., and Dangl, J.L. (1997). A novel zinc finger protein is encoded by the Arabidopsis *Lsd1* gene and functions as a negative regulator of plant cell death. Cell 88, 685-694.

Freialdenhoven, A., Peterhansel, C., Kurth, J., Kreuzaler, F., and Schulze-Lefert, P. (1996). Identification of genes required for the function of non-race-specific mlo resistance to powdery mildew in barley. Plant Cell 8, 5-14.

Gray, J., Close, P.S., Briggs, S.P., and Johal, G.S. (1997) A novel suppressor of cell death in plants encoded by the *Lls1* gene of maize. Cell 89, 25-31.

Hammond-Kosack, K.E., and Jones, J.D.G. (1996). Resistance gene-dependent plant defense responses. Plant Cell 8, 1773-1791.

Kim, K.H., Heu, M.H., Park, S.Z., and Koh, H.J. (1991). New mutants for endosperm and embryo characters in rice. Korean J. Crop Sci. 36, 197-203.

Lewis, N.G., and Yamamoto, E. (1990). Lignin: occurrence, biogenesis and biodegradation. Annu. Rev. Plant Physiol. Plant Mol. Biol. 41, 455-496.

Mittler, R., Shulaev, V., and Lam, E. (1995). Coordinated activation of programmed cell death and defense mechanisms in transgenic tobacco plants expressing a bacterial proton pump. Plant Cell 7, 29-42.

Nagarathna, K.C., Shetty, S.A., and Shetty, H.S. (1993). Phenylalanine ammonia lyase activity in pearl millet seedlings and its relation to downy mildew disease resistance. J. Exp. Bot. 44, 1291-1296.

Peterhansel, C.. Freialdenhoven, A., Kurth, J., Kolsch, R., and Schulze-lefert, P. (1997). Interaction analyses of genes required for resistance responses to powdery mildew in barley reveal distinct pathways leading to leaf cell death. Plant Cell 9, 1397-1409.
Rao, M.V., Paliyath, G., Ormrod, D.P., Murr, D.P., and Watkins, C.B. (1997). Influence of salicylic acid on H_2O_2 production, oxidative stress, and H2O2-metabolizing enzymes. Plant Physiol. 115, 137-149.

Willekens, H., Chamnongpol, S., Davey, M., Schraudner, M., Langebartels, C., Montagu, M.V., Inze, D., and Camp, W.V. (1997). Catalase is a sink for H_2O_2 and is indispensible for stress defence in C3 plants. EMBO J. 16, 4806-4816.

Wolter, M., Hollricher, K., Salamini, F., and Schulzee-Lefert, P. (1993). The mlo resistance alleles to powdery mildew infection in barley trigger a developmentally controlled defence mimic phenotype. Mol. Gen. Genet. 239, 122-128.

FUNGITOXIC RESPONSES OF RICE LEAVES AND CALLI AS RELATED TO BLAST RESISTANCE

A.A. AVER'YANOV, T.D. PASECHNIK, V.P. LAPIKOVA,
Research Institute of Phytopathology
p/o B.Vyazemy, Moscow region 143050, Russia

L.M. GAIVORONSKAYA
Russian People's Friendship University
Moscow 117198, Russia

1. Abstract

The toxicity of rice cell exometabolites to blast spores was studied. The cultivars represented the host susceptibility or resistance (complete, partial or mixed). Diffusates collected from intact leaves of the susceptible cultivar, 1 or 2 days after inoculation, was found to inhibit spore germination weakly whereas this ability was significantly stronger in all resistant cultivars. Active oxygen (AO) species (H_2O_2 or $\cdot OH$) may be involved in the toxicity since the latter was abolished by appropriate scavengers. The contribution of radical O_2^- to the toxicity was large in completely resistant cultivars and absent in the partially resistant ones. Cultivars with mixed type of resistance were intermediate in this respect. So, the leaf diffusate fungitoxicity was likely indicative of the resistance. Analogous antifungal responses were found in rice callus cultures, with the similar specificity to fungus strains as in intact plants from which the cultures were derived. Diffusates of spore-inoculated calli were weakly if any fungitoxic in compatible combinations but strongly toxic in incompatible ones. The reaction depended on Ca^{2+} ions. During the first hours after inoculation, the toxicity of callus diffusates was mediated by AO and then by other substances formed as a result of AO production. Therefore, rice cells in a callus culture retain fungitoxic responses characteristic of the varietal resistance and peculiar to intact plants.

2. Introduction

"Active oxygen (AO) species" usually mean superoxide O_2^- and hydroxyl $\cdot OH$ free radicals, hydrogen peroxide H_2O_2 and singlet oxygen $^1O_2$. Many facts witness that these compounds produced by plants are involved in mechanisms of disease resistance both as triggers and direct toxicants (Wojtaszek, 1997).

D. Tharreau et al. (eds.), Advances in Rice Blast Research, 86–92.
© 2000 *Kluwer Academic Publishers. Printed in the Netherlands.*

In infected plants, most assays of AO were carried out in the extracellular space (Baker and Orlandi, 1995). Similarly, we found the rise in chemically tested superoxide production on a surface or in diffusates of blast-infected rice leaf segments. Simultaneously, leaf diffusates acquired AO-mediated toxicity to the fungus (Lapikova et al., 1994). Both indexes correlated positively each other. They were significantly higher in completely resistant than in susceptible cultivars 1-2 days after inoculation. It was suggested that AO-dependent fungitoxicity of diffusates of infected leaves may contribute to the inheritable resistance to blast and may be considered as a pre-symptomatic marker of this resistance.

The aims of the present work were (1) to follow relations between fungitoxicity of rice leaf diffusates of intact plants (rather than leaf segments) and inheritable blast resistance including the partial one and (2) to reveal the role of AO in the toxic effect. As well, we studied callus cultures of different rice cultivars to find out (3) whether isolated cells retain the ability to the resistance-related fungitoxic response and (4) are AO species involved in this response.

3. Material and Methods

Intact rice plants at the age of 4 leaves were inoculated alternatively by sprayed or fixed drops of spore suspension. With the first method, the disease incidence was evaluated as a relative area of 4th leaves covered with compatible-type lesions in 6-8 days. With the second method, 4th leaves of each seedling were immobilised in an inoculation chamber, and infective drops, fixed by holders, were applied to the upper leaf surface without any wounding (Pasechnik et al., 1995). Disease lesions of each type, according to Latterel et al. (1965), were counted and represented as a percentage of a total number of infective drops.

Fixed infective drops were collected 1 - 2 days after inoculation and liberated from inoculum spores to yield a diffusate of infected leaves. That of healthy leaves was prepared by an incubation of water droplets under the same conditions. A new spore population (a test-organism) of the same strain was mixed with a diffusate. The relative inhibition of their germination (against the germination in water) was the measure of a diffusate fungitoxicity. To assess the involvement of AO species in this action, appropriate antioxidant reagents were added to diffusates along with spores (Aver'yanov and Lapikova, 1988).

Rice callus cultures were maintained on the Murashige and Skoog medium for one week after the isolation from primary calli. Pieces of callus (5 mg) were submerged into 100 _l of a spore suspension inoculum (water in a healthy counterpart) in 96-well tissue culture plate. At different time intervals, callus diffusates were prepared, and their fungitoxicity was tested like those of leaves.

4. Results

4.1. INTACT LEAVES

At first we compared the two inoculation techniques and found close similarity of visible symptoms of resistance (or susceptibility). Therefore, the method of fixed drops, necessary to prepare diffusates, reproduces the normal plant-pathogen interactions without artefacts.

The cv. Sha-tiao-tsao was susceptible to all strains tested and manifested predominant compatible-type lesions. Cvs. Zenith and Tadukan were completely resistant and symptomless. Other cultivars produced small and rare compatible lesions

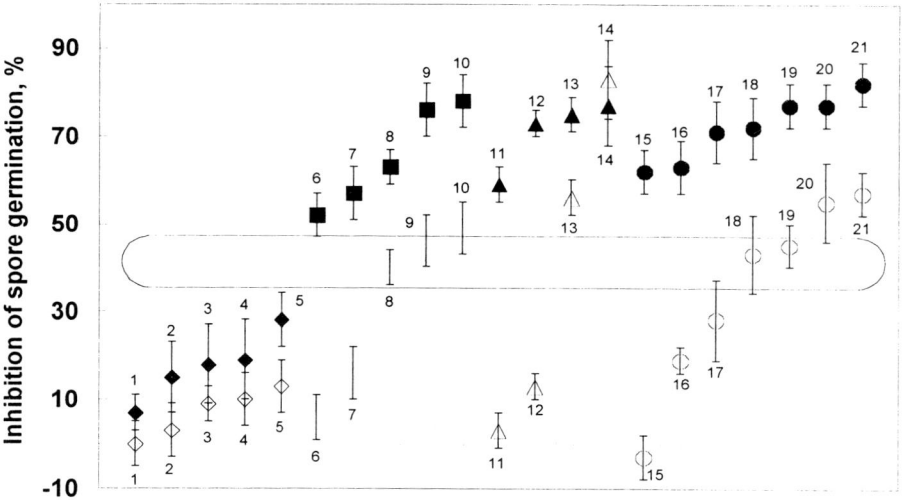

Figure 1. Fungitoxicity of diffusates of healthy (empty symbols) and infected (filled symbols) leaves. Maximum values of the first 2 post-inoculation days are given.

Numbers near symbols indicate cultivar / strain combinations. Susceptibility: (1) Sha-tiao-tsao / F67-57; (2) Sha-tiao-tsao / KK-37; (3) Sha-tiao-tsao / H5-3; (4) Sha-tiao-tsao / KP-57; (5) Sha-tiao-tsao / KP-69. Complete resistance: (6) Zenith / KK-37; (7) Tadukan / H5-3; (8) Zenith / H5-3; (9) Tadukan / KP-69; (10) Tadukan / KK-37. Partial resistance: (11) Norin 18 / H5-3; (12) Shimokita / KK-37; (13) Shimokita / H5-3; (14) Shimokita / KP57. Mixed resistance: (15) CNAB kweichow / KK-37; (16) China 1039 / H5-3; (17) China 1039 / KP-57; (18) CNAB kweichow / KP-57; (19) CNAB kweichow / F67-57; (20) China 1039 / KK-37; (21) CNAB kweichow / KP-69. Full arrest of spore germination corresponds to 100% inhibition, their germination in water is defined to be inhibited by 0%. Negative values mean a stimulation of germination as compared to water. Means ± SD from 5 series of 100 spores are represented.

which we ascribed to the partial resistance. Cvs. Norin 18 and Shimokita contained genes of vertical resistance but not to the strains H5-3 and KP-57, respectively; we qualified these combinations as a "pure" partial resistance. The case of cvs. CNAB kweichow and China 1039 was considered as an example of a "mixed" (complete and partial) resistance. The mixed character of symptoms was the especially evident in the combination of CNAB kweichow with KP-57 where abundant necrotic spots, typical of complete resistance, appeared.

Like the leaf-segment model, intact plants produced diffusates capable of inhibition of blast spore germination (*Figure 1*). This ability of healthy plants did not depend on their varietal resistance against the fungal strains tested. Meanwhile, after inoculation, the fungitoxicity usually increased. Its level in one or two days was significantly higher in resistant than in susceptible cultivars (that is depicted as a gap between the groups of cultivars in the plot). Therefore, fungitoxic exometabolites of infected leaves may be constituents or, at least, markers of blast resistance. Interestingly, this seems to be valid for cultivars with partial resistance, as well.

The involvement of AO, namely H_2O_2 and $\cdot OH$, in the toxic effects was clearly indicated by complete or strong protection of spores by addition of catalase or sodium formate (hydroxyl scavenger) to diffusates. However, superoxide dismutase (SOD) was not a universal antidote. It protected spores well from poisoning by the diffusates of completely resistant cvs. Zenith and Tadukan but was absolutely ineffective with partially resistant Norin 18 and Shimokita. This suggests that the former (but not the least) pair of cultivars prodices superoxide in fungitoxic doses. Effects of some other factors was also found to distinguish between two kinds of resistance. Thus, another hydroxyl scavenger, thiourea, behaved like SOD. The diffusates of completely resistant cultivars lost their toxicity after boiling but those of partially resistant remained unchanged. Effects of light or of histidine on the toxicity also depended on the type of resistance.

Results with mixed resistance cultivars were not so unequivocal and, sometimes, intermediate as to the degree of spore protection by antioxidant reagents. Thus, the substantial effect of SOD, reminiscent of complete resistance, was found in the combination of CNAB kweichow and KP-57 associated with massive necrosis.

So, fungitoxic responses of blast-infected rice leaves appear to contribute to an inheritable resistance to this disease. In general, the responses are mediated by AO, and the role of the particular AO species may depend on the resistance genotype.

4.2. CALLUS CULTURES

To elucidate whether the antifungal responses and their host-pathogen specificity occur at the cellular level we investigated the rice callus culture. The experimental strategy was the same as with intact plants but only the completely resistant cultivars were tested. We compared levels of callus diffusate toxicity and the data on resistance of whole plants (Kiyosawa, 1976; Correa-Victoria and Zeigler, 1994; D.Tharreau, personal communication) from which callus cultures were derived.

It was found that callus diffusates may also be fungitoxic (*Figure 2*). Without infection, this index did not characterise the resistance. Inoculated calli increased their fungitoxicity so that all susceptible and all resistant cultivars differed significantly in 18 h. The only exception was the compatible combination of Ishikare Shiroke and Ken 54-20 where the toxicity exceeded the level of susceptible cultivars. However, this instance represents the moderate susceptibility (Kiyosawa, 1976) which is not typical compatibility. Furthermore, the wound inoculation was used in the cited paper of Kiyosawa. The more natural spray inoculation described recently (Hayashi et al., 1998) revealed the resistant reaction in this combination.

Figure 2. Fungitoxicity of diffusates of healthy (empty symbols) and infected (filled symbols) calli in 18 h after inoculation.

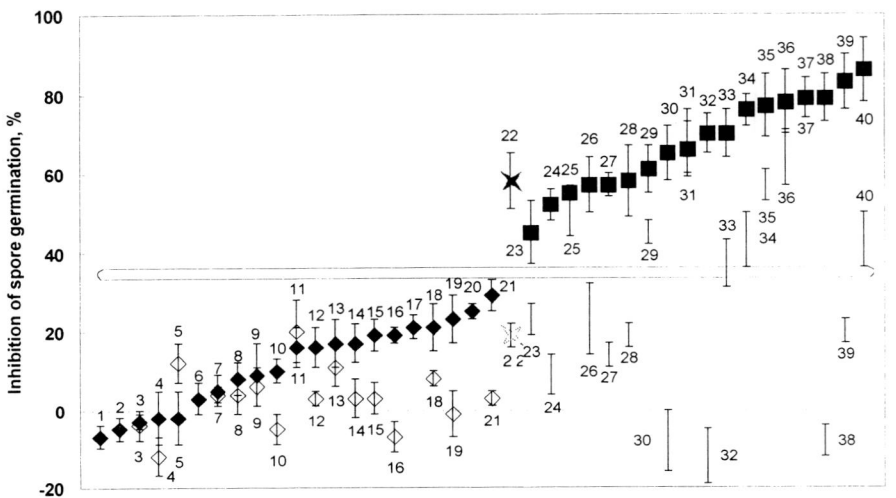

Numbers near symbols indicate cultivar / strain combinations. Susceptibility: (1) Shin 2 / Ken 54-20; (2) Aishi Asahi / Ken 54-20; (3) Maratelli / Ken 54-20; (4) Aishi Asahi / Kyu 82-359A; (5) Shin 2 / Ina 168; (6) Shin 2 / BR 26; (7) Zenith / PH-31; (8) Maratelli / PH-31; (9) Maratelli / Kyu 82-359A; (10) Sha-tiao-tsao / KK-37; (11) Maratelli / Ina168; (12) Maratelli / H5-3; (13) Aishi Asahi / PH-31; (14) Ishikare Shiroke / PH-31; (15) Fukunishiki / PH-31; (16) Sha-tiao-tsao / KP-57; (17) Maratelli / BR 26; (18) Sha-tiao-tsao / H5-3; (19) Ishikare Shiroke / H5-3; (20) Fukunishiki / BR 26; (21) Shin 2 / H5-3. Moderate susceptibility: (22) Ishikare Shiroke / Ken 54-20. Resistance: (23) Tsuinake / Ken 54-20; (24) Zenith / Ken 54-20; (25) Fukunishiki / Ina168; (26) Ishikare Shiroke / Kyu 82-359A; (27) Aishi Asahi / Ina168; (28) Ishikare Shiroke / Ina168; (29) Fukunishiki / Ken 54-20; (30) Fukunishiki / Kyu 82-359A; (31) Shin 2 / Kyu 82-359A; (32) Zenith / Kyu 82-359A; (33) Tsuinake / Ina168; (34) Tsuinake / H5-3; (35) Zenith / Ina168; (36) Tsuinake / Kyu 82-359A; (37) Zenith / H5-3; (38) Zenith / KK-37; (39) Fukunishiki /H5-3; (40) Zenith / KP-57.

It must be noted that the high toxicity of callus diffusate was the sign of not only resistance of particular cultivars but of incompatibility of particular host/parasite combinations. This was the most evident in the interactions of cvs. Shin 2 and Aishi asahi with strains Ina 168 and Kyu 82-359A.

The induction of antifungal responses involves presumably calcium ions. Thus, callus of cv. Zenith inoculated by the strain Kyu 82-359A in the presence of 0.1 mM of EGTA (the chelator of this ion) or $LaCl_3$ (the blocker of Ca^{2+} channels in plasma membrane) did not increase the toxicity of its diffusate.

TABLE 1. Time-dependent changes of callus diffusate fungitoxicity and of its sensitivity to antioxidant reagents. The table represents the inhibition of spore germination in diffusates collected at times indicated.

Additions to	Time after inoculation			
diffusates	6 h	9 h	18 h	24 h
cv. Fukunishiki / strain Kyu 82-395A				
None	—	—	65 ± 7	90 ± 5
Catalase, 100 _g/ml	—	—	15 ± 8	73 ± 7
SOD, 100 _g/ml	—	—	10 ± 8	43 ± 7
Mannitol, 10 mM	—	—	-1 ± 7	84 ± 5
Thiourea, 0.5 mM	—	—	18 ± 7	82 ± 7
cv Tsuinake / strain Kyu 82-395A				
None	46 ± 5	68 ± 5	78 ± 7	86 ± 5
Catalase, 100 _g/ml	13 ± 6	43 ± 5	68 ± 6	82 ± 7
SOD, 100 _g/ml	8 ± 6	15 ± 7	68 ± 7	75 ± 7
Mannitol, 10 mM	27 ± 3	23 ± 7	72 ± 9	65 ± 4
Thiourea, 0.5 mM	17 ± 6	34 ± 7	82 ± 5	83 ± 7
cv. Zenith / strain H5-3				
None	49 ± 7	67 ± 7	79 ± 5	—
Catalase, 100 _g/ml	26 ± 6	2 ± 8	67 ± 7	—
SOD, 100 _g/ml	18 ± 7	-7 ± 8	67 ± 7	—
Mannitol, 10 mM	-3 ± 9	-10 ± 7	72 ± 4	—
Thiourea, 0.5 mM	6 ± 5	-1 ± 7	74 ± 6	—

In some incompatible combinations, callus diffusates possessed the fungitoxicity which, in contrast to leaves, was quite insensitive to antioxidants and so independent of AO. We suggested that this state was a temporary and followed the time-course of the response between 6 and 24 h after an inoculation (Table 1). It was found that the toxicity levels grew gradually during this period. The signs of AO involvement in the diffusate toxicity, found in 18 h, disappeared later, in the couple Fukunishiki / Kyu 82-359A. Meanwhile, if this AO-dependence was absent in 18 h it was revealed before, in combinations of cv. Tsuinake with Kyu 82-359A and of Zenith with H5-3. Therefore, the AO-dependent toxicity was associated with an early period of host-parasite interaction, and this period was different in different combinations.

The delayed accumulation of antifungal exometabilites other than AO may result or not from the previous oxidative burst. To unravel this possibility we attempted to prevent the lately-expressed fungitoxicity insensitive to antioxidants by addition of these reagents earlier. In the combination Tsuinake / Kyu 82-359A, we administrated ·OH-scavengers mannitol or formate into an inoculum. This treatment diminished the toxicity of diffusates collected in 6; 9 and 18 h either. So, the generation of AO was necessary for the forthcoming appearance of other toxic compounds.

Therefore, AO species underlie, directly or indirectly, the cell fungitoxic response which is related to the blast resistance.

5. Discussion

Our data show that rice leaves of blast-resistant cultivars respond to an infection by the evolution of exometabolites harmful to the parasite. Besides the inhibition of spore germination, the same compounds may hinder the fungal development at later stages (Pasechnik *et al.*, 1997). This ability of leaves is indicative of the inheritable resistance. The partial resistance, rather obscure as to its physiological mechanisms, may also depend on this phenomenon.

The fungitoxic action of drop diffusates of plant tissues is usually attributed to phytoalexins. In our experiments, the rapidity of the fungitoxic response and its sensitivity to antioxidant agents makes AO more probable candidate at least at an early period of host-pathogen interaction. Nevertheless, phytoalexins and, perhaps, products of lipid peroxidation may play the same role later.

The fungitoxic response reflecting the host-pathogen specificity is obviously the cellular character since it is peculiar to undifferentiated cultured rice cells. This reaction might be of practical interest as a marker of blast resistance readable several days before visible symptoms appear. Furthermore, cell cultures do not produce typical symptoms at all, and the approach described may be useful in search for resistant cell clones.

Authors thank Drs. J.-L. Notteghem and D. Tharreau for the kindly provided plant and fungal material.

6. References

Aver'yanov, A.A. and Lapikova, V.P. (1988) Fungitoxicity determined by active forms of oxygen in excretions of rice leaves, *Soviet Plant Physiology* **35**, 873-881.

Baker, C.J. and Orlandi, E.W. (1995) Active oxygen in plant pathogenesis, in: *Annu.Rev.Phytopathol*, vol.33, Annual Reviews Inc., Palo Alto, pp. 299-321.

Correa-Victoria, F.J. and Zeigler, R.S. (1994) Pathogenic variability in *Pyricularia grisea* at a rice blast "hot spot" breeding site in eastern Colombia, *Plant Disease* **77**, 1029-1035.

Hayashi, N., Ando, I., and Imbe, T. (1998) Identification of a new resistance gene to a Chinese blast fungus isolate in the Japanese rice cultivar Aichi Asahi, *Phytopathology* **88**, 822-827.

Kiyosawa, S. (1976) Method for testing and gene analyses of blast resistance of rice varieties, *Oryza* **13**, 1-32.

Lapikova, V.P., Aver'yanov, A.A., Petelina, G.G., Kolomiets, T.M., and Kovalenko, E.D. (1994) Fungitoxicity of leaf excretions as related to varietal rice resistance to blast, *Russian J. of Plant Physiol.* **41**, 108-113.

Latterel, F.M., Marchetti, M.A., and Grove, B.R. (1965) Co-ordination of effort to establish an international system for race identification in *Piricularia oryzae*, in: R.F. Chandler (ed.), *The rice blast disease. Proc. of a Symp. at the Internat. Rice Research Institute*, The Johns Hopkins Press, Baltimore, pp. 257-274.

Pasechnik, T.D., Aver'yanov, A.A., Lapikova, V.P., Kovalenko, E.D., and Chernova, N.A. (1995) Fungitoxicity of rice leaf excretion due to activated oxygen: contribution to partial resistance to blast disease, *Russian J. of Plant Physiol.* **42**, 330-337.

Pasechnik, T.D., Lapikova, V.P., and Aver'yanov, A.A. (1997) Inhibition of pre-penetration development of blast fungus during the infection of resistant rice cultivars, *Eur. J. of Plant Pathol.* **103**, 747-750.

Wojtaszek, P. (1997) Oxidative burst: an early plant response to pathogen infection, *Biochem. J.* **322**, 681-692.

INDUCED RESISTANCE AGAINST RICE BLAST

H.K. MANANDHAR[1], H.J. LYNGS JØRGENSEN[2], S.B. MATHUR[3] and
V. SMEDEGAARD–PETERSEN[2]
[1]Plant Pathology Division, Nepal Agricultural Research Council, P.O. Box
1126, Khumaltar, Lalitpur, Nepal; [2]Plant Pathology Section, Department
of Plant Biology, The Royal Veterinary and Agricultural University,
Thorvaldsensvej 40, DK–1871 Frederiksberg C, Denmark; [3]Danish
Government Institute of Seed Pathology for Developing Countries,
Ryvangs Alle 78, DK–2900 Hellerup, Denmark.

1. Introduction

Induced disease resistance which may be either local or systemic throughout the plant
is the process of active resistance dependent on the host plant's physical or chemical
barriers, activated by biotic or abiotic agents (Kloepper *et al.*, 1992). Induced resistance
has been known since the beginning of the century (Chester, 1933) and has been
referred to as acquired resistance, acquired immunity, immunization, or vaccination.

It is only for the past two to three decades that induced resistance has been
extensively investigated in laboratories around the world with emphasis on providing
environmentally sound means of disease control. Induced resistance has been reported
in more than 25 crops, including rice (Tuzun and Kuć, 1991).

The objective of this review is to give an overview of the various agents used to
induce resistance against *P. oryzae* in rice and summarize the results presented on
mechanisms of protection i.e. the changes in the host related to increased resistance
occurring after application of the agents.

2. Reports on Induced Resistance against Rice Blast by Biotic Agents

There are several reports where protection of rice by various biotic agents have been
claimed or suggested to be due to induced resistance.

2.1. AVIRULENT *PYRICULARIA ORYZAE*

Kiyosawa and Fujimaki (1967) and Ohata and Kozaka (1967) independently reported
control of blast by pre–inoculation with avirulent *P. oryzae*. Other reports have
confirmed the ability of avirulent *P. oryzae* to protect rice against blast (Park and Kim,
1983; Iwano, 1987; Fujita *et al.*, 1990; Arase and Fujita, 1992; Manandhar *et al.*,

D. Tharreau et al. (eds.), Advances in Rice Blast Research, 93–104.

1998b). Also pre–inoculation with mutants of *P. oryzae,* generated by exposure to UV radiation, could control blast (Shen *et al.,* 1990). The mutants were avirulent or mildly virulent to the test cultivar of rice. In all cases, the protection was local. Only Park and Kim (1983) reported reduced lesion size up to 3–cm above the preinoculation site.

The degree of protection may vary with the lesion type incited by the inducer isolates. Thus, Fujita *et al.* (1990) found the highest degree of protection (>80%) with avirulent isolates inciting quite small eye–shaped spots with brown margins.

Lesions resulting from an inoculum mixture of virulent and avirulent *P. oryzae* were smaller than the ones caused by the virulent isolate alone (Ohata and Kozaka, 1967; Iwano, 1987). Manandhar *et al.* (1998b) also found reduction in disease severity when coinoculating avirulent and virulent *P. oryzae* (1:1 mixture). The virulent and avirulent isolates were not antagonistic to each other in dual culture experiments. In general, however, the longer the period (up to 72 h) between inoculation with avirulent and virulent *P. oryzae*, the higher degree of protection was found (Iwano, 1987; Arase and Fujita, 1992; Manandhar *et al.* 1998b).

In field experiments, pre–inoculation of rice plants with an avirulent isolate of *P. oryzae* at the tillering and heading stages reduced leaf and panicle blast, respectively (Iwano, 1987), but no yield data were given. Manandhar *et al.* (1998b) performed three field experiments at two locations of Nepal, and found that avirulent *P. oryzae* significantly reduced the disease incidence when applied at both booting and heading stages. However, grain yield and thousand grain weight were not significantly increased in any of the experiments.

2.2. OTHER ORGANISMS

In Japan, Hemmi *et al.* (1936) reported that *Bipolaris oryzae* suppressed both the germination and germ tube elongation of *P. oryzae* when rice plants were inoculated with both pathogens together. Similar observations were made by Aoki (1937). No anatagonism between the two pathogens was observed in dual culture experiments (Andersen *et al.,* 1947), indicating that there was no direct inhibitory effect by *B. oryzae. Bipolaris sorokiniana,* a non–rice pathogen, protected rice both locally and systemically against blast (Manandhar *et al.,* 1998b), the local effect was stronger than the systemic. No antagonistic activity was found between *B. sorokiniana* and *P. oryzae* in dual culture experiments. *B. sorokiniana* reduced the disease under both greenhouse and field conditions and furthermore increased the grain yield significantly in the field (Manandhar *et al.,* 1998b).

Thieron *et al.* (1995) used the bacterium *Clavibacter michiganensis* ssp. *tesselarius* (non–pathogenic to rice) to protect rice against *P. oryzae* and found more than 70% suppression of the pathogen in the leaves inoculated with the bacterium. Smith and Métraux (1991) reported systemic protection of rice against *P. oryzae* after inoculation with *Pseudomonas syringae* pv. *syringae,* a bacterium causing hypersensitive reactions in rice. Thus, in systemically protected leaves, the blast lesions were reduced in number and size by 85% and 50%, respectively. Reimmann *et al.* (1995), however, could not obtain any systemic protection against blast when they tested the same rice cultivar with the same bacterial strain.

2.3. FUNGAL EXTRACTS AND TOXINS

Seed treatment with culture extracts of *P. oryzae* was found to protect rice against the same fungus (Watanabe, 1951). Later, Watanabe (1957) detected the presence of a plant hormone resembling β–indole acetic acid in the extract, and suggested that seed treatment with the hormone might be beneficial for germination, growth, yield, and resistance against blast. Also, root treatment with mycelial extract of *Cephalothecium* sp. reduced blast infection (Yoshii, 1950).

Ouyang *et al.* (1987) found reduction in blast severity after treatment with toxins of *P. oryzae*. A glycoprotein elicitor isolated from mycelium of *P. oryzae* was also found to reduce blast when applied directly to the leaves by infiltration or press injury (Schaffrath *et al.*, 1995; Thieron *et al.*, 1995).

3. Reports on Induction of Resistance against Blast by Abiotic Agents

3.1. CHEMICALS

Several chemicals have been studied for their ability to protect rice against *P. oryzae*. Of these, di–ethyl sulphate (Gangadharan and Mathur, 1976), the phenanthroline complex of cobalt (Nikolaev *et al.*, 1991), 2–chloroisonicotinamides and their derivatives (Yoshida, 1992; Seguchi *et al.*, 1992), and fatty acids such as hydroxy– and epoxy–octadecadienoic acids (Song *et al.*, 1994) were reported to induce resistance in rice against *P. oryzae*. Other chemicals reported as resistance inducers include cycloheximide, p–chloromercuribenzoate, sodium malonate, 2,4–dichlorophenoxyacetic acid (Sengupta and Sinha, 1987; Sinha, 1990), plant hormones such as auxins (2,4–D, NAA, IAA), ethylene or their precursors (Matsumoto *et al.*, 1980; Iwata *et al.*, 1981) as well as some amino acid derivatives such as *N*–benzoyl amino acids, *N*–phenylacetyl amino acids (Takano *et al.*, 1972) and *N*–lauroyl-L–valine (Homma *et al.*, 1973). Similarly, some fungicides such as dichlorocyclopropane (WL 28325®; 2,2–dichloro–3,3–dimethyl cyclopropane carboxylic acid) (Langcake and Wickins, 1975), probenazole (Oryzemate®; 3–allyloxy–1,2–benzisothiazole–1,1–dioxide) (Watanabe *et al.*, 1979; Iwatata *et al.*, 1980) and tricyclazole (Beam®; 5–methyl–1,2,4–triazolo(3,4–b)–benzothiazole) (Nikolaev *et al.*, 1994) are known to protect rice plants systemically by activating defence mechanisms.

Arimoto *et al.* (1976) reported induced resistance by soaking rice seeds in a dodecyl DL–alaninate hydrochloride solution. The protection was detected 20 days after seeding, reached a maximum after 30 days and persisted for 75 days. Furthermore, Arimoto *et al.* (1991) reported generational succession of DL–alanine dodecylester HCl–induced protection against blast. Blast was decreased by 75% under greenhouse conditions and by 50% under field conditions on plants grown from seeds soaked in a 500 ppm solution. A reduction in blast severity of 61% in greenhouse and 52% in field was observed on plants grown from seeds harvested from the originally treated plants (second generation). When the treatment was repeated on alternate generations, the protective effect reached 97% in the fouth generation.

Treatment with salicylic acid was found to protect rice seedlings systemically against blast (Cai and Zheng, 1996). Manandhar *et al*. (1998a) also found reduced blast infection after treatment with salicylic acid. In greenhouse experiments, salicylic acid gave good protection when applied as foliar spray or soil drench, but not as seed treatment. In field experiments, foliar sprays at vegetative stages, reduced neck blast incidence. However, grain yield was increased significantly in only one of two experiments.

Kamata (1955) reported that the resistance of rice against *P. oryzae* was increased by high concentrations of iron, manganese, zinc and copper in the leaf. Seed treatment with cupric chloride or ferric chloride provided good control of blast under different cropping conditions (Sengupta and Sinha, 1987). Ragab and Afifi (1988) observed reduced blast infection and increased grain yield after spraying with iron. Manandhar *et al*. (1998a) found ferric chloride and di–potassium hydrogen phosphate effective in reducing leaf blast under greenhouse conditions. The efficacy of the chemicals depended on the method of application. Ferric chloride was effective when used as soil drench or foliar spray. Di–potassium hydrogen phosphate was only effective when applied as foliar spray. Neither of the chemicals were effective when used as a seed treatment. In field experiments, two or more foliar applications of ferric chloride at vegetative growth stages provided the best control of neck blast and increase in grain yield. Di–potassium hydrogen phosphate behaved in a similar way in one of the two experiments.

3.2. MISCELLANEOUS AGENTS

Guo and Luo (1994) reported induced resistance in rice against both leaf and neck blast by seed treatment or foliar spray with a medicinal drug, Baizhi.

Exposure of whole rice plants to UV–light for 15 min prior to inoculation with *P. oryzae* reduced leaf blast severity by more than 70%. Also the initial appearance of disease was delayed at least by 16 h in UV–irradiated compared with untreated control plants (H.K. Manandhar, unpublished results).

4. Cytological Events and Changes Associated with Induced Resistance

For the control agents reviewed above, it has been suggested that induced resistance was responsible for inhibiting *P. oryzae*. However, it is necessary to investigate the mechanisms by which a control agent operates before it is possible to claim that induced resistance is involved in the protection. This should start with microscopic examination of the pathogen with and without application of the control agent in order to determine where and when it is inhibited and whether visible defence reactions are activated. Furthermore, studies should be performed to determine whether the control agent has any direct toxic effect on *P. oryzae*, and whether defence–related genes are expressed. The latter type of investigation may be combined with biochemical studies determining the timing and magnitude of accumulation of antimicrobial compounds, eg. phytoalexins and defence–related enzymes. However, only in certain cases, such investigations have been performed (eg. in the case of probenazole and dichlorocyclopropane).

Penetration and intracellular hyphal growth of a virulent isolate of *P. oryzae* was inhibited in detached rice leaf sheaths pre–inoculated with non–host species of *Pyricularia* (Arase and Fujita, 1992). Manandhar (1996) observed less cytoplasmic granulation in rice leaves inoculated only with virulent *P. oryzae* than in leaves pre–inoculated with either *B. sorokiniana* or avirulent *P. oryzae* before virulent *P. oryzae*. Cytoplasmic granulation is a commonly observed phenomenon preceeding hypersensitive reactions of rice against *P. oryzae* (Tomita and Yamanaka, 1983).

Rapid lignification has been demonstrated in rice leaves pre–treated with salicylic acid (Cai and Zheng, 1997) and various other inducers, including probenazole, dichlorocyclopropane, 2,6–dichloroisonicotinic acid and a glycoprotein elicitor isolated from mycelium of *P. oryzae* (Thieron *et al.*, 1995).

On plants pre–treated with dodecyl DL–alaninate hydrochloride, germination, appresorial formation and penetration by *P. oryzae* were not different from untreated control plants, but the invading hyphae were stopped in epidermal and adjoining cells without subsequent penetration (Arimoto *et al.*, 1991). On the other hand, probenazole was found to act against penetration by *P. oryzae* into the leaf cells and also against subsequent extension of the invading mycelium into adjacent cells (Watanabe *et al.*, 1979). On plants pre–treated with dichlorocyclopropane, an increased proportion of small brown hypersensitive flecks appeared as a resistant reaction in response to challenge inoculation (Langcake and Wickins, 1975). The penetrated cells were degenerated and filled with brown melanin–like material.

5. Biochemical Changes Associated with Induced Resistance

5.1. IMPORTANT ENZYMES INVOLVED IN BIOSYSTHESIS OF ANTI-MICROBIAL COMPOUNDS

Phenylalanine ammonia–lyase (PAL), coniferyl alcohol dehydrogenase (CAD) and peroxidase (PO) are some major enzymes involved in catalyzing steps in the phenyl propanoid pathway, leading to lignification and biosynthesis of signaling and antimicrobial compounds (Bowels, 1990; Hammond–Kosack and Jones, 1996). PAL is, for example, a key enzyme in production of phytoalexins (Dixon and Paiva, 1995). Similarly, lipoxygenase (LOX) is involved in the biosynthesis of several hydroxy– and epoxy–fatty acids which have direct inhibitory effect against *P. oryzae* (Mori *et al.*, 1987; Namai *et al.*, 1993). LOX may also generate signal molecule such as jasmonic acid (Hammond–Kosack and Jones, 1996). In addition to these enzymes, there are several other enzymes (eg. phospholipase, esterase) involved in induced defence reactions (Sekizawa *et al.*, 1983).

Activities of PAL (Zhang *et al.*, 1987; Ouyang *et al.*, 1987), PO (Matsuyama, 1983), and LOX (Namai *et al.*, 1990; Ohata *et al.*, 1991) were found to be significantly higher in response to infection attempts by an avirulent than by a virulent *P. oryzae*. Thieron *et al.* (1995) reported increased activities of PAL, CAD and PO in leaves treated with *Clavibacter michiganensis* ssp. *tesselarius* or a glycoprotein elicitor from *P. oryzae*. Likewise, Manandhar *et al.* (1999) found earlier and higher accumulation of peroxidase transcript in *B. sorokiniana*–inoculated and UV–light–irradiated leaves (6–12 h after treatment) than in leaves inoculated with *P. oryzae*. In all cases, the activities were measured in response to inducer inoculation alone. Smith and Métraux (1991), who found systemic protection of rice by pre–inoculation with *Pseudomonas syringae* pv. *syringae*, observed increased activities of PAL, CAD and POX in the leaves pre-inoculated with the bacterium, but not in the systemically protected leaves.

PAL acitivity was enhanced in probenazole–treated (Iwata *et al.*, 1980), *N*–cyanometyl–2–chloroisonicotinamide–treated (Yoshida, 1992) and salicylic acid–treated (Cai and Zheng, 1997) rice plants. Increased PO activity was detected in rice plants treated with the resistance inducing fungicides probenazole (Watanabe *et al.*, 1979), and dichlorocyclopropane (Langcake and Wickins, 1975). Similarly, plants treated with *N*–cyanomethyl–2–chloroisonicotinamide had higher LOX activity (Yoshida, 1992). In all cases, the enhancement was especially high when the plants were challenge inoculated with *P. oryzae* after the inducer treatment.

Treatment of rice plants with probenazole, dichlorocyclopropane or 2,6–dichloroisonicotinic acid (INA) alone could increase the lipoxygenase activity (Thieron *et al.*, 1995). Treatment with INA also resulted in higher levels of jasmonic acid (Schweizer *et al.*, 1995). Also foliar spray with di–potassium hydrogen phosphate, salicylic acid and ferric chloride caused proxidase transcript to accumulate within 24 hours (Manandhar *et al.*, 1996).

From mycelium of *P. oryzae*, Iwata *et al.* (1982) isolated a water soluble glycoprotein which elicitated PO activity in rice leaves. Schaffrath *et al.* (1995) also isolated a glycoprotein elicitor isolated from the mycelium of *P. oryzae* and they suggested that it might be involved in the recognition of rice by *P. oryzae*. Xiao *et al.* (1994) suggested that extracellular glycoprotein(s) bind germ tubes and appressoria to the substrate, and at least a part of these glycoprotein(s) are further involved in sensing and transmitting information about induction of cellular differentiation.

5.3. PHYTOALEXINS

Unspecified phytoalexins (Uehara, 1958; Cha *et al.*, 1982) and specific phytoalexins such as oryzalexins A to F (Akatsuka *et al.*, 1985; Sekido *et al.*, 1986; Kato *et al.*, 1993, 1994), momilactones A and B (Cartwright *et al.*, 1981) and a flavonoid, sakuranetin (Kodama *et al.*, 1992) have been isolated from both blast–infected and UV–irradiated rice leaves. Generally, higher concentrations of the phytoalexins were found in incompatible than in compatible interactions (Uehara, 1958; Cha *et al.*, 1982).

Heavy metals and cell wall extracts of *P. oryzae* and UV–irradiation stimulated the accumulation of oryzalexins and momilactones in rice (Kodama *et al.*, 1988). Likewise, production of momilactones in response to infection with *P. oryzae* was markedly enhanced by prior treatment with dichlorocyclopropane (Cartwright *et al.*, 1980). Also several other chemicals reported to induce resistance against blast (eg. cupric chloride, ferric chloride, cycloheximide, p–chloromercuribenzoate and sodium malonate) are known to be phytoalexin inducers (Sengupta and Sinha, 1987, Sinha, 1990).

5.4. PATHOGENESIS-RELATED PROTEINS

Pathogenesis–related (PR) proteins are host–encoded polypeptides that accumulate as a result of infection by pathogens or application of elicitors (Eyal and Fluhr, 1991). Although several PR–proteins show antimicrobial activities *in vitro*, their role in the defence of plants remains unclear. However, a coordinated activity of several PR–proteins may be necessary for resistance (Hammond–Kosack and Jones, 1996).

Several PR–proteins have been shown to be activated in rice plants in response to *P. oryzae* or other agents. Du and Wang (1992) observed increased activities of PR–2 (β–1,3 glucanase) and PR–3 (chitinase) in rice after infection by *P. oryzae* or treatment with a culture extract of the fungus. Accumulation of both PR–2 and PR–3 proteins was increased in the leaves pre–inoculated with *P. syringae* pv. *syringae* (Smith and Métraux, 1991), but not in the systemically protected leaves. Manandhar *et al.* (1999) found transcript accumulation for PR–1a, PR–2, PR–3, PR–4, and PR–5 (thaumatin–like) in response to treatment with *P. oryzae*, the non–rice pathogen *B. sorokiniana* or UV–irradiation. The accumulation was earlier (12 h after inoculation) and at a higher level for *B. sorokiniana* and UV–treatment than for *P. oryzae*.

PR–protein accumulation has also been observed for abiotic inducers. Thus, activities of PR–proteins, including chitinase were found to be increased by dichlorocyclopropane (Thieron *et al.*, 1995) and salicylic acid (Cai and Zheng, 1997). Likewise, Manandhar *et al.* (1996) found that foliar application of salicylic acid and di–potassium hydrogen phosphate caused accumualation of transcripts for PR–1a, PR–2, and PR–3 within 24 h in rice. In addition, they also found that salicylic acid was able to cause accumulation of PR–4 transcript. A probenazole–inducible gene for an intracellular PR–protein in rice has been cloned (Midoh and Iwata, 1996).

5.5. ACTIVE OXYGEN SPECIES

The production of reactive oxygen species probably plays a key role in plant defence against pathogens in many incompatible interactions (Hammond–Kosack and Jones, 1996). Also in resistance of rice against *P. oryzae*, the possible involvement of oxygen radicals has been suggested in several reports (Aver'yanov *et al.*, 1987, 1993; Lapikova *et al.* 1994). Thus, germination of spores of *P. oryzae* was inhibited by a diffusate from leaves of a blast–resistant rice cultivar (Aver'yanov and Lapikova, 1989). The diffusates from blast–infected resistant cultivars were more toxic than those from blast–infected susceptible cultivars (Lapikova *et al.*, 1994). The fungitoxicity of the diffusate was concluded to be due to active oxygen species such as O_2^{-1}, HO, and H_2O_2 since the the toxic effect was prevented by addition of superoxide dismutase, catalase, HO scavengers, and $^1O_2$ quenchers.

Resistance against *P. oryzae* induced by heat (Aver'yanov *et al.*, 1993) and by the phenanthroline complex of cobalt (Nikolaev *et al.*, 1991) was related to generation of oxygen radicals. Oxygen radicals were also found to be involved in the antiblast activity of fungicides such as tricyclazole, fthalide and probenazole (Nikolaev *et al.*, 1994).

6. Practical Application of Induced Resistance

Induced resistance using microorganisms has been demonstrated successfully in rice against blast under laboratory and greenhouse conditions. The potential of induced resistance, however, has not been utilized in practical disease control. One of the reasons for this is the very limited number of investigations that have been conducted under field conditions to determine the efficacy and stability of induced resistance. This review found so far only two reports (Iwano, 1987 and Manandhar et al., 1998b) in which the inducer organisms were evaluated against rice blast under field conditions. In general, there are several other issues to be considered when using microorganisms as resistance inducers. They include economics of the application technology, possible harmful effects of the inducers to nearby crops, and sensitivity of inducer to field environment. Manandhar et al. (1998b) discussed some of the issues and indicated the possibility of using B. sorokiniana in rice against blast but they did not calculate the cost benefit ratio. Some problems in the development of a control agent based on induced resistance have been discussed by Jørgensen and Smedegaard-Petersen (1994).

Microbial metabolites, extracts or elicitors may also be used as resistance inducers and such inducers may be less sensitive to environmental stress than microorganisms. A number of successful studies have been made in rice using various fungal extracts or toxins against blast, but not a single example is available where they have been used in practice. It could be simply due to unreasonable production cost of such inducers.

An overview of present review shows that the inducing agents with the potentially greatest chance of success are chemicals which do not have direct toxic effect against pathogen or environment. Such inducers may be applied like conventional pesticides with existing equipment. Some fungicides having resistance inducing effects against rice blast are already in use, eg. probenazole and tricyclazole, and others may be evolved later. For example, the chemical DL-alanine dodecylester HCl which has shown generational effect in reducing rice blast (Arimoto et al., 1991) could be a promising candidate for real induced resistance. Chemicals such as phosphates used for the induction of resistance in cucumber (Gottstein and Kuć, 1989) and maize (Reuveni et al., 1994) have shown significant effect in rice against blast (Manandhar et al., 1998a). Such chemicals are rather inexpensive, easily obtainable and non-toxic, and may also serve as fertilizers. Furthermore, several studies suggest that use of micronutrients, especially iron may be helpful in inducing resistance against rice blast.

Finally, the use of induced resistance in a crop like rice has certain advantages, eg. seeds or seedlings of rice can be treated in several ways before transplantation.

References

Akatsuka, T., Kodama, O., Sekido, H., Kono, Y., and Takeuchi, S. (1985) Novel phytoalexins (oryzalexins A, B and C) isolated from rice blast leaves with Pyricularia oryzae: 1. Isolation, characterization and biological activities of oryzalexins, Agricultural and Biological Chemistry 49, 1689–1694.
Andersen, A.L., Henry, B.W., and Tullis, E.C. (1947) Factors affecting infectivity, spread, and persistence of Piricularia oryzae Cav., Phytopathology 37, 94–110.

Aoki, K. (1937) Physiological studies on the conidial germination of *Piricularia oryzae* and *Ophiobolus miyabeanus*, *Forschungen auf dem Gebiet der Pflanzenkrankheiten* 3, 147–176.

Arase, S. and Fujita, K. (1992) Induction of inaccessibility to *Pyricularia oryzae* by pre-inoculation of *P. grisea* in rice leaf-sheath cells, *Journal of Phytopathology* 134, 97–102.

Arimoto, Y., Homma, Y., Ohtsu, N., and Misato, T. (1976) Studies on chemically induced resistance of plants to diseases. I. The effect of a soaking of rice seed in dodecyl DL-alaninate hydrochloride on seedling infection by *Pyricularia oryzae*, *Annals of the Phytopathological Society of Japan* 42, 397–400.

Arimoto, Y., Homma, Y., Yoshino, R., and Saito, S. (1991) Generational succession of DL-alanine dodecylester HCl-induced resistance to blast disease in rice plants, *Annals of the Phytopathological Society of Japan* 57, 522–525.

Aver'yanov, A.A. and Lapikova, V.P. (1989) Fungitoxicity determined by active forms of oxygen in excretions of rice leaves, *Soviet Plant Physiology* 35, 873–881.

Aver'yanov, A.A., Lapikova, V.P., and Djawakhia, V.G. (1993) Active oxygen mediates heat-induced resistance of rice plants to blast disease, *Plant Science* 92, 27–34.

Aver'yanov, A.A., Lapikova, V.P., Umnov, A.M., and Dzhavakhiya, V.G. (1987) Generation of superoxide radical by rice leaves in relation to blast resistance, *Soviet Plant Physiology* 34, 301–306.

Bowles, D.J. (1990) Defense-related proteins in higher plants, *Annual Review of Biochemistry* 59, 873–907.

Cai, X.Z. and Zheng, Z. (1996) Effect of exogenous salicylic acid on resistance of rice seedlings to blast, *Chinese Rice Research Newsletter* 4(3), 8–9.

Cai, X.Z. and Zheng, Z. (1997) Biochemical mechanisms of salicylic acid-induced resistance in rice seedling to blast, *Acta Phytopathologica Sinica* 27, 231–236.

Cartwright, D.W., Langcake, P., and Ride, J.P. (1980) Phytoalexin production in rice and its enhancement by a dichlorocyclopropane fungicide, *Physiological Plant Pathology* 17, 259–267.

Cartwright, D.W., Langcake, P., Pryce, R.J., Leworthy, D.P., and Ride, J.P. (1981) Isolation and characterization of two phytoalexins from rice as momilactones A and B, *Phytochemistry* 20, 535–537.

Cha, J.S., Cho, B.H., and Kim, K.C. (1982) Differences of phytoalexin-like activities in compatible and incompatible combinations of rice cultivars and races of *Pyricularia oryzae*, and decrease of those activities by heavy application of nitrogen, *Korean Journal of Plant Protection* 21, 27–33.

Chester, K.S. (1933) The problem of acquired physiological immunity in plants, *The Quarterly Review of Biology* 8(2), 129–154 and 275–324.

Dixon, R.A. and Paiva, N.L. (1995) Stress-induced phenylpropanoid metabolism, *Plant Cell* 7, 1085–1097.

Du, L.C. and Wang, J. (1992) Activities and distribution of chitinase and β-1,3-glucanase in rice induced by *Pyricularia oryzae*, *Acta Phytopathologica Sinica* 18, 29–36.

Eyal, Y. and Fluhr, R. (1991) Cellular and molecular biology of pathogenesis related proteins, *Oxford Surveys of Plant Molecular and Cell Biology* 7, 223–254.

Fujita, Y., Sonoda, R., and Yaegashi, H. (1990) Leaf blast suppression by pre-inoculation of some incompatible lesion-type isolates of *Pyricularia oryzae*, *Annals of the Phytopathological Society of Japan* 56, 273–275.

Gangadharan, C. and Mathur, S.C. (1976) Di-ethyl sulphate induced blast resistance in rice variety Mtu 17, *Science and Culture* 42, 226.

Gottstein, H.D. and Kuć, J. (1989) Induction of systemic resistance in cucumber by phosphates, *Phytopathology* 79, 176–179.

Guo, J.R. and Luo, K. (1994) Studies on induced effects of rice varieties to blast disease, *Journal of Hunan Agricultural College* 20, 66–70.

Hammond-Kosack, K.E. and Jones, D.G. (1996) Resistance gene-dependent plant defense responses, *The Plant Cell* 8, 1773–1791.

Hemmi, T., Ikeya, D., and Inoue, Y. (1936) Influence of *Ophiobolus miyabeanus* on the penetration of *Piricularia oryzae*, in the host body, *Agriculture and Horticulture* 11, 953–964.

Homma, Y., Shida, T., and Misato, T. (1973) Studies on the control of plant diseases by amino acid derivatives (1) Effect of N-lauroyl-L-valine on rice blast, *Annals of the Phytopathological Society of Japan* 39, 90–98.

Iwano, M. (1987) Suppression of rice blast infection by incompatible strain of *Pyricularia oryzae* Cav., *Bulletin of Tohoku National Agricultural Experimental Station* No. 75, 27–39.

Iwata, M., Iwamatsu, H., Suzuki, Y., Watanabe, T., and Sekizawa, Y. (1982) Inducers of host peroxidase from rice blast fungus, *Annals of the Phytpthological Society of Japan* **48**, 267–274.

Iwata, M., Sekizawa, Y., Iwamatsu, H., Suzuki, Y., and Watanabe, T. (1981) Effects of plant hormones on peroxidase activity in rice leaf and incidence of rice blast, *Annals of the Phytopathological Society of Japan* **47**, 646–653.

Iwata, M., Suzuki, Y., Watanabe, T., Mase, S., and Sekizawa, Y. (1980) Effect of probenazole on the activities of enzymes related to the resistant reaction in rice plant, *Annals of the Phytopathological Society of Japan* **46**, 297–306.

Jørgensen, H.J.L. and Smedegaard-Petersen, V. (1994) Constraints in the implementation of induced resistance against necrotrophic pathogens of cereals, in P. Lepoivre (ed.), *Biological Control of Fruit and Foliar Disease*, Proc. Workshop CEC Prog. "Competitiveness of Agriculture and Management of Agricultural Resources", 1993, European Commission, Directorate-General for Agriculture, Working document for the European Commission reference F.II.3-SJ/0010, pp. 77–85.

Kamata, E. (1955) Relationship between iron, manganese, zinc and copper contents in rice plant and its resistance to the wilting disease (*Piricularia oryzae*), *Proceedings of the Crop Science Society of Japan* **23**, 281–282.

Kato, H., Kodama, O., and Akatsuka, T. (1993) Oryzalexin E, a diterpene phytoalexin from UV–irradiated rice leaves, *Phytochemistry* **33**, 79–81.

Kato, H., Kodama, O., and Akatsuka, T. (1994) Oryzalexin F, a diterpene phytoalexin from UV–irradiated rice leaves, *Phytochemistry* **36**, 299–301.

Kiyosawa, S. and Fujimaki, H. (1967) Studies on mixture inoculation of *Pyricularia oryzae* on rice. I. Effects of mixture inoculation and concentration on the formation of susceptible lesions in the injection inoculation, *Bulletin of the National Institute of Agricultural Sciences, Tokyo*, **D 17**, 1–19.

Kloepper, J.W., Tuzun, S., and Kuć, J.A. (1992) Proposed definations related to induced disease resistance, *Biocontrol Science and Technology* **2**, 349–351.

Kodama, O., Miyakawa, J., Akatsuka, T., and Kiyosawa, S. (1992) Sakuranetin, a flavanone phytoalexin from ultraviolet–irradiated rice leaves, *Phytochemistry* **31**, 3807–3809.

Kodama, O., Suzuki, T., Miyakawa, J., and Akatsuka, T. (1988) Ultraviolet-induced accumulation of phytoalexins in rice leaves, *Agricultural and Biological Chemistry* **52**, 2469–2473.

Langcake, P. and Wickens, S.G.A. (1975) Studies on the dichlorocyclopropanes on the host–parasite relationship in the rice blast disease, *Physiological Plant Pathology* **7**, 113–126.

Lapikova, V.P., Aver'yanov, A.A., Petelina, G.G., Kolomiets, T.M., and Kovalenko, E.D. (1994) Fungitoxicity of leaf excretions as related to varietal rice resistance to blast, *Russian Journal of Plant Physiology* **41**, 108–113.

Manandhar, H.K. (1996) *Rice Blast Disease: Seed Transmission and Induced Resistance*, Ph.D. Thesis, The Royal Veterinary and Agricultural University, Copenhagen, Denmark, pp. 110.

Manandhar, H.K., Jørgensen, H.J.L., Mathur, S.B., and Smedegaard-Petersen, V. (1998a) Resistance to rice blast induced by ferric chloride, di-potassium hydrogen phosphate and salicylic acid, *Crop Protection* **17**, 323–329.

Manandhar, H.K., Jørgensen, H.J.L., Mathur, S.B., and Smedegaard-Petersen, V. (1998b) Suppression of rice blast by preinoculation with avirulent *Pyricularia oryzae* and the nonrice pathogen *Bipolaris sorokiniana*, *Phytopathology* **88**, 735–739.

Manandhar, H.K., Mathur, S.B., Smedegaard-Petersen, V. and Thordal-Christensen, H. (1999) Accumulation of transcripts for pathogenesis-related proteins and peroxidase in rice triggered by *Pyricularia oryzae*, *Bipolaris sorokiniana* and UV-light, *Physiological and Molecular Plant Pathology* (in press).

Matsumoto, K., Suzuki, Y., Mase, S., Watanabe, T., and Sekizawa, Y. (1980) On the relationship between plant hormones and rice blast resistance, *Annals of the Phytopathological Society of Japan* **46**, 307–314.

Matsuyama, N. (1983) Time course alteration of lipid peroxidation and the activities of superoxide dismutase, catalase and peroxidase in blast-infected rice leaves, *Annals of the Phytopathological Society of Japan* **49**, 270–273.

Midoh, N. and Iwata, M. (1996) Cloning and characterization of a probenazole-inducible gene for an intracellular pathogenesis-related protein in rice, *Plant and Cell Physiology* **37**, 9–18.

Mori, A., Enoki, N., Shinozuka, K., Nishino, C., and Fukushima, M. (1987) Antifungal activity of fatty acids against *Pyricularia oryzae* related to antifungal constituents of *Miscanthus sinensis*, *Agricultural and Biological Chemistry* **51**, 3403–3405.

Namai, T., Kato, T., Yamaguchi, Y., and Hirukawa, T. (1993) Anti-rice blast activity and resistance induction of C–18 oxygenated fatty acids, *Bioscience, Biotechnology and Biochemistry* **57**, 611–613.

Namai, T., Kato, T., Yamaguchi, Y., and Togashi, J. (1990) Time–course alteration of lipoxygenase activity in blast–infected rice leaves, *Annals of the Phytopathological Society of Japan* **56**, 26–32.

Nikolaev, O.N. and Aver'yanov, A.A. (1991) Involvement of superoxide radical in the mechanism of fungicidic action of phthalide and probenazol, *Soviet Plant Physiology* **38**, 375–381.

Nikolaev, O.N., Aver'yanov, A.A., Lapikova, V.P., and Dzhavakhiya, V.G. (1994) Possible involvement of reactive oxygen species in action of some anti–blast fungicides, *Pesticide, Biochemistry and Physiology* **50**, 219–228.

Nikolaev, O.N., Novodarova, G.N., Kolosova, E.M., Aver'yanov, A.A., Dzhavakhiya, V.G., and Vol'pin, M.E. (1991) Induction of resistance of rice plants to blast under the influence of a phenanthroline complex of cobalt and its possible mechanism, *Doklady, Botanical Sciences* **316–318**, 26–28.

Ohta, H., Shida, K., Peng, Y.L., Furusawa, I., Shishiyama, J., Aibara, S., and Morita, Y. (1991) A lipoxygenase pathway is activated in rice after infection with rice blast fungus *Magnaporthe grisea*, *Plant Physiology* **97**, 94–98.

Ohata, K. and Kozaka, T. (1967) Interaction between two races of *Piricularia oryzae* in lesion formation in rice plants and accumulation of fluorescent compounds associated with infection, *Bulletin of the National Institute of Agricultural Sciences, Tokyo*, C 21, 111–135.

Ouyang, G.C., Ying, C.Y., Zhu, M.H., and Xue, Y.L. (1987) Induction of disease resistance by spores and toxins of *Pyricularia oryzae* in rice and its relation to the phenylpropane pathway, *Plant Physiology Communications* **4**, 40–42.

Park, S.K. and Kim, K.C. (1983) Effects of mixing and reciprocal inoculation with compatible and incompatible races of *Pyricularia oryzae* on the enlargement of disease lesions of rice blast, *Korean Journal of Plant Protection* **22**, 300–306.

Ragab, M.M. and Afifi, M.A. (1988) Effect of microelements zinc, manganese and iron on blast disease of rice, in *Abstracts of Papers*, 5th International Congress of Plant Pathology, Kyoto, Japan, p. 251.

Reimmann, C., Hofmann, C., Maugh, F., and Dudler, R. (1995) Characterization of a rice gene induced by *Pseudomonas syringae* pv. *syringae*: requirement for the bacterial *lemA* gene function, *Physiological and Molecular Plant Pathology* **46**, 71–81.

Reuveni, R., Agapov, V., and Reuveni, M. (1994) Foliar spray of phosphates induces growth increase and systemic resistance to *Puccinia sorghi* in maize, *Plant Pathology* **43**, 245–250.

Schaffarth, U., Scheinpflug, H., and Reisener, H.J. (1995) An elicitor from *Pyricularia oryzae* induces resistance responses in rice: isolation, characterization and physiological properties, *Physiological and Molecular Plant Pathology* **46**, 293–307.

Schweizer, P., Buchala, A., and Métraux, J.P. (1995) The octadecanoic pathway mediates defence responses against pathogen attack in rice plants, *Journal of Cell Biochemistry* **21**, 490 (Abstract).

Seguchi, K., Kurotaki, M., Sekido, S., and Yamaguchi, I. (1992) Action mechanism of *N*-cyanomethyl-2-chloroisonicotinamide in controlling rice blast disease, *Journal of Pesticide Science* **17**, 107–113.

Sekido, H., Endo, T., Suga, R., Kodama, O., Akatsuka, T., Kono, Y., and Takeuchi, S. (1986) Oryzalexin D (3,7-dihydroxy-(+)-sabdaracopimaradiene), a new phytoalexin isolated from blast infected rice leaves, *Journal of Pesticides Science*, **11**, 369–372.

Sekido, H. and Akatsuka, T. (1987) Mode of action of oryzalexin D against *Pyricularia oryzae*, *Agricultural and Biological Chemistry* **51**, 1967–1971.

Sekizawa, Y., Komiya, I., Iwata, M., Watanabe, T., Murata, S., and Umemura, K. (1983) A kinetic survey on an activation of enzyme participating in host–defence mechanism in rice leaf infected with blast fungus, *Journal of Pesticide Science* **8**, 121–123.

Sengupta, T.K. and Sinha, A.K. (1987) Phytoalexin inducer chemicals for control of blast (Bl) in West Bengal, *International Rice Research Newsletter* **12**(2), 29–30.

Shen, Y., Huang, D.N., Qiu, D.W., Fan, Z.F., Wang, J.X., and Yuan, X.P. (1990) Exploration of induced resistance in rice plants by buff pigment mutants of *Pyricularia oryzae*, *Chinese Journal of Rice Science* **4**(3), 139–142.

Sinha, A.K. (1990) Basic research on induced resistance for crop disease management, in P. Vidhyasekaran (ed.), *Basic Research for Crop Disease Management*, Daya Publishing House, Delhi, pp. 87–101.

Smith, J.A. and Métraux, J.P. (1991) *Pseudomonas syringae* pv. *syringae* induces systemic resistance to *Pyricularia oryzae* in rice, *Physiological and Molecular Plant Pathology* **39**, 451–461.

Song, F.M., Ge, X.C., Zheng, Z., Wu, W.L., and Wu, Y.L. (1994) Effect of two octadecadienoic acids on rice resistance to blast at seedling stage, *Chinese Journal of Rice Science* **8**(3), 162–168.

Takano, S., Suzuki, T., and Sahashi, Y. (1972) Inhibitory actions upon the rice blast fungi by amino acids, *N*–benzoyl amino acids and N–phenacetyl amino acids, *Journal of the Agricultural Chemical Society of Japan* **46**, 309–312.

Thieron, M., Schaffrath, U., Reisener, H.J., and Scheinpflug, H. (1995) Systemic acquired resistance in rice: studies on the mode of action of diverse substances inducing resistance in rice to *Pyricularia oryzae*, *Mededelingen Faculteit Landbouwkundige en Toegepaste Biologische Wetenschappen, Universiteit Gent.* **60**(2b), 421–429.

Tomita, H. and Yamanaka, S. (1983) Studies on the resistance reaction in the rice blast disease caused by *Pyricularia oryzae* Cavara. I. The pathological changes in the early infection stage of the inner epidermal cells of leaf sheath, *Annals of the Phytopathological Society of Japan* **49**, 514–521.

Tuzun, S. and Kuć, J. (1991) Plant immunization: an alternative to pesticides for control of plant diseases in the greenhouse and field, in J. Bay–Petersen (ed.), *Biological Control of Plant Diseases*, FFTC Book Series No. 42, Taiwan, pp. 30–40.

Uehara, K. (1958) On the production of phytoalexin by the host plant as a result of interaction between the rice plant and the blast fungus (*Piricularia oryzae* Cav.), *Annals of the Phytopathological Society of Japan* **23**, 127–130.

Watanabe, T. (1951) Studies on the vaccine therapy of the blast disease of rice plants. 6. The effects of various vaccines of rice blast fungus to the development of the rice plant and the causal fungus, *Shokubutsu Byogai Kenkyu, Kyoto* **4**, 55–63.

Watanabe, T. (1957) Studies on the vaccine therapy of the blast disease of rice plants. X. The detection of the plant hormone, *Annals of the Phytopathological Society of Japan* **22**, 143–147.

Watanabe, T., Sekizawa, Y., Shimura, M., Suzuki, Y., Matsumoto, K., Iwata, M., and Mase, S. (1979) Effects of probenazole (Oryzemate® on rice plants with reference to controlling rice blast, *Journal of Pesticide Science* **4**, 53–59.

Xiao, J.Z., Ohshima, A., Kamakura, T., Ishiyama, T., and Yamaguchi, I. (1994) Extracellular glycoprotein(s) associated with cellular differentialtion in *Magnaporthe grisea*, *Molecular Plant Microbe Interactions* **7**, 639–644.

Yoshi, K. (1950) Studies on *Cephalothecium* as a means of artificial immunization of agricultural crops. II. On the effect of treatment by the dried mycelium powder of *Cephalothecium* on the development of leaf blast in rice seedlings, *Annals of the Phytopathological Society of Japan* **14**, 9–10.

Yoshida, H. (1992) Antiblast activity and mode of action of 2-chloroisonicotinamides, *Journal of Pesticide Science* **17**, S241–S249.

Zhang, J.T., Duan, G.M., and Yu, Z.Y. (1987) Relationship between phenylalanine ammonia–lyase (PAL) activity and resistance to rice blast, *Plant Physiology Communications* **6**, 34–37.

BREEDING STRATEGY FOR RICE BLAST RESISTANCE IN EGYPT

I.R. AIDY, A.O. BASTAWISI and M.R. SEHLY

Rice Research & Training Center, Sakha, 33717, Kafr El-Sheikh, Egypt

Abstract

Rice blast disease caused by the fungus *Pyricularia grisea* is one of the most biotic stresses constrains in rice productivity in Egypt. Breeding for blast resistance are our concern in varietal improvement program. Since the Egyptian cultivars were showing susceptibility. Understanding gene type, mode of inheritance and stability of the resistant varieties are essential for transfer the blast resistance to popular cultivars. The previous studies for exotic and local rice varieties under Egyptian conditions showed that inheritance of blast resistance was dominant in most cases and one to three major genes controlled the mode of resistance.

Our strategy for breeding to blast resistance are: 1) gene accumulation resulted of introducing two leading rice cultivars, Sakha 101, and Sakha 102; 2) gene pyramiding resulted of introducing the high yielding rice variety Giza 178; 3) gene rotation where resistance cultivars are concentrated in hot spots in Nile Delta. Multiline varieties is another strategy but still under evaluation. The level of partial resistance is under our concern in all strategies.

The stability of resistance of the twenty Egyptian rice cultivars to leaf blast caused by *Pyricularia grisea* was studied using three stability parameters viz. Low mean disease score, small regression coefficient and least deviation values from regression. Cultivars like IR 28, Sakha 101, Giza 178, Giza 175, Sakha 102, and Giza 181, showed low mean disease score (1.71,1.80,1.84,1.88 respectively),and small regression coefficient values ranged between (0.03 and 0.26) as well as least deviation from regression. Therefor these cultivars were found to show stable resistance to leaf blast and can be used as new sources for blast resistance in breeding program.

1. Introduction

Rice blast disease, Caused by *Pyricularia grisea* (Cooke) Sacc. (*Pyricularia oryzae* Cavara) (teleomorph : *Magnaporth grisea*) is one of the most serious biotic stress constraint to rice productivity in Egypt (Aidy *et al*, 1994). Two phases of infection are known; leaf blast during the vegetative stage of rice plant

D. Tharreau et al. (eds.), Advances in Rice Blast Research, 105–111.

development and panicle infection during reproductive stage. This latter effect usually have more economic importance since it is direct on yield and quality reductions (Khush , 1977; Ou, 1985; Roumen, 1992; Surek & Beser, 1997). Breeding for developing blast resistant varieties are our concern in varietal improvement program. Since the Egyptian cultivars were showing susceptibility. However, the resistance is broken down in short time 2-6 years (Kiyosawa, 1965) with cultivars released as resistant showing susceptibility after few years of widespread cultivation.

Understanding gene type, mode of inheritance and stability of the resistant cultivars are essential for transfer the blast resistance to popular cultivars. The previous studies for exotic and local rice varieties under Egyptian conditions showed that inheritance of blast resistance was dominant in most cases and one to three major genes controlled the mode of resistance (Omar *et al* , 1970; El - Azizi, 1972, Maximos, 1974; Balal *et al*, 1977; Aidy 19874; Maximos *et al*, 1985) and El - Malky 1997).

The objectives of this study are :

I) Emphasize the breeding strategy for blast resistance in Egypt.

II) Clarify varietal stability under Egyptian conditions.

I - Breeding strategy for blast resistance :

Three main strategies are utilizing under Egyptian conditions :

a) Gene accumulation :

Nahda was a cultivated variety released in 1953 with a good level of resistance to blast, after the spread of the variety resistance was broken- down. Two resistant parents were crossed to Nahda: the American variety Calady 40 produced the variety Giza 171 and the Japanese variety Kinmaze to produce the resistant variety Giza 172. Both Giza 171 and Giza 172 were released 1977 and cultivated in a large area until the resistant

genes were broken-down. In 1980's another resistant genes were applied when resistant gene from local cross CZ 242   (Nagina / Norin 25) were introduced to the variety Giza 172 to introduce Giza 176 where its resistance was broken-down.   Another resistant gene was introduced from the variety Milyang 79 to introduce the variety Sakha 101 which has a good level of resistance to blast and released in 1997. Another variety, Sakha 102, was released in 1997 with the same strategy as shown in the figure (1)

Fig. 1 : Gene accumulation scheme for rice varieties Sakha 101 and Sakha 102

1- Sakha 101 :

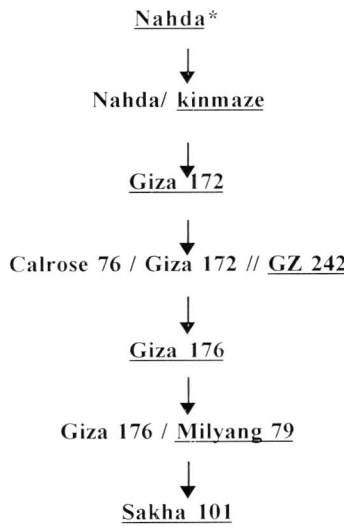

Nahda*

↓

Nahda/ kinmaze

↓

Giza 172

↓

Calrose 76 / Giza 172 // GZ 242

↓

Giza 176

↓

Giza 176 / Milyang 79

↓

Sakha 101

2- Sakha 102 :

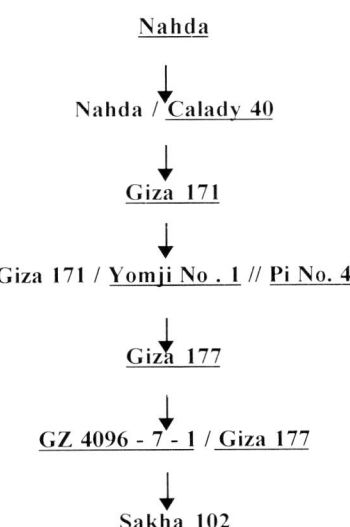

Nahda

↓

Nahda / Calady 40

↓

Giza 171

↓

Giza 171 / Yomji No . 1 // Pi No. 4

↓

Giza 177

↓

GZ 4096 - 7 - 1 / Giza 177

↓

Sakha 102

• The line under the variety represents its resistance at releasing

b) **Gene Pyramiding** :

Where two or more resistant genes could be incorporated in one variety . This strategy were applied in Egypt to produce the variety Giza 175, where three resistant parents (IR 28, IR 1541 - 6, and Giza 180) were incorporated to the variety Giza 14, followed by the variety Giza 178 where fourth source of resistance were applied from the variety Milyang 49, as the scheme showing below :

IR 28 / **IR 1541 - 6** // **Giza 180** / **Giza 14**

↓

Giza 175 / **Milyang 49**

↓

Giza 178

c) Gene rotation :
Since rice area concentrated in North Delta gene rotation has a small role in our strategy. However in the hot spots of blast mainly rice areas where the poultry farms are concentrated the resistant varieties are recommended .

d) Others :
Under the studying utilizing isogenic lines. Besides all the previous strategies, partial resistance always is under our concern that is why we study the stability of Egyptian rice varieties for blast resistance to maintain a high level of partial resistance in our new released varieties .

II. STABILITY OF RESISTANCE TO LEAF BLAST DISEASE IN SOME EGYPTIAN RICE CULTIVARS :

The extensive studies on the genetic complete blast resistance (major genes) revealed that there are several gene (s) against *Pyricularia oryzae* in rice (Khush, 1977; Balal *et al,* 1977; Aidy, 1984 and Maximos *et al*, 1985). However, complete resistance controlled by major genes is quite in stable (Kiyosawa, 1981). It is due to the presence of a large number of races of pathogen and blast fungus which are highly variable as well as genetic potential for pathogenic variation were noticed and reported by many investigators and in several countries (Padmanabhan *et al*, 1970; Abdel- Hak, 1981; El. Refaei *et al*, 1986 and Shely *et al*, 1990). Therefore, this study was undertaken to test the stability of resistance to leaf blast disease *(Pyricularia grisea)* in some Egyptian rice cultivars in five years at three locations.

2. Materials and methods

Twenty Egyptian rice cultivars were evaluated for reaction to leaf blast in the uniform Blast Nursery (Ou, 1985 and Sehly *et al*, 1990) for five successive seasons during 1993 - 1997 at three locations Viz., Sakha, Gemmiza and Zarzoura. All normal agronomic practices were followed and randomized complete block design with three replications was used . A highly susceptible rice cultivar Sabieny was used as an infector .Reaction of leaf blast was estimated after 30 days of seeding following the standard scole as recommended by IRRI (1993). The method outlined by Eberhart and Russel (1966) was used to analyze the date for determining the resistance stability of different cultivars .

3. Results and discussion

The analysis of variance date for the twenty rice cultivars exposed to *Pyricularia grisea* in the uniform blast nursery at three locations, namely, Sakha, Gemmiza and Zarzora, which were located in north medial and west of delta for five years (1993- 1997) constituting 15 different environments are presented in Table 1. The data revealed the existence of highly significant differences among the evaluated genotypes (G), environments (E) and G x E interactions, indicating the presence of variation between the tested cultivars and years for leaf blast reaction.   So, it's clarified that each cultivar responded characteristically to change in the environment. similar results were reported by Bhardwaj and Singh (1983) and Kumar *et al* (1996).

Table 1: Analysis of variance for twenty rice cultivars exposed to *Pyricularia grisea* in blast nurseries at three locations for five years .

Source of variation	of	Ms
Reps . in Environments	30	0.95**
Genotypes (G)	19	200.15**
Genotypes x Environment :		
Interaction (G E)	266	2.25**
Environments	14	16013**
Error	570	0.22

** **significant at 1% level .**

The data of leaf blast reaction showed that a highly significant difference among the twenty rice cutivars for mean disease score. Eight cultivars, Yabani Lulu, Nahda, Giza159, Giza14, Giza172, Giza171, Giza170 and Reiho were susceptible. The disease score for these cultivars were 6.86, 6.82, 6.53, 6.28, 6.04, 6.00, 5.75 and 5.55 respectively. Whereas nine cultivars viz, IR 28 Sakha 101, Giza 178, Giza175, Sakha102, Giza181, Giza180 GZ1368-5-4 and Giza177 were resistant with mean disease score ranged between 1.71 (IR 28) and 2.11 (Giza 177). However the cultivars Agami and Nabatat Asmer were moderately resistant, while Giza 176 was moderately susceptible (Table 2). The estimates of the three stability parameters viz, mean disease score, regression coefficients (bi) and deviation from regression (S^2d) for the twenty rice cultivars are given in table 2. IR 28, Sakha 101, Giza 178, Giza 175,   Sakha 102 and Giza 181 were more stable in their resistance than other cultivars since the regression coefficient value for these cultivars were least among the resistant cultivars and deviation from regression were also quite small, followed by Giza 180, GZ 1368-5-4 and Giza 177 with a slightly higher disease score and slightly higher regression than the cultivars mentioned before but S^2d almost equal to zero. For the susceptible and moderately resistant or susceptible cultivars, it has high mean disease score with S^2d significantly greater than zero. Therefore these cultivars were less stable in their response to *Pyricularia grisea* (Table 2)

Table 2 : Mean leaf blast score and stability parameters for twenty Egyptian rice cultivars under 15 different environments.

Cultivars	Mean	by	S^2d
Nabatat Asmer	3.55	0.395	1.292
Yabani Lulu	6.86**	2.352**	0.919
Agami	2.60	0.156	0.658
Giza 14	6.28**	1.458**	0.944
Nahda	6.82**	1.992**	0.502
Giza 170	5.75**	1.925**	0.673
Giza 159	6.53**	2.346**	1.714
Giza 171	6.00**	1.647**	0.867
Giza 172	6.04**	1.624**	0.032
Giza 180	1.93	0.066	0.003
Reiho	5.55**	2.233**	0.118
IR 28	1.71	0.170	0.104
GZ 1368-5-4	1.97	0.322	0.058
Giza 181	1.88	0.036	0.005
Giza 175	1.84	0.234**	0.044
Giza 176	4.82	2.400	0.614
Giza 177	2.11	0.464	0.063
Giza 178	1.81	0.129	0.055
Sakha 101	1.80	0.265	0.090
Sakha 102	1.88	0.104	0.002

** **significant at 1% level.**

Thus, from the above stability of resistance of some Egyptian rice cultivars results it appeared that the cultivars IR28, Sakha 101, Giza 178, Giza 175 and Sakha 102 had stable resistance to leaf blast disease and they are widely adapted for general cultivation under different environment conditions. However, the presence of various races in different epidemiological regions may erode the stability of resistance these of cultivars if these races happen to be specifically virulent on them. Similar rsesults were reported by Bhardwaij and Singh (1983), Kumar (1996) and Guimaraes *et al* (1998) .

References

Abdel - Hak, T.M. 1981. Rice diseases and assessment of their compact in Egypt Proceedings First National rice Institute Conference, Feb. 21- 25 : 114 - 121.

Aidy, I.R. 1984. A study on the genetic behavior of resistance to rice blast and brown spot diseases in rice. Ph.D. Thesis, Fac. Agric., Ain. Shams Univ., pp 32- 47.

Aidy, I.R ; E.A. Draz and M.R. Sehly. 1994. Rice varietal resistance to blast disease under different test conditions. Proc. 6th Conf. Agron., AL. Azhar Univ., Cairo, Egypt, Vol. 1, Sept 1994: 223 - 230.

Balal, M.S ; A.K. A. Selim; S. H. Hassanien, and M. A. Maximos. 1977. Inheritance of resistance to leaf and neck blast in rice. Egyptian Journal of Genetic and Cytology 6 (2) : 332- 341.

Bhardwaj , C.L. and B.M. Singh . 1983 . The stability of resistance *to Pyricularia Oryzae* in rice . Indian Phytopath . 36 (1) : 422 - 426 .

Eberhart, S. A. and W. L. Russel. 1966. Stability parameters for comparing varieties. Crop. Sci. 6: 36 - 40.

EL-Azizi, A. M. 1972. Genetic behavior of resistance to blast disease, *Pyricularia Oryzae* Cav. in rice and some other economic characters. Ph.D. Thesis, Fac. Agric. , Ain Shams Univ. , pp. 35 - 51.

EL-Malky, M. M. 1997. Studies on some genetic characters in rice using tissue culture techniques. M.Sc. Thesis, Fac. Agric., Shibin El-Kom Univ. , pp. 74 - 102.

EL-Refaei, M. I. ; M. M. Ragab and M. A. Afifi, 1986. Studies on blast disease of rice in Egypt caused by *Pyricularia Oryzae* Cave. 2. Physiological races and varietal resistance. Egyptian Society of Applied Microbiology. Proc. VI. Conf. Microbiol., Cairo.

Guimaraes, E.P: M.C Amezquita and F. Correa - Victoria . 1998 Determination of minimum number of growing seasons for assessment of disease resistance stability in rice . Crop Sci. Vol . 38 : 67 - 71 .

International Rice Research Institute (IRRI) 1993. Standard evaluation system for rice, 3rd Edition July 1993, 17 - 18.

Khush, G. S. 1977. Disease and insect resistance in rice. Advances in Agronomy Vol. 29 : 265 - 335.

Kiyosawa, S. 1965. Ecological analysis on breakdown of resistance in resistant varieties and breeding counterplan against it. (In Japanese), Nogoyo Gijutsu (Agr. Tech.). 20: 465-470.

Kiyosawa, S. 1981. Gene analysis for blast resistance. Oryzae .18 : 196 - 203.

Kumar, S.S ; S.V. Kumar and V.N. Reddy . 1996. Phenotypic stability in rainfed rice (Oryza sativa) Indian Journal of Agricultural Sciences . 66 (12) 705 – 707 .

Maximos, M. A. 1974. Inheritance of resistance to Leaf blast *disease Pyricularia Oryzae* Cav. in rice and its relation to neck blast and some other economic characters. Ph. D. Thesis, Fac. Agric. , Ain. Shams Univ. , pp 40- 166 .

Maximos, M. A; A. A. Tayel; R. A. El. Adawy and I. R. Aidy. 1985. Genetic analysis of resistance to blast disease in rice (*Pyricularia Oryzae* Cav .). Annals Agric . Sci., Fac. Agric ., Ain Shams Univ., Cairo, Egypt Vol. 30 (1) : 383 - 398 .

Omar, A. M. ; S. H. Hassanien; A. K. A. Selim, and M. A. Maximos. 1970. Genetic behavior of field reaction to blast disease of rice in U. A. R. Ain Shams Univ. Fac. Agric. , Cairo, Research Bull. No. 567 : 1- 12.

Ou, S. H. 1985. Rice diseases, 2nd edition. Commonwealth Mycological Institute, Kew, Surrey, England. 330 pp.

Padmanabhan, S. Y. ; N. K. Chakrabarti; S. C. Mathur and J. Veeraaghovon. 1970. Identification of pathogenic races of *P. Oryzae* in India. Phytopathology, 60 : 1574 - 1577.

Roumen, E. C. 1992. Effect of leaf age on components of partial resistance in rice to leaf blast. Euphytica 63 : 271-279.

Sehly, M. R. ; Z. H. Osman ; H. A. Mohamed and A. O Bastawisi. 1990. Reaction of Some rice entries to *P. Oryzae*_ Cav. and race picture in 1988 . The Sixth Congress of Phytopathology. Cairo. Mach 1990 : 159 - 170.

Surek, H. and N. Beser. 1997. Effect of blast disease on rice yield. International Rice Research Notes. Vol. 22. (1) : 25- 26.

BREEDING FOR BLAST RESISTANCE IN RICE IN WEST AFRICA

B. N. SINGH[1], M. P. JONES[1], S. N. FOMBA[2], Y. SERE[1], A. A. SY[1],
K. AKATOR[1], P. NGNINBEYIE[3], AND S. W. AHN[4]
[1] *WARDA / ADRAO, Bouake, Côte d'Ivoire;* [2] *Rice Research Station,
Rokupr, Sierra Leone;* [3] *IRA, Dscahng, Cameroon;* [4] *IRRI, Manila,
Philippines*

1. Introduction

Rice is grown in an area of 4.3 million ha in 17 WARDA member countries of West
Africa (FAO, 1994). The total annual production is 7.4 million tons of paddy with an
average yield of 1.7 tons ha^{-1}, which is the lowest compared to other rice growing
regions of world .The rice import in WARDA member states in 1995 was 2.43 million
tons with a value of 684 million US$ (WARDA Data Base, 1998). The average yield in
the region is low due to presence of many biophysical constraints, which reduces the
yield potential of the varieties. Rice blast (*Pyricularia grisea*) is one of the major biotic
constraints affecting rice production. It is a widespread and most serious disease in all
the rice growing ecologies (Awoderu, 1974; Bidaux, 1978; John *et al.*, 1985; Fomba and
Taylor, 1994). Seedling blast occurs in nursery stage in transplanted rice, leaf blasts
under direct seeded uplands and rainfed lowlands, and panicle blasts in all the ecologies.

The most effective and economical way to control blast is the use of resistant
varieties (Grill and Khush, 1979; Ou, 1985). However, resistant varieties do not
necessarily remain resistant for a long time, and often succumb to new races of blasts
due to selection pressure in pathogen populations. Breeding for stable and durable
resistance to blasts has been one of the major breeding objectives of the rice breeder and
pathologists. Efforts are underway to develop and select high yielding lines with blast
resistance, and tolerance to different biotic and abiotic constraints in the region through
various breeding strategies.

RICE GROWING ENVIRONMENTS (RGE) AND BLAST

The WARDA (1997) has classified RGE in 6 types based on its production potential.
These are uplands, rainfed lowlands, irrigated sahel, irrigated humid and sub-humid,
mangrove swamp, and deep water. The per cent area in different ecologies are, 40% for

D. Tharreau et al. (eds.), Advances in Rice Blast Research, 112–128.

uplands, 38% for rainfed lowlands, 3% irrigated sahel, 9% for irrigated humid and moist zone, 4% mangrove swamp, and 6 % in deep water. Productivity is maximum in irrigated sahel, but the production potential is maximum in rainfed lowlands (inland valley swamps and floodplains) due to its better soil quality and hydrology. Based on climate, length of growing period, annual rainfall, altitude, rainfall pattern, major rainfed crops and vegetation, the agroecological zones (AEZ) have been further classified as Sahel, Sudan savanna, northern Guinea savanna, southern Guinea savanna, derived and coastal savanna, humid forest, and mid altitude moist savanna (Jagtap, 1995; Singh *et al.*, 1997). Leaf blast is common in all RGE and all the AEZ, except irrigated sahel. It is more common in uplands, followed by rainfed lowland, irrigated mangrove swamp, and deep water.

RICE BLAST IN UPLAND RICE

Around 1.7 million ha area in west Africa is under upland rice with an average yield of 1 t ha^{-1}. Nigeria, Côte d'Ivoire, Guinea, Sierra Leone and Togo are the major upland rice growing countries. In Nigeria, early released varieties like FARO 1 (BG 79), FARO 3 (Agbede), and Ofada were observed as susceptible to blast (Awoderu, 1974). Olufowote (1977) reported 15-100 % loss in grain yield in FARO 11 (OS 6). Most of the tropical upland *japonica* varieties like Moroberekan, and Iguape Cateto in Côte d'Ivoire; and LAC 23 in Liberia and Guinea are tall types. Rok16 (Ngovie) released in 1978 in Sierra Leone, is a widely grown upland *indica* variety and durable resistant to blast (Table 1). Farmers prefer it due to awns,which reduces the bird damage. As these varieties are being grown under low input systems,and slash and burn practices disease severity is low in farmer's field. Ex-china in Nigeria, Peking in Gambia and Guinea Bissau, and Rok 3 in Sierra Leone are widely grown varieties by farmers, but susceptible to leaf blast. In Nigeria, ITA150 (FARO 46) shows durable resistance to blast. Its area has been on increase since 1987 and in 1998 around 50 % of upland rice area of total 500,000 ha in the country is under this variety (S.Fagade, personal communication). It is being grown from humid forest in south to moist savanna and dry savanna in north. For a variety like ITA 150 to be grown in a large area, in additional to blast resistance(Immune type), other traits like grain type (long bold), easy threshability, early maturity (90 days), intermediate plant height (110cm), tolerance to drought and acid uplands, resistance to leaf scald, intermediate amylose content and stable yield are important traits (IITA, 1984; Singh *et al.*,1997; Imolehin *et al.*,1997a). ITA257 (FARO 45) was equally high yielding, early maturity, drought tolerant, and immune to blast, but its area declined over the years due to its hard threshing grains, medium bold grain type, and shorter plant height. Some of the recently released varieties like FARO 47 (ITA117), FARO 48 (ITA301), FARO 49 (ITA 315), in Nigeria; FKR 33 in Burkina Faso, WAB 56-50, and WAB 56-104 in Côte d'Ivoire are resistant to blasts (Imolehin *et al.*, 1997a; Sie *et al.*, 1997). Its durability will be monitored in future.

TABLE 1: Resistant and susceptible rice varieties in different countries of West Africa

Countries	Susceptible varieties	Resistant line/ varieties	Hot spot locations
Benin	DS290, IR442, IR8, IR20, IR22, TN1, Gambiaka	ADNY11, ITA222, 11365, ITA212,NIARIS85-12, INARIS 88,ITA 304	Oueme valley, Moussourou,Save,
Burkina Faso	IR 1529 , ITA 306,C 74	FKR 5 (IRAT 144), FKR 33, ITA123, IR64, FKR44	Karfiguela, Farko-Ba
Cameroon	IR 46 (RL)	CICA8, ITA300, BKN7033, WITA4,	Mbo Plains, Bokle(near Garoua)
Côte d'Ivoire	Bouake 189	WITA1, WITA3,WITA 7;	M'be,Gagnoa,Man, Korhogo,Tombokro
Chad	IR46(RL)	Tox 728-1,WITA 4	Mala
Ghana	IR5, IR8, IR442, GR18	GR 19, GR 21, Sikamo	Nyankpala
Gambia	Peking	ADNY11, IR54, DJ11-509, IET3137, WITA2	Sapu
Guinea	IR5	Suakoko8, CK4, CK73, CK41, CK44	Kilissi
Guinea Bissau	Peking	Rok5, RD 15	Contuboel
Mali	Gambiaka	BG90-2, Kogoni91-1	Sikasso, Baguineda
Mauritania	-	Jaya, IR28,ITA306	-
Niger	-	IR1529-680-3, IR54	-
Nigeria	OS6, FARO8, ITA 222, Ex-China, ITA 212, ITA306, BG90-2	ITA150,ITA257,ITA301, ITA 315, Sipi 692033, WITA 4, WITA1	Ikenne, Ibadan, Badeggi, Onne, kadawa
Senegal	IR8, IR442, IKP,	DJ 11-509,SE 319-G, WITA2, IR 13240-108-2-2-3, ITA 306	Djibelor,Sefa
Sierra Leone	CCA, Rok 3, Kuatic Kundir	Suakoko 8, Mahsuri,, Rok14, Rok5, WAR 1 ADNY11, Rok 16WAR 77,	Rokupr
Togo	C 74	ITA 222, TGR 1, ITA 212, IR 841,WITA 4	Adeta

RICE BLAST IN RAINFED LOWLAND AND IRRIGATED RICE

Around 1.6 million ha area in west Africa is under this ecology, with an average yield of 1.4 tons per ha. Countries like Nigeria, Senegal, Sierra leone, Ghana, Benin, Burkina Faso, Chad and Togo. FARO 8 (MAS 2401), an introduction from Indonesia is most popular variety for rainfed lowlands in Nigeria. It was released in 1963 and till 1965, it was resistant, but in 1966 it was observed as susceptible (Awoderu and Ojomo,1981). In Burkina Faso and Togo, earlier C74 was found good for rainfed lowland, but it was observed as susceptible to leaf and panicle blasts (Poisson, 1978; Akator, 1981).

After the development of semi-dwarf rice from IRRI, many of them were released in 70's in West Africa. Some of the early introductions like IR8, IR442 , IR20, IR22 and IR 5 were released in Nigeria, Ghana, Benin, Cameroon, Cote d'Ivoire, Liberia and others. In the initial years, they had high yield, but soon succumbed to leaf blast and were withdrawn from cultivation (Awoderu and Esuruso,1975; Carpenter, 1977; Bidaux,1978; Vodouhe *et al.*, 1981). Other initially blast resistant varieties recommended from time to time, too succumbed to blast later. Farmers not differentiate the varieties for rainfed and irrigated lowland ecologies and often plant irrigated varieties in rainfed lowlands. FARO 29 (BG 90-2) released in 1984 in Nigeria was susceptible to blast by 1990. However, BG90-2 is the single most widely grown variety (60,000 ha) in Mali and other Sahelian counties, where blast is not a major problem. FARO36 (ITA 222) was observed susceptible to leaf blast in 1988 in northern Nigeria, two years after its release (FACU, 1988). In northern Cameroon and Chad, IR 46 is the most popular variety for irrigated rice. When farmers plant it under rainfed lowland drought prone conditions, it is severely affected by leaf blast. During monitoring team visit in 1992, severe leaf blast infection was observed in IR 46 at Bokle research site, near Garoua (WARDA, 1992a). Recently, FARO35 (ITA212), and FARO37 (ITA306) was observed as susceptible to leaf and panicle blasts in farmers field in Kaduna state, Nigeria in 1995 and 1996 (Singh *et al.*,1997). In 1997 wet season, around 10,000 ha was affected by panicle blasts in Jigawa state, northern Nigeria (Singh, 1997).ITA 212 has blast resistance gene from Tetep and initially it had moderately resistant reaction to blast (IITA, 1980). Mahsuri, the most popular rainfed lowland variety in south and Southeast Asia is released as Rok 25 in Sierra Leone was observed in 1997 as susceptible to leaf blast in Côte d'Ivoire. In first lowland rice breeding Task Force meeting in 1992, leaf and panicle blast was rated as premier production constraints in rainfed lowland ecologies in West Africa.(WARDA, 1992b). Two of the recently released varieties in Nigeria, viz, FARO44 (Sipi692033) and FARO50 (ITA 230) are resistant to leaf blast (Imolehin *et al.*,1997b). Some of the other released varieties in past, like ITA 123 in Burkina Faso, ADNY 11 in Benin Republic and Sierra Leone, and Suakoko 8 in Liberia and Sierra Leone are resistant to blast (Table 1). The recently released varieties like WITA1 and WITA 3 in Côte d'Ivoire, and Cisadane in Nigeria are resistant to blast (Singh, 1998). The durability of their resistance will be monitored.

RICE BLAST IN MANGROVE SWAMP AND DEEP-WATER RICE

Mangrove swamp rice is grown in around 200,000 ha in West Africa. It is one of the major rice growing ecology in Sierra Leone, Guinea Bissau, Senegal, Guinea and Gambia. Transplanting is a more common practice in its cultivation. Some of the popular varieties are Rok5, Rok10, Kuatic Kundur, WAR1, and WAR 77. Seedling blast is a major problem in nursery grown under upland conditions. Kuatic Kundur, a popular variety with farmers is severely affected by seedling blast. Deep-water rice is grown in around 250,000 ha mainly in Mali, Guinea Bissau, Guinea, Niger and Nigeria. Leaf blast is common in vegetative stage, but due to taller plant type damage is not severe. Gambiaka variety in Benin Republic and Mali is susceptible to seedling blast in early stage (Table 1).

Rice blast epidemics in irrigated rice have been reported in Japan (Fukunaga, 1965), Korea (Crill *et al.*,1982; Kim,1994), Egypt (Sehli and Balal, 1994), and China (Shen and Lin, 1994). In West Africa, such epidemic has not occurred on a large area due to diversity of rice growing environments, and large number of varieties, both *japonica* and *indicas*, grown by farmers. Yield losses are severe at individual plot levels only. Tropical upland *Japonicas* seems to be more adapted and resistant to blast than *indicas*.

ECOTYPE AND LOCATION SPECIFICITY OF BLAST RESISTANCE

Varietal differences in blast resistance have been observed in West Africa and it varies from one country to another, and also within the country. FARO8, a popular rainfed lowland variety in Nigeria, is resistant to leaf blast in Côte d'Ivoire and highly susceptible in Nigeria. Similarly, ITA 306 is susceptible to leaf blast in rainfed lowlands in Nigeria and resistant in Côte d'Ivoire and Burkina Faso (WARDA, 1993). IR 46, ITA 212, ITA 306, ITA 222 grown in rainfed lowlands are severely affected by leaf and panicle blasts than irrigated lowlands. This reaction of location and ecotype specificity suggests that there is a need to recommend the varieties for specific ecologies and for a country. Ou (1963) reported differences in varietal reactions of same varieties grown in Taiwan, Philippines and Vietnam. Taichung Native 1, which is first high yielding·semi-dwarf *indica* variety developed in Taiwan, China was observed as highly susceptible in Vietnam and Philippines, and was highly resistant in Taiwan. Similar observations were observed with respect to *japonica* varieties like Taichung 65, Taichung 176, and Kaohsiung 24 (Ou, 1963).

BLAST REACTION TO DIFFERENTIAL VARIETIES

Leaf blast reactions to 8 international differentials (Ling and Ou, 1969) have been evaluated by Awoderu (1970) in Nigeria, Mbodj *et al.* (1988) in Senegal ,Casamance, and Awoderu (1990) in Côte d'Ivoire. Awoderu (1970) reported 11 Nigerian and 11 inter-national races of blast in Nigeria. Nigerian race NG 5 and NG 10, and international race IA 65 and IB 1 were found highly virulent. All the 11 differential varieties were susceptible to NG5. Amongst the international differentials, only Raminad Str.3 was resistant to IB1 and Zenith was resistant to IA 65. Mbodj *et al.*(1988) in their evaluation at Sefa, Thiar and Djibelor, found out that only Raminad Str.3 was resistant at all the three locations, and other 7 differential lines were susceptible at Sefa.The reaction varied at other two locations. Zenith and Usen were observed as resistant at Thiar, but susceptible at Djibelor, and NP 125 resistant at Djibelor but susceptible at Thiar. Awoderu (1990) in Côte d'Ivoire in his study with blast differentials observed that there were four different types of internationals blast races at Bouake (IF 1), Man (IB 15),

TABLE 2: Varietal reaction of 25 lines for partial blast resistance at M'be, Côte d'Ivoire, 1997 wet season

Entry No	Designation	Origin	Variety group[1]	Ecotype[2]	Seedling vigor*	Leaf blast[3] 42DAS	DFLA[4]	Panicle blast*
1	Arborio	Italy	J	IL	9	9	14	-
2	B 40	Indonesia	I	IL	7	8	14	5
3	BR 27	Bangladesh	I	U	3	8	16	1
4	Caiapo	Brazil	I/J	U	1	5, PR	18	3
5	Drago	Italy	J	IL	7	9	16	5
6	Gigante Vercelli	Italy	J	IL	5	8	14	5
7	GZ5379-22-2	Egypt	J	IL	5	6, PR	17	3
8	IR 36	IRRI	I	IL	3	3	18	3
9	IR 50	IRRI	I	IL	3	7	17	7
10	IR 64	IRRI	I	IL	1	6,PR	17	5
11	ITA 257	IITA	I/J	U	1	6,PR	17	3
12	Javae	Brazil	I	IL	1	0	>40	5
13	Li-Jiang-Xin-Tuan-Hei-Gu	China	J	IL	9	9	13	
14	Mahsuri	Malaysia	I/J	RL	3	8	15	5
15	Makmur	Malaysia	I	IL	3	3	31	1
16	Metica 1	Colombia	I	IL	7	8	14	5
17	MTU 1001	India	I	IL	3	7	16	1
18	MTU 1003	India	I	IL	3	6,PR	17	5
19	Oryzica Llanos 4	Colombia	I	IL	3	1	25	1
20	Seribu Gantang	Malaysia	I	IL	1	1	22	1
21	Suweon 349	korea	J	IL	3	7	17	3
22	Swarna	India	I	IL	7	9	15	5
23	Tainung 70	Taiwan China	J	IL	5	1	25	3
24	Moroberekan, PR check	Cote d'Ivoire	J	U	1	6,PR	16	1
25	IR 5, S check	IRRI	I	IL	7	8	13	3

Mean (Diseased leaf area at 42DAS):48.1%; LSD 23.1; CV:28.1% ; R^2 :0.92

*SESscale,IRRI1996

1I:*Indica*;J:*Japonica*;I/J:*Indica-Japonica*cross

2 U: Upland; RL: Rainfed lowland; I :Irrigated lowland

3 SES Scale, IRRI ;PR: Partially resistant lines

4 FLA: Days to first lesion appearance

Odienne (IA 11), and Tombokro (IA 92). Only Dular variety was resistant at all the four locations, and others were susceptible at one or other locations. CI 8970 was another variety, which was resistant at three out of 4 locations. Variations in reaction to leaf and panicle or neck blast have been reported in past. Zenith differential variety, observed as moderately resistant to leaf blast was observed as highly susceptible to neck blast (Jacquot, 1978). Screening at IITA, Ibadan has shown that differential varieties like Fukinishiki, Toride 1, Tetep, Raminad Str 3, Pia Kan tao, and NP 125 were resistant to moderately resistant (IITA, 1983). During 1997 wet season, a set of blast differentials were screened at IITA. Raminad Str 3 was observed as resistant; NP 125 and Dular as moderately resistant; and Usen, and Kanto 51 as highly susceptible. WARDA is developing a set of differentials to characterise the different blast populations in the west Africa. Pathological and molecular characterisation of blast populations diversity from west Africa is being carried out in collaboration between WARDA and HRI/NRI/CABI/IMI,U.K. Some of the hot spot locations for blast screening in West Africa are: Ikenne, Onne, Ibadan for uplands, and Kadawa for panicle blasts in Nigeria; Mbe, Man, Gagnoa, and Korhogo in Côte d'Ivoire; Adeta in Togo; Mbo and Bokle in Cameroon; Rokupr in Sierra Leone; Karfiguela in Burkina Faso.

BLAST REACTIONS TO DIFFERENT RESISTANT GENES

Kiyosawa (1972, 1984) has reported 13 monogenically blast resistant genes at 9 loci in *japonica* varieties. Bidaux (1977), and Mbodj *et al.* (1988) have studied resistance of these genes in west Africa. Bidaux (1977) in his study at Bouake Côte d'Ivoire, with 14 varieties observed that Pi-ta 2 gene had maximum number of 30 days for retardation of the initiation of the epidemic in variety Pi no 4. Mbodj (1988) also observed that Pi-ta 2 gene in variety Pi no 4 was only resistant variety evaluated at 3 locations in Casamance, Senegal. All other 13 varieties were susceptible at one or other locations.

The *indica* blast gene differentials from IRRI were screened for their leaf blast reactions at IITA Ibadan, Nigeria. The study was carries out in 1994 and 1995, and 1997 wet seasons. Out of the five lines evaluated , only Pi-2 gene from C101 A51 had 0 score (Immune), while C101 LAC (Pi-1), C104 PKT (Pi-3), C101 PKT (Pi-4a), and C105 TTP 4L23 (Pi-ta) were highly susceptible and ineffective. In leaf blast screening at M'be, Côte d'Ivoire in 1997 wet season, C101A51 and C101LAC had partial resistance score of 5 and Pi-ta was observed as highly susceptible (Table 2). Out of 6 RIL lines evaluated during 1994 and 1995 at Ibadan, only RIL 249 (Score 0), and RIL 545(score 3) were effective. Other RIL's like RIL 10, 23,29, and 276 were susceptible. At Mbe, the gene combinations Pi-1 and Pi-2 in BL 121 and BL122 were resistant, while combinations Pi-1 and Pi-4 in BL 141, and gene combinations Pi-2 and Pi-4 in BL 241 and 245 had partial resistance (Table 2).

BREEDING STRATEGIES FOR BLAST RESISTANCE

As blast is one of the major disease of all the ecologies, breeding for resistance to blast is the major breeding objective at WARDA. Different breeding methods, such as bulk and pedigree methods after hybridization, anther culture, wide hybridization between glaberrima and sativa and backcross methods are followed. Selection for blast resistance at WARDA on-station and off-site hot spot locations, and through upland, lowland, irrigated, integrated pest management (IPM), and mangrove swamp Task Forces are the major activities. African Blast screening Nursery (ABSN) is also distributed to the NARS in Africa to select the location specific blast resistant lines.

SCREENING OF RESISTANT DONORS

Screening of resistant donors for hybridization is one of the major activities being carried through ARBN and Task Forces. In lowland rice breeding Task Force, a special nursery as "leaf and panicle blast screening nursery (LPBSN) was constituted in 1992 to select donors for hybridization. In IPM Task Force, lines are being selected through blast characterisation project at hot spot locations. Some of the improved blast resistant donors used for lowland improvement are ITA123, ITA239, ITA 302, ITA324, ITA414, ITA416, Tox 3118-47-1-1-2-3, Tox 3226-5-2-2-2, TCA80-4, IR36, IR72, WITA1, WITA 3 and others. For uplands,WAB 56-50,WAB 56-104,ITA 257,LAC23,IRAT 10, Moroberekan and other are being used.

HYBRIDIZATION AND SELECTION

In lowland breeding programme, the breeding activities were carried out at IITA, Ibadan, Nigeria till 1997. Single, three way, and double crosses were made to incorporate the resistance to blast, iron toxicity, rice yellow mottle virus, gall midge,drought and high yield from different donors. Parents are selected based on the multiple resistance. But several times donor parents have only one or two desirable traits. At Mbe, backcrosses and anther culture techniques have been used in crosses between *sativa* with *glaberrima*. At Ibadan, blast screening were carried out from F_2 , in an injector row technique. Mixed seeds of 4 to 5 highly susceptible lines were seeded in screenhouse, two weeks earlier than test entries. 250 gm seed of each cross in F_2 is grown in vertical rows and are subjected to mixed field isolates of blast. The blast resistant plants of 25 to 30 days old were transplanted in field for further selections. Individual plants were harvested in F_2 and the lines from F_3 onwards (pedigree lines) were handled as per pedigree method. Each 10th entry is used as susceptible or resistant check for comparison. At one time in a screenhouse 3,600 pedigree lines were grown, and in three screenhouses 10,800 lines were screened. The screening was carried out every four months, and annually over 30,000 pedigree lines were being screened. Scoring as per IRRI, SES method is carried out. The lines having 0,1,3,and 5 scores were uprooted and transplanted in field for further evaluation. Selections were made between and within progeny rows. In this method, 40 % of lines were discarded in seedbeds and remaining 60% are transplanted in field. The screening is carried out till a line is fixed. t varies from F4 to F9 generations from different cosses. The screening is carried out now at Mbe. Fixed lines are evaluated in observational nursery (ON) for various biotic and abiotic stresses. Selected lines are further evaluated in observational ,replicated and advanced yield trials.

GENETIC MALE STERILE FACILITATED RECURRENT SELECTION (GMSFRS)

By using six genetic male sterile lines from IRRI, crosses were made with 11 donors in 1984 in first cycle, and 8 new donors in third cycle. The random mating population was thus developed and screened for their blast resistance (Singh *et al.*,1996).Seeds from male sterile plants were harvested in subsequent cycles and after five cycles of random mating, male fertile plants were selected as per pedigree method. The fixed lines are being evaluated for their yield potential and resistance to several biotic and abiotic stresses. Two lines viz., RF 85C-C1-1-37-1-2-2-3, and Tox 85C-C5-85-1 were promising under irrigated and rainfed lowland conditions(WARDA,1995). During 1997 wet season 6,021 lines derived from GMSFRS were evaluated for their blast resistance and other agronomic traits, and 570 lines were selected for further evaluation.

INTERSPECIFIC PROGENIES FOR BLAST RESISTANCE

Oryza glaberrima, the African rice is known to have a very high degree of blast resistance (Awoderu and Ojomo, 1981). Earlier attempts to incorporate resistance from it to *O.sativa* have not been successful due to high degree of sterility in F2. Recent success at WARDA has generated a large number of lines with fertile segregants (Jones *et al.*, 1996). Around 400 Interspecific progenies were evaluated during 1996-97, and majority of them had excellent resistance to blast. Some of the promising high yielding lines with blast resistance are WAB 450-I-B-P-106-HB, WAB 450-I-B-P-160-HB and WAB 450-I-B-P-28-HB (WARDA, 1996).

SELECTION OF LINES WITH PARTIAL RESISTANCE (PR)

Advanced breeding lines, and IRBN from IRRI, Philippines is regularly screened for their blast characerization. During 1997 wet season at WARDA main station at M'be, Côte d'Ivoire, two IRBN nurseries were evaluated for their leaf and panicle blast reactions. In first set, 25 lines from 11 countries were grown in lattice design 5 x 5, with Moroberekan and IR 5, as PR and susceptible checks, respectively. In second set, 81 lines (inclusive of 25 above lines) were evaluated in 9 x 9 simple lattice. The design was as per multi-environment evaluation of rice blast (IRRI, 1997). Co39 and Martelli were used as spreader, and IDSA 6 as a barrier, resistant check. Entries were seeded in upland field,using sprinkler irrigation for favourable disease development. Data were recorded on days to first susceptible lesion appearance (DFLA) from days after seeding (DAS), seedling vigour, diseased leaf area % at every 5 days interval, 15 days after sowing till 42 days. Seedling and panicle blasts reactions at harvest (50 panicles) were converted to SES scale (IRRI, 1996). The scale was improved to differentiate between scores 4 to 9. In scale 9, the middle top leaf was infected; in scale 8 the central leaf was not infected; in scale 7 top 1-2 leaves had no infection, and in scale 4 to 6, top 2-3 leaves were free from infection. The disease appeared first after 12 days of sowing. The slow blasting lines, based on the r-value(rate of disease development) in scale 4 to 6 were classified as partial resistant (Bidaux, 1978). For panicle blast infection, the entries from first replication were transplanted at 42 DAS. In set 1and 2, four lines, viz. Javae, Alianka, Milyang 55, and Tetep were observed as Immune; another 10 lines a score of 1 and 3, 22 lines had partial resistance (score 4, 5 and 6), and rest 56 % of lines were susceptible to highly susceptible (Table 2& 3). IR 50 was only lines observed as susceptible to panicle blast. The maximum DLA at 42 DAS was 40%, as of Moroberekan, PR check. Most of the *Japonica* varieties were susceptible to highly susceptible. The high seedling vigor was also observed as an imporatnt trait of all PR lines. Two to three top leaves of PR lines were free from leaf blast infection. The DFLA in susceptible and highly susceptible lines was 13 to 16 DAS, while in PR lines at 17-18 DAS. The resistant and PR lines will be further evaluated for utilisation in breeding programme.

GERMPLASM EXCHANGE AND SELECTION IN SEGREGATING POPULATIONS THROUGH TASK FORCES

In first meeting of the Lowland rice breeding Task Force, leaf blast was ranked as number one major constraints for rice production in rainfed lowland and fourth in irrigated lowland ecosystem (WARDA, 1992b). Leaf and panicle blast screening nursery (LPBSN), and selections from the segregating generations (F_2 and F_3 bulks) activities were undertaken to screen and select the lines at hot spot locations. NARS in Cameroon, Ghana, Benin Burkina Faso, Sierra Leone, and Togo were provided small grant to carryout the activities. Selections were made in the segregating populations for use within the country and distribution to other NARS. Selected lines from LPBSN were used by national programs in their yield trials, and further in crossing programme by WARDA. Many times the selections were not only based on blast resistance, but resistance to other location specific biotic and abiotic stresses, plant height and grain yield.

MOLECULAR MAPPING

QTL mapping studies are being carried out at M'be, Bouake with populations derived from crosses between WAB 56-50 and WAB 56-104 with *O. glaberrima* like CG 14, CG 20 and IG 10.

TABLE 3: Reaction of lines for their reaction to leaf and panicle blasts under natural infestation at Mbe, Côte d'Ivoire during 1997 wet season

Entry No	Designation	Origin	Variety group[1]	Ecotype[2]	Seedling vigor*	Leaf blast[3] 42DAS	DFLA[4]	Panicle blast*
1	ADT 30	India	I	IL	9	9	14	-
2	Aichi Asahi	Japan	J	IL	9	9	16	-
3	Alianka	CIAT	I	IL	2	0	>42	0
4	Araguaia	Brazil	J	U	1	7	17	3
5	Ariete	Italy	J	I	9	9	16	0
6	Bahagia	Malaysia	I	I	2	7	19	0
7	Balilla	Italy	J	U	9	9	16	1
8	BL 121	IRRI	I		2	1	25	0
9	BL 122	IRRI	I		2	1	22	1
10	BL 141	IRRI	I		1	6 PR	18	3
11	BL 142	IRRI	I		4	6 PR	17	3
12	BL 241	IRRI	I		3	5 PR	19	3
13	BL 245	IRRI	I		2	4 PR	18	1
14	BPT 1235	India	I	RL	9	9	15	3
15	BR IRGA 410	CIAT	I	IL	7	9	16	1
16	BR 7	Bangladesh	I	U	1	7	19	3
17	Carreon	Philippines	I		1	1	19	1
18	Colombia 1	Colombia	I	U	5	7	16	0
19	CO 39	India	I	IL	9	9	14	0
20	C101A51	IRRI	I		3	5 PR	18	1
21	C101LAC	IRRI	I		1	5 PR	19	5
22	C101PKT	IRRI	I		9	9	14	-
23	C105TTP 4L23	IRRI	I		5	8	16	3
24	Diana	Italy	J	IL	9	9	16	-
25	Elio	Italy	J	IL	9	9	19	1
26	Giza 159	Egypt	J	IL	9	9	17	1
27	Giza 176	Egypt	J	IL	9	9	18	-
28	IRAT 144	BurkinaFaso	J	U	1	8	17	1
29	IRI 372	Korea	J	IL	7	6 PR	18	-
30	IR 24	IRRI	I	RL	6	7	16	1
31	IR 42	IRRI	I	IL	7	9	16	0
32	IR 74	IRRI	I	IL	3	4 PR	19	1
33	ITA 212	IITA	I	IL	3	6 PR	20	1
34	Kanto 51	Japan	J	IL	9	8	16	1
35	K 59	Japan	J	IL	9	9	16	1
36	Loto	Italy	J	IL	9	9	14	-
37	Milyang 15	Korea	J	IL	9	9	18	-

38	Milyang 55	Korea	I	IL	1	0	>40	3
39	Milyang 80	Korea	J	IL	5	6 PR	18	5
40	Milyang 82	Korea	I	IL	2	6 PR	25	5
41	Nonganbyeo	Korea	J	IL	7	8	15	1
42	NP 125	India	I	RL	3	6 PR	18	3
43	OM 997-6	Vietnam	I	IL	2	6 PR	19	1
44	Oryzica 1	Colombia	I	IL	9	9	15	3
45	Peta	Indonesia	I	RL	7	9	18	0
46	Reiho	Japan	J	IL	5	6 PR	19	1
47	Setanjung	Malaysia	I/J	U	5	8	17	1
48	Shin 2	Japan	J	U	7	8	15	3
49	Suweon 303	Korea	J	IL	4	7	15	1
50	Ta-Poo-Cho-Z	Taiwan, China		IL	1	1	20	5
51	Taikeng 7	Taiwan, China	J	IL	9	7	17	0
52	Tetep	Vietnam	I	RL	1	0	>40	0
53	Toride 1	Japan	J	IL	6	6 PR	19	0
54	UPLRi 5	Philippines	I	U	1	5 PR	17	1
55	Usen	Japan	J	IL	7	9	14	1
56	Xiangzi 3150	China	I	IL	1	1	25	0
57	Zenith	USA	J	IL	5	8	18	3

Mean (Diseased leaf area at 42 DAS): 22.97 % ; LSD 18.57; CV: 22.97%; R^2: 0.96

*, 1,2 and 3 as of Table 1

Conclusions

Rice blast damage still occurs in most of the rice producing countries in the region and released varieties succumbs to change in virulence of blast population. For effective and efficient management of rice blast disease through host plant resistance, following strategies are being undertaken to develop stable and durable blast resistant varieties.

I. The screening in segregating generations are being carried out in specific ecologies such as upland, rainfed lowland, irrigated and mangrove swamp. For screening ,the resistant plants and lines are selected after seedling stage, vegetative and panicle blast screening. Segregating populations are also screened at hot spot locations in the region.

II. Genetic male sterile facilitated recurrent selection is being carried out to recombine and pyramid blast resistant genes though random mating populations.

III. Blast resistant genes from *O.glaberrima* are being introgressed through backcrossing in *O.sativa* and interspecifics are being produced with better resistance and higher yields.

IV. Another culture lines between *Japonica/indica*, *indica/indica* and *sativa/glaberrima* are being generated to reduce the breeding cycles and incorporate the resistance genes.

V. Elite breeding lines are tested in multilocational blast nursery through ARBN and LPBSN at hot spot locations. Promising lines in advance yield trials are characerized for their blast resistance as per IRRI SES scale. Javae, Oryzica Llanos 4, Seribu Gantang, Tainung 70, Alianka, Milyang 55, Ta-Poo-Cho-Z, and Xiangzi 3150 are some of the highly resistant donors.

VI. To develop durable resistant varieties, blast resistant donors and elite lines are constantly screened for their multiple resistance to various biotic and abiotic constraints. For upland ecologies, drought, leaf scald , sheath rot and panicle blast, tolerance to acid uplands, threshability ,and grain quality are important traits; while for lowlands drought, leaf scald, iron toxicity, RYMV and ARGM are the important traits in West Africa.

VII. FARO 46 (ITA 150) in Nigeria is a new durable blast resistant cultivar for upland ecology. New blast resistant varieties such as WAB 56-50, WAB 56-104 and FKR 33 for uplands; and Cisadane, WITA 1, and WITA 3 have been released recently. Their durability of blast resistance will be monitored in future.

VIII. Pathological and molecular characterisation of rice blast pathogen population in West Africa is under progress in collaboration with HRI/NRI/CABI-IMI ,U.K. This will help in better understanding of population diversity in Africa as compared to Asian or other populations. A set of west African blast differentials lines are being developed to characterise the blast populations diversity. International blast differentials, and Kiyosawa and IRRI gene differentials are being used to monitor the resistance source and genes. Tetep, Carreon, are some of the resistant Pi-2 gene in C101A51 shows immune reaction to blast races at Ibadan, while both Pi-1 in C101LAC and Pi-2 showed partial resistance to blast at M'be, Côte d'Ivoire. BL 121 and BL 122 gene combinations (Pi-1 and Pi-2) were found as resistant and its resistance will be monitored in multilocational trials. This will be further used in hybridization programme.

References

Akator, S.K., 1981. Methods to combat blast in Togo. pp 35-42. In: Proceedings of the symposium on rice resistance to blast. Montpellier, France 18-21 March, 1981.

Awoderu, V.A., (1970). Identification of races of *Pyricularia oryzae* in Nigeria. Plant Dis.Rep.,54:520-523.

Awoderu, V.A., (1974). Rice diseases in Nigeria.PANS.20:416-424.

Awoderu, V.A., (1990). Evaluation of several rice varieties for horizontal resistance to leaf blast and neck rot in Ivory Coast. *J. Basic Microbiol.* 30:649-654.

Awoderu, V.A. and. Esuruoso, O.F., (1974). Reduction in grain yield of two rice varieties infected by the rice blast in Nigeria. *Nig. Agric. J.* 11:170-173.

Awoderu, V.A. and Ojomo O.A., (1981). Identification of sources of resistance to blast and variability of *Pyricularia oryzae*. pp 269-281. Proc. of the symposium on rice resistance to blast. Montpellier, France 18-21 March, 1981.

Bidaux, J.M., (1978). Screening for horizontal resistance to rice blast in Africa. pp 154-174. In: Rice in Africa. Academic Press, London.

Carpenter, A.J., (1977). Crop losses affecting rice in Liberia. In: *Plant Protection for the Rice* Crop., Seminar Proceedings N°4. WARDA Liberia. p 160-173.

Crill, J.P. and Khush G.S., (1982). Effective and stable control of rice blast with monogenic resistance. In: Evolution of the gene rotation concept for rice blast control p 87-102. International Rice Research Institute, Manila, Philippines..

Crill, J.P., Ham, Y.S., and. Beachell, H.M, (1982). The rice blast disease in Korea and its control with race prediction and gene rotation. In Evolution of the gene rotation concept for rice blast control. pp 123-130. International Rice Research Institute, Manila , Philippines.

FACU, (1988). Problems and prospects for rice production in northern Nigeria. Study Tour Report. September, 1988. 33 pp. Federal Agricultural Co-ordinating Unit, Abuja, Nigeria.

FAO, (1994). Food and Agricultural Organization Production Year Book for 1994: 48:70-71.

Fukunaga, K., (1965). Fungicide development for blast control. pp 409-414. In: Rice Blast Disease. IRRI, Philippines. John Hopkins Press, Baltimore.

Fomba, S.N. and Taylor, D.R., (1994). Rice blast in West Africa. Its Nature and Control: pp 343-356. In: Rice Blast Disease. Ed: R.S. Zeigler, S.A. Leong and P.S.Teng. CAB International, U.K. and IRRI, Philippines.

Imolehin, E.D., Ukwungwu, M.N., Kehinde, J.K., Maji, A.T. Akinremi, J.A., SinghB.N., and Oladimeji, O., (1997a). FARO 45 to FARO 49: two early maturing and three medium maturing upland rice varieties released in Nigeria. *Intl. Rice Research Notes*: 22 (3): p. 24-25.

Imolehin, E.D., Ukwungwu, M.N., Kehinde, J.K., T.Maji, A., Akinremi, J.A., Singh, B.N., and Oladimeji, O., (1997 b). FARO 44 and FARO 50, two irrigated lowland rice varieties for Nigeria. *Intl. Rice Res. Notes*; 22 (3): p. 21-22.

IITA, (1980). Annual Report for 1980. IITA, Ibadan, Nigeria.

IITA,. (1983). Cereals Improvement programme. Annual Report 1983. pp 19-20. International Institute of Tropical Agriculture. Ibadan, Nigeria.

IITA., (1984). Annual Report 1984. International Institute for Tropical Agriculture, Ibadan, Nigeria.

IRRI., (1996). Standard Evaluation System for Rice. 4th Edition. July 1996. IRRI, Manila, Philippines. pp. 52.

IRRI., (1997). Multi-environment Evaluation of Rice Blast Second IRBN- Supplement 1997). IRRI, Manila, Philippines.

Jacquot, M., (1978). Varietal improvement programme for pluvial rice in Francophone Africa. pp. 117-129. In: Rice in Africa: Ed: I.W.Buddenhagen and C.J.Persley. Academic Press.

Jagtap, S.S., (1995).Environmental characerization of the moist lowland savanna of Africa.Potential and constraints for crop production. Pro. Intl. workshop , Cotonou, Republic of Benin, 19-23 September, 1994. IITA, Ibadan, Nigeria.

John, V.T., Alam, M.S. and Thottappilly, G., (1985). Diseases and insect pests of wetland rice in tropical Africa. In The wetlands and rice in sub-Saharan Africa. Ed.A.S.R. Jou and J.A. Lowe. IITA, Ibadan, Nigeria. pp. 141-150.

Jones, M.P., Dingkuhn, M., Aluko G.K. and Mande, S., (1996). Using backcrossing and doubled haploid breeding to generate weed competitive rices from *O. sativa* x *O.glaberrima* Steud genepools.pp 61-79. In: Interspecific Hybridization: Progress and Prospects: WARDA 1996.

Kim, C.K., (1994). Blast management in high input, high yield potential, temperate rice ecosystems.pp. 451-463. In: Rice Blast Disease. Ed: R.S. Zeigler, S.A. Leong, P.S.Teng.IRRI & CABI.

Kiyosawa, S., (1972). Genetics of blast resistance. In: *Rice Breeding*, pp. 203-225. IRRI,Philippines.

Kiyosawa, S., (1984). Establishment of differential varieties for pathogenicity test of rice blast fungus.Rice Genet.Newletter. 1; 95-97.

Ling, K.C.and Ou., S.H., (1969). Standardization of international races numbers of *Pyricularia oryzae*.Phytopath.59:339-342.

Mbodj, Y., Gaye, S, .Diaw,.S and Faye, A., (1988). La pyriculariose du riz en Casamance, Senegal - Variabilite du pathogene. L'Agronomie Tropicale 43 (2): 106-110.

Notteghem, J.L.(1986). Blast resistance methodologies in the Ivory Coast, 1972-85.pp. 305-316. In:Progress in Upland Rice Research.IRRI. Manila, Philippines.

Olufowote, J.O., (1977). Farox 56/30, a promising upland rice variety for Nigerian farmers.Nig.J.Genet 1: 76-88.

Ou, S.H., (1963). Varietal reactions of rice to blast. In The Rice Blast Disease. pp 223-234. International Rice Research institute. John Hopkins Press, Baltimore, USA.

Ou, S.H., (1985).rice Diseases.2nd edition.CMI,Kew, U.K. 380 pp.

Poisson, C., (1978). Rice in Upper Volta.pp 347-348. In Rice in Africa. Ed. I.W. Buddenhagen and G.J. Peresley. Academic Press.

Sehly, M.R. and Bala, M.S., 1994. Resistance to rice blast disease in Egypt.pp. 597. In: Rice Blast Disease. Ed: R.S. Zeigler, S.A. Leong , P.S.Teng. IRRI & CABI.

Sere, Y. (1981). Rice blast prevention in Upper Volta. (In) Proceedings of the symposium on rice resistance to blast. Montpellier France 18-21 march, 1981.pp. 51-65.

Shen, M., and Lin, J.Y. 1994. The economic impact of rice blast disease in China. pp. 321-331. In: rice Blast Disease. Ed: R.S. Zeigler, S.A. Leong, P.S.Teng.IRRI & CABI.

Sie, M., Sere Y.,and Sonou, A., (1997). FKR 33: a popular upland variety in Burkina Faso.*Int. Rice Res. Notes* 22 (1) : 32.

Singh, B.N.(1997). Trip Report: Niger Republic and Nigeria, 13-27 October, 1997. WARDA, Bouake, Côte d'Ivoire.

Singh, B.N., (1998).advances in varietal improvements for lowland ecologies in west Africa. Paper presented in the annual meeting of the Breeding Task Force, 01-02 April, 1998. WARDA, Bouake.

Singh, B.N., Masajo, T.M. and Oladimeji, O.A., (1996). Use of genetic male sterile facilitated recurrent selection for blast resistance in rice. *Int.Rice Res.Notes*: 21 (1): 28-29.

Singh, B.N., Fagade, S., Ukwungwu, M.N., Williams, C., Jagtap, S.S., Oladimeji, O., Efisue, A. and Okhidievbie, O., (1997). Rice growing environments and biophysical constraints in different agroecological zones of Nigeria. *Met. J.2* (1):35-44.

Vodouhe, S.R., Djegui, N. and Amadji, F., (1981). Impact of blast on rice cultivation in the People's republic of Benin. (In) Proceedings of the symposium on rice resistance to blast. Montpellier France 18-21 march, 1981.pp .27-33.

WARDA (1992a). Proceedings of the first meeting of the lowland rice breeding Task force. January 20-21, 1992. Ibadan, Nigeria. WARDA Task force meeting Series 322 pp.

WARDA, (1992b). Report of the monitoring tour to Nigeria and Cameroon.October 15-21, 1992. WARDA Task Force meeting series 10.pp 21.

WARDA., (1996). Annual Report. 1996. West Africa Rice Development Association, Bouake, Côte d'Ivoire.

WARDA(1997).WARDA MediumTerm Plan,1998-2000.WARDA, Bouake.pp 81.

RESISTANCE TO BLAST IN HYBRID RICE

E. P. GUIMARAES, A. S. PRABHU, M. C. FILIPPI, and A. S. SILVA
Embrapa Arroz e Feijão
Caixa Postal 179, 75375-000 Santo Antônio de Goiás, GO, Brasil

1. Introduction

Rice blast caused by *Pyricularia grisea* (Cooke) Sacc., is one of the most important diseases both in irrigated and upland rice ecosystems, in Brazil. Low durability of the blast resistance has been a common factor for all cultivars released so far to farmers. Therefore, durable resistance has been the main goal of our national rice breeding program. The National Research Center for Rice and Beans (Embrapa Arroz e Feijao), has been using different approaches to increase the durability of the cultivars for blast resistance. The normal procedures in the breeding projects consist, crosses between disease resistant sources and commercial rice cultivars widely adapted to local conditions, pedigree selection in hot spot sites and construction of a broad based population including well known sources of durable resistance using recurrent selection procedure (Prabhu and Guimarães, 1990; Filippi and Prabhu, 1997). Recently, we have embarked on hybrid rice breeding for irrigated rice as a supplement to the ongoing breeding projects. High yielding lines improved by conventional breeding for pest resistance, grain quality, among other desirable agronomic traits, are being used as parents of the hybrid breeding project. Incorporation of blast resistance is one of the major objectives, because the hybrids are prone to disease outbreaks due to intensive cultural practices such as high fertilizer rates and the hypervariability of *P.grisea*.

It is well known that additive effects are important to determine partial resistance. According to Bonman (1992) the durable resistance of some of the rice cultivars is associated with polygenic partial resistance that shows no evidence of race specific major gene resistance. Partial resistance sensu Parlevliet (1988) is quantitative resistance based on minor genes whose effects are small and cannot be discerned individually. Notteghem (1985) showed that additive effects play a major role in partial resistance in upland rice. Nevertheless, working with hybrid rice combinations, Veillet et al. (1996) detected positive heterosis for all components of partial blast resistance.

The strategy followed for hybrid development is the one described by Lin and Yuan (1980) and used in China since 1976. Three lines are required, a `A` line, a genetic-cytoplasmatic male-sterile germplasm, a `B` line, capable of maintaining the sterility of the `A` line, and one `R` line, responsible to restore the fertility of the `A` line. Seeds of the male-sterile line are produced by crossing `A` and `B` lines. Hybrid is the F_1 seed obtained when `A` and `R` lines are combined.

D. Tharreau et al. (eds.), Advances in Rice Blast Research, 129–136.
© 2000 *Kluwer Academic Publishers. Printed in the Netherlands.*

In general, blast resistance is controlled by one to four dominant genes depending upon the race (Kiyosawa, 1981; Mackill and Bonman, 1992; Filippi and Prabhu, 1996). In some cultivars resistance is also conferred by incompletely dominant or recessive genes (Yu et al., 1987; Oka and Lin, 1957). Because the commercial product of the hybrids is the F_1, any single cross involving one resistant parent with major gene resistance, in majority of the crosses the hybrids are expected to exhibit resistant reaction. This feature facilitates the hybrid breeder's job, but it does not solve the problem of the durability of the blast resistance. Partial resistance is quantitative based on minor genes and is inherited polygenically (Lin, 1986; Roumen, 1993).

Development of durable blast resistance for disease prone environments require evaluation and selection of lines in hot spot screening sites exposing them to diverse pathogen populations (Correa-Victoria and Zeigler, 1993). This procedure would permit identification of lines with broad spectrum, major gene resistance. However, the selection for partial resistance is possible when major race–specific genes are absent and require controlled greenhouse tests. The partial resistance of the cultivar cannot be measured in absolute terms. It is always a relative measure and is often compared with a standard susceptible cultivar (Parlevliet, 1988). In the case of hybrids the standards checks are their parents.

The level of partial resistance was assessed to leaf blast in cultivars by inoculating with a specific isolate of *P. grisea* and measuring components of resistance such as infection efficiency, lesion size, sporulation capacity, etc. (Villareal et al., 1981; Yeh and Bonman, 1986; Roumen, 1993). Therefore, it is necessary to identify a race virulent to both parents and hybrids possessing major genes.

Thus this work was done, mainly to evaluate the performance of several hybrids, male-sterile (A), and restorer (R) lines under heavy disease pressure in rice blast nursery for vertical resistance and to determine the level of partial resistance of some hybrids developed by "Embrapa Arroz e Feijão" in Brazil.

2. Blast Nursery Test

A standard method for testing rice germplasm for leaf blast resistance in the uniform blast nursery was initially proposed during the Rice Symposium held at IRRI, Philippines, in 1963 (Ou, 1963). In Brazil, the National Blast Nursery (VNB), coordinated by Embrapa, as a part of the National Rice Research Program, is a similar trial conducted annually at seven to eight locations in different parts of the country.

A set of 79 'R' lines were included in VNB 1993 and conducted at eight environments in Brazil. The disease assessment was made using a 0-9 scale (International Rice Research Institute, 1988). The restorer lines evaluated in the VNB showed a wide range of reactions ranging from highly resistant to highly susceptible, but the majority of them exhibited resistant reaction (Table 1). Out of the 79 restorer lines 46.8% showed resistant reaction (0-3). The effectiveness of the resistance gene varied in different locations depending upon the climatic conditions and the prevailing pathogen population. For example, the 'R' line CNA 5741 was resistant in four locations and susceptible in the other four sites. Line CNA 5557 was resistant across all environments, nevertheless, there was not a single line that was susceptible in all trials.

The disease pressure was lowest at Pindorama and Jaciara among the test sites. The superior performance of these 'R' lines is expected as they were subjected to a series of tests for resistance both in the blast nurseries and in the field during the process of germplasm development of irrigated rice. These results insure the availability of potential resistant male parent sources for the hybrid project.

TABLE 1. Leaf blast reaction in some 'R' lines evaluated at eight locations in VNB 1993, in Brazil

'R' line (CNA)	Pelotas	Goiania-1	Goiania-2	Jaciara	Lucas	Cachoeirinha	Pinda	Pindorama
5746	4	5	4	2	2	2	3	4
4995	4	5	4	3	1	3	3	1
5524	3	2	2	2	1	1	1	1
5731	3	5	3	3	2	2	2	1
5709	4	1	1	2	3	4	2	1
5213	4	1	1	3	3	4	1	1
3882	4	5	1	2	3	4	4	1
3461	4	4	1	2	1	2	1	1
5557	3	1	4	1	2	1	1	1
5741	3	5	9	1	2	9	8	1
4900	4	5	9	1	9	5	8	1
3989	4	5	9	2	7	9	8	1
6444	4	1	2	1	1	4	1	1
7532	4	1	1	1	1	1	4	1
Average	3.7	3.3	3.6	1.9	2.7	3.6	3.3	1.2
Local check	9	6	9	7	5	9	4	1

Total number of 'R' lines testes = 79

3. Determination of Level of Partial Resistance in the Hybrids

The quantification of partial resistance in the hybrids is complicated because of the presence of major genes. The first step was to identify one race or lineage that would be capable of defeating all resistance genes present in the test material. Inoculation tests were conducted under controlled greenhouse conditions. Six hybrids, their parents, two maintainers and CO 39 (susceptible check) were utilized to identify the virulent race. Race identification was previously done based on the standard eight international differentials.

The test material was planted in plastic trays (15 x 30 cm) containing 6 kg of soil fertilized with NPK (5g of 5-30-15 + Zn, 3g of ammonium sulfate per 6 kg of soil). Sixteen cultivars were sown (10 to 12 seeds) in 5-cm rows in one tray.

Isolates of *P. grisea* were collected from irrigated rice cultivars, from field trials in the experimental stations of Embrapa Arroz e Feijão, Goiania, Brazil. Single conidial isolates were established from sporulating lesions and stock cultivars were maintained on filter paper disks in sterilized butterpaper bags at $4°C \pm 1°C$ in the refrigerator. For sporulation the isolates were grown on oatmeal agar in Petri dishes and incubated at $25°C$ for 7 days. Inoculum was prepared as described earlier (Prabhu et al., 1992) and concentration was adjusted to 3×10^5 conidia per ml. Twenty two day old plants were inoculated by spraying with aqueous spore suspension on the leaves, until run-off, using an atomizer connected to an air compressor. The plants after inoculation were incubated in moist chamber for 24 hours and later they were maintained in the greenhouse under high humidity (70-90%) and high temperature (26 to $30°C$). Disease assessment was made 7 to 9 days after inoculation. The disease ratings of 0 to 3 were considered incompatible or resistant and 4 to 9 compatible or susceptible in a SES scale of 0-9 (International Rice Research Institute, 1988). Inoculation experiments were repeated three times and the consistent uniform reactions were taken into consideration.

Differential genotype by isolate interaction is evident from the results in Table 2. In half of the cases where one of the parents showed resistant reaction to a given race the F_1 hybrid was resistant indicating the dominant nature of the resistance, in the other half the genetic control seems to be due to recessive genes. The isolate Tetep L^3 pertaining to race IB-9 and lineage BZ-8 was selected for further study because of its virulence to both hybrids as well as their parents and maintainers. The host from which the isolate originated determined the virulence pattern. The isolates from Tetep, Metica and BR-IRGA 409 showed relatively high frequency of virulence to the test material.

TABLE 2. Disease reaction of hybrids and their parents to nine isolates of *Pyricularia grisea*

Parents and hybrid	Isolates								
	Cica 8	Aliança 2[4]	Metica 2[4]	Tetep L[3]	BR-IRGA 409 L2[1]	BR-IRGA 409 L1[1]	Oryzica Llanos 5	CO 39 L1[1]	C101[3] -PKT
0461/0Z7	R/S	R/R	S/S	S/S	S/S	S/S	R/R	R/S	R/S
H37	R	R	S	S	S	S	R	S	S
0461/0ZD	R/S	R/S	S/S	S/S	S/S	S/S	R/R	R/S	R/S
H38	R	S	S	S	S	S	R	R	S
0461/0ZG	R/S	R/S	S/S	S/S	S/S	S/S	R/S	R/S	R/S
H39	S	R	S	S	S	S	R	S	S
0461/11P	R/R	R/S	S/S	S/S	S/S	S/S	R/S	R/S	R/R
H200	R	S	S	S	S	S	S	S	R
046H/0Z D	R/S	R/S	R/S	S/S	S/S	R/S	R/R	R/S	R/S
H16	R	R	S	S	S	R	R	R	R
046H/16 B	R/R	R/R	R/R	S/S	S/R	R/R	R/R	R/R	R/R
H326	R	R	R	S	R	R	R	R	R
CO39	S	S	S	S	S	S	S	S	S
IR36	R	R	R	R	R	R	R	R	R

CO 39 and IR36 are susceptible and resistant checks, respectively
R = Resistant and S = Susceptible

Partial resistance was analyzed in two 'A' lines (046I and 046H), represented by their maintainers (#2RE and #2RF), four 'R' lines (0Z7, 0ZD, 0ZG and 11P), and five hybrids (H16, H37, H38, H39, and H200). One hundred seeds of each test material were sown in separate trays. The layout of the experiment was randomized block design with three replications. The same inoculation and evaluation procedures described above were employed.

The components of resistance studied include, number of sporulating lesions per unit area of leaf (SLN), percentage of infected leaves (PIL) and percentage of leaf area affected (PLA). The sporulating lesion number was determined based on 5.0 cm leaf area of 20 seedlings per tray, seven days after inoculation. The number of infected leaves with at least one sporulating lesion per seedling were counted in a sample of 50 seedlings per genotype five days after inoculation. At nine days after inoculation, the percentage of leaf area infected was measured in a sample of 30 seedlings per replicate, selected at random. The sporulation capacity was measured according to the disc method described by Roumen et al. (1992). For the variance analysis the data were transformed to arc sen $\sqrt{x}$ or $\sqrt{x+1}$. The results are presented in Table 3.

TABLE 3. Sporulating lesion number (SLN), number of infected leaves (NIL), and percentage of leaf area infected (PLA) of hybrids, parents and maintainers inoculated with the isolate Tetep L[3] (race IB-9) of *Pyricularia grosea*

Genotypes	SLN (per cm²)	NIL (%)	PLA (%)
11P[*]	2.85[a]	32.19a	33.17 b
CO 39 (Susceptible check)	2.71[a]	43.90ab	77.83a
0ZD[*]	1.72ab	19.02 b	26.17 b
0ZG[*]	1.71ab	27.61 b	16.43 b
0Z7[*]	1.24ab	24.47 b	29.80 b
H16[**]	0.83ab	26.40 b	29.13 b
#2RE[***]	0.65ab	27.22 b	22.33 b
H37[**]	0.62ab	24.08 b	14.80 b
H38[**]	0.61ab	26.76 b	14.23 b
046H[****]	0.58ab	21.90 b	10.67 b
H39[**]	0.52ab	22.26 b	30.50 b
H200[**]	0.46ab	19.39 b	10.93 b
046I[****]	0.19 b	18.11 b	9.10 b
#2RF[***]	0.14 b	21.89 b	11.00 b

The mean lesion number per cm² was transformed to $\sqrt{x+0.5}$ and percentage leaf area infected to arc sin $\sqrt{x}$. Average followed by the same letter do not differ statistically at 5% level using Tukey test
• Restorer; ** Hybrid; *** 'B' line; and **** 'A' line

There were no statistically significant difference among the hybrids as well as between the hybrids and their parents for the three components utilized for partial resistance measurement. The differences in sporulation capacity were not observed among the hybrids and their parents. However, considering the lesion number, the hybrids showed intermediate levels of partial resistance compared to their parents (Figure 1). The hybrids H200, H37, H38, H39 and H16, in decreasing order, showed superior performance. Yeh and Bonman (1986) and Roumen (1993) consider that the sporulating lesion number is the most important component of partial resistance and the genetic analysis showed that the inheritance of resistance to lesion number is determined by several minor genes.

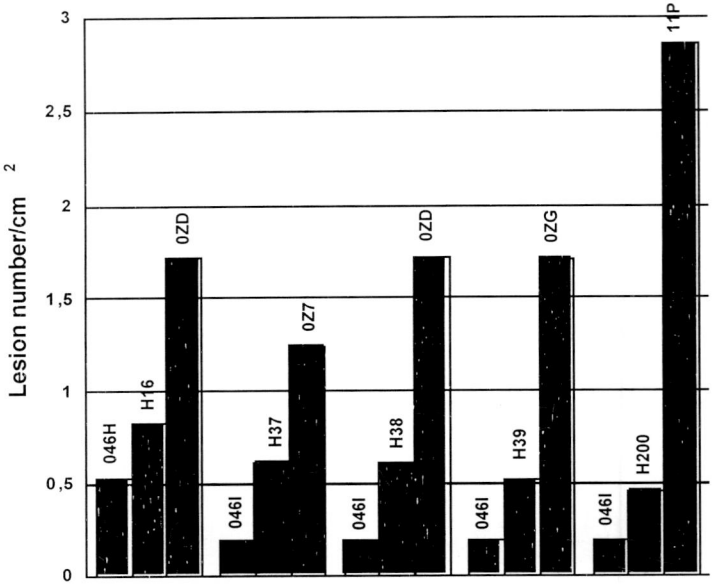

FIGURE 1.Lesion number per cm^2 of four hybrids (H16, H37, H38, H39 and H200) and their parents (046H, 046I, 0Z7, 0ZD, 0ZG, and 11P) in artificial inoculation tests with race IB-9 of *Pyricularia grisea*

The results showed that significant levels of partial resistance is accumulated in lines with major gene resistance even though they were not screened for it in the breeding process. In the irrigated ecosystem, partial resistance will be more effective with controlled irrigation and other cultural practices than under upland conditions. Flooding increases the resistance of rice cultivar to blast through nutritional changes in the host plant tissue (Kim, 1986).

However, it is important to combine both vertical and partial resistance by prior selection of the desirable parents both in the blast nursery and greenhouse testing procedures. The multilocation tests of germplasm in rice blast nursery, provide an indirect estimate of partial resistance. The disease severity index obtained from IRBN results was positively correlated with the partial resistance levels observed under field conditions (Ahn, 1994; Prabhu et al., 1997). There is always a positive correlation between major and minor gene resistance (Ahn and Ou, 1982). Further, the use of molecular markers linked to resistance genes facilitate the selection of restorers both for complete and partial resistance (McCouch et al., 1994). The chance that a matching virulent race or pathotype overcome the major gene resistance of F$_1$ hybrid is much less than in the monoculture of a resistant commercial rice cultivar grown in extensive areas. However, when the effectiveness of major resistance gene is lost with time the hybrids should exhibit adequate levels of partial resistance.

4 References

Ahn, S.W. (1994) International collaboration on breeding for resistance to rice blast, in R.S. Zeigler, S.A. Leong, P.S.Teng (eds). *Rice Blast Disease*, CAB International, Wallingford, pp.137-154.

Ahn, S.W. and Ou, S.H. (1982) Quantitative resistance of rice to blast disease, *Phytopathology* **72**, 279-282.

Bonman, J.M. (1992) Durable resistance to rice blast disease – environmental influences, *Euphytica* **63**, 115-123.

Correa-Victoria, F. and Zeigler, R.S. (1993) Pathogenic variability in *Pyricularia grisea* at a rice blast "hot spot" breeding site in eastern Colombia, *Plant Disease* 77, 1029-1035.

Filippi, M.C. and Prabhu, A.S. (1996) Inheritance of blast resistance in rice to two *Pyricularia grisea* races, IB-1 and IB-9, *Brazilian Journal of Genetics* **19**, 599-604.

Filippi, M.C. and Prabhu, A.S. (1997) Selección recurrente.para resistencia parcial a *Pyricularia grisea* Sacc. en arroz en Brasil, in E.P. Guimarães (ed.), *Selección Recurrente en Arroz*, CIAT, Cali, pp. 217-226.

International Rice Research Institute (1988) *Standard Evaluation System for Rice*, International Rice Research Institute, Los Baños, pp. 14-15.

Kim, C.H. (1986) Effect of water management on the etiology and epidemiology of rice blast caused by *Pyricularia oryzae* Cav. PhD Thesis, Louisiana State University, Louisiana, U.S.A.

Kiyosawa, S. (1981) Genetic analysis of blast resistance, *Oryza* **18**, 196-203.

Lin, S.C. (1986) Genetic analysis of minor gene resistance to blast in japonica rice, in *Rice Genetics*, International Rice Research Institute, Los Baños, pp. 451-469.

Lin, S.C. and Yuan, L.P. (1980) Hybrid rice in China, in *Innovative Approaches to Rice Breeding, Selected Papers from the 1979 International Rice Research Conference*, International Rice Research Institute, Los Baños, pp. 35-51.

McCouch, S.R., Nelson, R.J., Thome, J., and R.S.Zeigler (1994) Mapping of blast resistance genes in rice, R.S. Zeigler, S.A . Leong, P.S.Teng (eds), CAB International, Wallingford, pp.167-186.

Mackill, D.J. and Bonman, J.M. (1992) Inheritance of blast resistance in near isogenic lines of rice, *Phytopathology* **82**, 746-749.

Notteghem, J.L. (1985) Définition d'une estratégie d'utilisation de la resistance par analyse génétique des relations hôte-parasite. Cas du couple riz-*Pyricularia oryzae*, *Agron Trop* **40**, 129-147.

Oka, H.I. and Lin, K.I. (1957) Genetic analysis of resistance to blast disease in rice (by biometrical genetic method), *Japanese Journal of Genetics* **32**, 20-27.

Ou, S.H. (1963) A proposal for an international program of research on the rice blast disease, in *The Rice blast disease*, Johns Hopkins, Maryland, pp. 441-446.

Parlevliet, J.E. (1988) Identification and evaluation of quantitative resistance, in K.J. Leonard and W.G. Fry (eds), *Plant Disease Epidemiology, Genetics, Resistance, and Management*, Vol.2, McGraw-Hill, New York, pp. 215-248.

Prabhu, A.S. and Guimarães, E.P. (1990) Estratégias de controle de brusone em arroz de sequeiro, *Summa Phytopatologica* **16**, 47-56.

Prabhu, A.S., Filippi, M.C., and Castro, N. (1992) Pathogenic variation among isolates of *Pyricularia oryzae* affecting rice, wheat and grasses in Brazil, *Tropical Pest Management* **38**, 367-371.

Prabhu, A.S., Ribeiro, A.S., Soave, J., Souza, N.S., Kempf, D., Filippi, M.C., Rangel, P.H.N., and Zimmermann, F.J.P. (1997) Vivero Nacional de Piricularia: Progreso, perspectivas y utilización como fuente de progenitores para la selección recurrente, in E.P. Guimarães (ed.), *Selección Recurrente en Arroz*, CIAT, Cali, pp. 217-226.

Roumen, E.C. (1993) *Partial resistance in rice to blast and how to select for it*, Thesis, Wageningen Agricultural University, Wageningen.

Roumen, E.C., Bonman, J.M., and Parlevliet, J.E. (1992) Leaf age related partial resistance to Pyricularia oryzae in tropical lowland rice cultivars as measured by the number of sporulating lesions, *Phytopathology* **82**, 1414-1417.

Veillet, S., Filippi, M.C., and Gallais, A. (1996) Combined genetic analysis of partial blast resistance in an upland rice population and recurrent selection for line and hybrid values, *Theor Appl Genet* **92**, 644-653.

Villareal, R.L., Nelson, R.R., Mackenzie, D.R., and Coffaman, W.R. (1981) Some components of slow blasting resistance in rice, *Phytopathology* **71**, 608-611.

Yeh, W.H., and Bonman, J.M. (1986) Assessment of partial resistance to *Pyricularia oryzae* in six rice cultivars, *Plant Pathology* **35**, 319-323.

Yu, Z.H., Mackill, D.J. and Bonman, J.M.(1987) Inheritance of resistance to blast in some traditional and improved rice cultivars, *Phytopathology* **77**, 323-326.

Acknowledgement: This work was supported by CNPq

EFFECTIVE CONTROL OF RICE BLAST DISEASE WITH 'SASANISHIKI' MULTILINE

S. KOIZUMI[1] AND T. TANI[2]

[1]*Department of Lowland Farming, Tohoku National Agricultural Experiment Station, Yotsuya, Omagari 014-0102, Japan*
[2]*Mountainous Region Agricultural Research Institute, Aichi-ken Agricultural Research Center, Inabu, Kitashitara, Aichi 441-2513, Japan*

1. Introduction

Blast caused by *Pyricularia grisea* (telemorph *Magnaporthe grisea*) (Rossman *et al.*, 1989) is the most destructive disease in Japan. Many rice cultivars with complete resistance have been developed to control the disease. However, the resistance in the cultivars has been broken down within several years after their release due to the increase in new races of rice bast fungus virulent to the resistance (Kiyosawa, 1974; Yamanaka and Yamaguchi, 1987). To prevent the breakdown of the resistance, the use of multilines proposed by Jensen (1952) and Borlaug (1959) was suggested, and reduction of blast development in mixtures of rice cultivars and near-isogenic lines with different complete resistance genes to blast has been reported (Koizumi, 1983; Shindo and Horino, 1989; Koizumi, 1994; Koizumi and Fuji, 1994; Koizumi *et al.*, 1996; Nakajima *et al.*, 1996a).

Under the circumstances, 'Sasanishiki' multiline was first registered as a rice multiline to control blast disease in Japan (Matsunaga, 1996). The multiline was released in 1995 and cultivated in 5,800 hectares of farmers' fields in Miyagi Prefecture of northern Japan in 1997. However, the effective and long lasting control method of blast with the multiline is not yet clarified. This study was conducted to clarify the effective control method of blast with the multiline.

2. 'Sasanishiki' Multiline and its Components

A leading rice cultivar of Japan, 'Sasanishiki', was used as the recurrent parent to develop the multiline. 'Sasanishiki' holds a complete resistance gene *Pi-a* to blast and level of partial resistance in it to the disease is low.

Nine near-isogenic lines (NILs) were developed from the recurrent parent by five to eight time backcrosses at Miyagi Prefectural Furukawa Agricultural Experimental Station. One (+ line) of the nine NILs lacks complete resistance genes effective to almost all Japanese strains of rice blast fungus, and each of the other NILs has a different complete resistance gene (*Pi-i*, *Pi-k*, *Pi-k^m*, *Pi-z*, *Pi-ta*, *Pi-ta²*, *Pi-z^i* and *Pi-b*) to blast in addition to the *Pi-a* gene derived from the recurrent parent, although the *Pi-a* gene in the *Pi-ta²*, *Pi-z^i* and *Pi-b* lines was not yet

D. Tharreau et al. (eds.), Advances in Rice Blast Research, 137–145.
© 2000 *Kluwer Academic Publishers. Printed in the Netherlands.*

confirmed since there are not blast fungus strains to check it in them (Matsunaga, 1996; Table 1).

Six (*Pi-k*, *Pi-k*^m, *Pi-z*, *Pi-ta*, *Pi-ta*² and *Pi-z*' lines) of the nine NILs are currently registered and four (*Pi-k*, *Pi-k*^m, *Pi-z* and *Pi-z*' lines) of the six are mixed and cultivated as a multiline named 'Sasanishiki BL'. Levels of partial resistance in all the NILs to blast are similar to that in 'Sasanishiki' (Matsunaga, 1996).

3. Mixture Trials

To clarify the effective control of blast with the multiline, we conducted paddy field trials in natural heavy blast epidemics from 1993 to 1996. For the trials, we made six square meter plots with three replications and 150 square meter plots with no replication. Seeds of the NILs and 'Sasanishiki' were uniformly mixed according to the decided proportions in the mixture plots before sowing, and rice seedlings were transplanted 30 × 15 cm apart by hand or machine. Blast severity was assessed by

Table 1. Near-isogenic lines developed from rice cultivar 'Sasanishiki' and their reaction to main races of *Pyricularia grisea* in trial fields

Near-isogenic line	Complete resistance genotype	Japanese race			
		007	037	077	107
Tohoku 1	+	S	S	S	S
Tohoku 2	*Pi-i Pi-a*	S	S	S	S
Tohoku 3	*Pi-k Pi-a*	R	S	S	R
Tohoku 4	*Pi-k*^m *Pi-a*	R	S	S	R
Tohoku 5	*Pi-z Pi-a*	R	R	S	R
Tohoku 6	*Pi-ta Pi-a*	R	R	R	S
Tohoku 7	*Pi-ta*²*	R	R	R	R
Tohoku 8	*Pi-z*'*	R	R	R	R
Tohoku 9	*Pi-b**	R	R	R	R
'Sasanishiki'	*Pi-a*	S	S	S	S

* *Pi-a* unknown.

R= resistant; S= susceptible.

the Asaga (1981) scale and RADPC (relative area under the disease progress curve) values (Bonman *et al.*, 1989) were calculated from the severity and the numbers of observation days.

3.1. COMPARISON OF BLAST CONTROL BETWEEN 'SASANISHIKI' MULTILINE AND FUNGICIDE TREATMENTS

3.1.1. *Mixtures of Susceptible 'Sasanishiki' and its Resistant Near-Isogenic Line*
'Sasanishiki' and its near-isogenic *Pi-z^t* line were used for the mixtures. 'Sasanishiki' was susceptible to all of the races of *Pyricularia grisea* distributed in our trial fields and the *Pi-z^t* line was resistant to all of them (Table 1). 'Sasanishiki' was mixed with the *Pi-z^t* line in proportions of 1 : 1 and 1 : 3, and blast severity in the mixtures was compared with that in pure stands of 'Sasanishiki' treated and untreated with blasticides (Table 2).

Table 2. Blast severity (RADPC [a]) in mixtures of rice cultivar 'Sasanishiki' and its near-isogenic lines, and in pure stands of 'Sasanishiki' treated and untreated with blasticides

Blast	Trial	'Sasanishiki' : *Pi-z^t* line		TML[b]	'Sasanishiki'	
		1 : 1	1 : 3		Blasticide-treated [c]	Non-treated
Leaf	1994			0.6 [d]	0.5	4.2
	1995-1	1.2	0.5	0.6	1.7	11.1
	1995-2	1.5		0.8	0.5	6.9
	1996-1	0.5	0.2	0.3	0.1	1.9
	1996-2	1.3		0.7	0.3	6.2
Panicle	1994			15.2	8.6	51.3
	1995-1	34.2	15.5	29.1	16.2	84.2
	1995-2	17.7		12.5	4.0	49.0
	1996-1	27.2	12.9	17.1	13.0	69.3
	1996-2	24.6		25.6	20.5	81.0

[a] Relative area under the disease progress curve.

[b] Mixture of equal proportion of 'Sasanishiki' and its nine near-isogenic lines.

[c] Probenazole granule (24g a.i./a) was submerged for leaf blast control and tricyclazole suspension
    c. (x 1000, 3g a.i./a) was sprayed twice for panicle blast control respectively.

[d] Underlined RADPC values in the mixtures are not significantly different from that in pure stand
    of 'Sasanishiki' treated with blastides in each trial.

The mixtures of the *Pi-z^t* line with 'Sasanishiki' inhibited leaf blast, and leaf blast severity in the mixtures including 50 and 75% *Pi-z^t* line was approximately equal to that in blasticide treated 'Sasanishiki' pure stand. However, panicle blast severity in the mixture of 50% *Pi-z^t* line was greater than that in the fungicide treated pure stand, although the severity in the mixture with 75% *Pi-z^t* line was statistically equal to that in the chemical applied pure stand (Table 2).

3.1.2. *Mixture of 'Sasanishiki' and its All Near-Isogenic Lines*
The mixture of equal proportion of 'Sasanishiki' and its nine NILs (TML) also reduced blast development. Leaf blast severity in TML was equal to that in the blasticide treated 'Sasanishiki' pure stand, although panicle blast severity in TML was greater than that in the chemical applied pure stand (Table 2). Most predominant Japanese race of rice blast fungus in TML was 007, which can attack three components ('Sasanishiki' , and + and *Pi-i* lines) of TML (Table 1).

3.1.3. *Other Mixtures*
We also examined blast severity in other four mixtures of respective equal proportions of 'Sasanishiki' and *Pi-i* line, *Pi-i* and *Pi-k* lines, *Pi-k*, *Pi-k*^{m}, *Pi-z* and *Pi-z*^{t} lines, and *Pi-ta*, *Pi-ta*^{2}, *Pi-z*^{t} and *Pi-b* lines.

Blast severity in the mixture of 'Sasanishiki' and *Pi-i* line was higher than that in the chemical treated pure stand of 'Sasanishiki' , and there was no or little blast reduction relative to mean of component pure stands in the mixture. The blast race 007 capable of attacking the two components (Table 1) predominated in the mixture.

On the other hand, blast severity in the mixtures of *Pi-k*, *Pi-k*^{m}, *Pi-z* and *Pi-z*^{t} lines, and *Pi-ta*, *Pi-ta*^{2}, *Pi-z*^{t} and *Pi-b* lines was equal to or lower than that in the chemical applied pure stand. This low severity is considered to be owing to few distribution of blast races virulent to the component lines in our trial fields.

In the mixture of *Pi-i* and *Pi-k* lines, blast severity was great in 1994 when the blast races (037 and 077) virulent to both of the components were predominant in the mixture (Table 1).

3.2. BLAST REDUCTION ON 'SASANISHIKI' MULTILINE

From data of blast severity in 23 mixtures of 12 different proportions of 'Sasanishiki ' and its NILs from 1993 to 1996, we calculated blast reduction in the mixtures to mean of component pure stands.

Mean of the reduction in our mixture trials was greater for leaf blast (50.2%) than for panicle blast (18.1%) as Mundt's (1994) summarization for rice cultivar mixtures. However, the mean value of panicle blast reduction in our trials was smaller than that of Mundt's, although the mean of leaf blast reduction from our data was not different from that of his summarization.

In our mixture trials, blast reduction to mean of component pure stands was great when the disease severity on susceptible lines was great and proportions of resistant lines in the mixtures were large. This suggests that barrier effect of resistant lines mainly caused the reduction in the mixtures, although Nakajima *et al.*(1996b) pointed out leaf blast depression due to induced resistance on the 'Sasanishiki' multiline.

3.3. RELATIONSHIP BETWEEN PANICLE BLAST SEVERITY AND YIELD
ON 'SASANISHIKI' MULTILINE

There were statistically significant negative correlations (P<0.01) between rough rice yields and percentages of spikelets diseased with panicle blast in all mixture trials from 1993 to 1996 as reported in pure stands of rice cultivars (Katsube and Koshimizu, 1970). Yield increased due to the mixtures, and the mean yield increase in all the mixtures relative to mean of component pure stands from 1993 to 1996 was 10.3%.

4. Blast Development on a Single Hill of 'Sasanishiki' in Pure Stands of Resistant Lines

To clarify the reason why the reduction of panicle blast in the mixtures was lower than that of leaf blast, we carried out this trial. A single hill of 'Sasanishiki' was planted in the center of each pure stand of resistant $Pi-z'$ and $Pi-b$ lines, and blast development due to natural infection on the single hill was examined and the development was compared with that in the pure stand of 'Sasanishiki'

Blast development was reduced by planting resistant lines around the single hill of 'Sasanishiki' , and the reduction was greater for leaf blast than for panicle blast (Figure 1). This result may indicate that level of auto-infection is high for panicle blast and barrier effect of resistant lines on the 'Sasanishiki' multiline is low for the disease.

MacKenzie (1979) indicated that the decrease in auto-infection increases the effect of multilines on cereal disease control using a mathematical model, and Mundt and Leonard (1985) and Wolfe (1985) pointed out that multilines and cultivar mixtures with small plants would be more effective for disease reduction due to low level of auto-infection. A low level of panicle blast reduction on the 'Sasanishiki' multiline may be partially due to a high level of auto-infection for panicle blast suggested by our study.

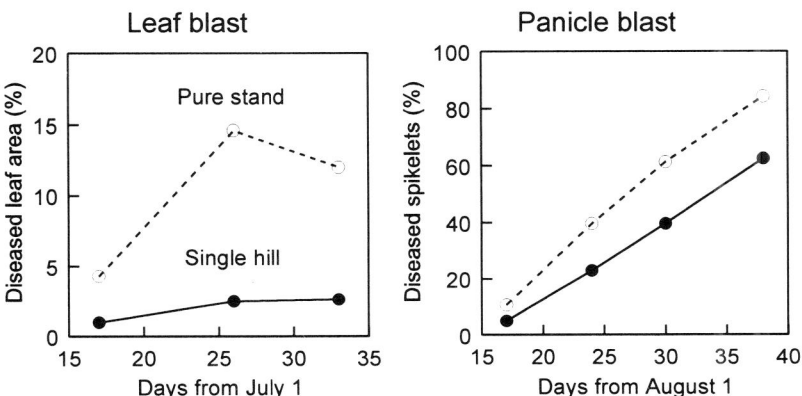

Figure 1. Blast development in pure stand of 'Sasanishiki' and on a single hill of 'Sasanishiki' in pure stands of resistant NILs.

5. Blast Races on 'Sasanishiki' Multiline

After monoconidial isolation of blast fungus from lesions in mixtures and pure stands of 'Sasanishiki' and its NILs, pathogenicity of the isolates was tested using Japanese differential rice cultivars (Yamada *et al.*, 1976), and their pathogenic races were identified.

5.1. CHANGE IN BLAST RACE FREQUENCY

Japanese race 007 was constantly predominant in the pure stands of 'Sasanishiki' , + and *Pi-i* lines and TML from 1993 to 1995 respectively, although kinds of isolated races were greater from TML than from the pure stands of 'Sasanishiki' , and + and *Pi-i* lines (Figure 2). The race 007 is virulent to 'Sasanishiki' , and + and *Pi-i* lines (Table 1).

On the other hand, at the second cropping of the mixture of *Pi-i* and *Pi-k* lines, blast fungus races, 037 and 077, which can attack both of the lines (Table 1), predominated (Figure 2). The races, 037 and 077, were also predominant in the mixture of equal proportion of *Pi-k*, *Pi-k'''*, *Pi-z* and *Pi-z'* lines from 1995 to 1996. The race 077 has virulence against three of the four lines and the race 037 can attack two of them (Table 1).

The 'super race' capable of attacking all components of TML was not isolated from every examined plot.

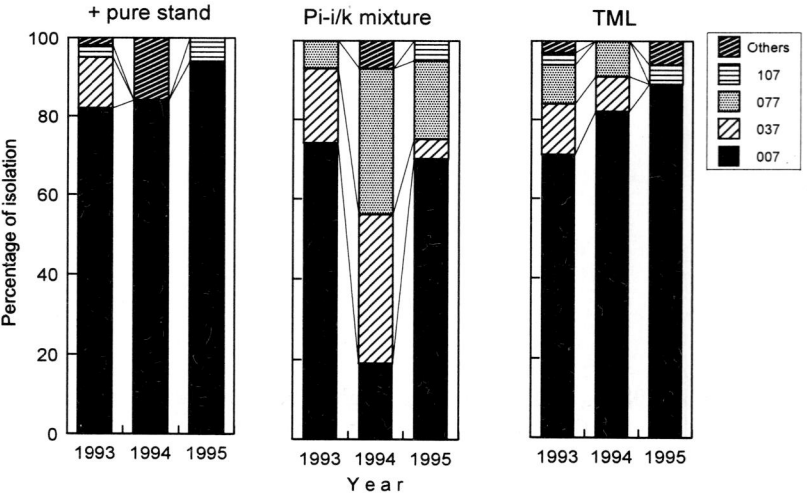

Figure 2. Japanese races of *Pyricularia grisea* isolated from pure stand of a NIL and mixtures of 'Sasanishiki' NILs. Pi-i/k mixture: mixture of equal proportion of *Pi-i* and *Pi-k* lines; TML: mixture of equal proportion of 'Sasanishiki' and nine NILs.

5.2. AVIRULENT BLAST RACES ISOLATED FROM PANICLE BLAST LESIONS

Table 3 represents percentages of avirulent *Pyricularia grisea* races isolated from panicle blast lesions of 'Sasanishiki' NILs. The avirulent races were isolated from panicle blast lesions especially on the *Pi-ta* and *Pi-z* lines. Moreover, on the *Pi-ta* and *Pi-z* lines, the avirulent races were isolated from neck node lesions as well as panicle branch and spikelet lesions. The isolation ratios of the avirulent races from the neck node lesions from the *Pi-ta* and *Pi-z* lines were 93 and 50 %, respectively.

Inoculation tests indicate that the avirulent races induce panicle blast lesions on the *Pt-ta* and *Pi-z* lines. Panicle blast due to the avirulent races also is considered to cause low level of panicle blast reduction on the 'Sasanishiki' multiline.

Table 3. Avirulent races of *Pyricularia grisea* isolated from panicle blast lesions of 'Sasanishiki' near-isogenic lines and panicle blast severity on them in the trial field

Near-isogenic line	Isolated avirulent races (%)	Diseased spikelets (%)
Pi-ta line (pure stand)	89	55
Pi-z line (pure stand)	45	12
Pi-k line mixed with *Pi-i* line	13	6
Pi-k line mixed with 'Sasanishiki'	19	5

6. Conclusions

Our trials indicated that increase in number of components and proportion of resistant components is necessary for effective control of rice blast disease with the 'Sasanishiki' multiline. However, available complete resistance genes in the multiline are limited, since several NILs are not effective in farmers' fields, although 14 complete resistance genes to blast were identified in Japan (Kiyosawa, 1974; Imbe and Matsumoto, 1985). Multilines with high levels of partial resistance to blast (especially panicle blast) should be urgently developed to effectively control rice blast disease with multines for long period.

REFERENCES

Asaga, K. (1981) A procedure for evaluating field resistance to blast in rice varieties, *J. Cent. Agric.*

Exp. Stn. 35, 51-138 (in Japanese, English summary).

Bonman, J. M., Estrada, B. A., and Bandong, J. M. (1989) Leaf and neck blast resistance in tropical lowland rice cultivars, *Plant Disease* 73, 388-390.

Borlaug, N. E. (1959) The use of multilineal or composite varieties to control airborne epidemic diseases of self-pollinated crop plants. in *Proc. 1st Int. Wheat Genet. Symp.*, Winnipeg, pp. 12-26.

Imbe, T., and Matsumoto, S. (1985) Inheritance of resistance of rice varieties to the blast fungus strains virulent to the variety 'Reiho', *Jpn. J. Breed.* 35:332-339.

Katsube, T., and Koshimizu, Y. (1970) Influence of blast disease on harvests in rice plant. 1. Effect of panicle infection on yield components and quality, *Bull. Tohoku Natl. Agric. Exp. Stn.* 39, 55-96 (in Japanese, English summary).

Kiyosawa, S. (1974) Studies on genetics and breeding of blast resistance in rice, *Misc. Publ. Bull. Natl. Agric. Sci., Ser. D* 1, 1-58 (in Japanese, English summary).

Koizumi, S (1983) Multiline cultivars as control measures against pathogenic races of rice blast fungus, *Shokubutu boeki* 37, 477-480, 548-551 (in Japanese).

Koizumi, S (1994) Effect of field resistance on leaf blast development in mixtures of susceptible and resistant rice cultivars, *Ann. Phytopathol. Soc. Japan* 60, 585-594.

Koizumi, S., and Fuji, S. (1994) Effect of mixtures of isogenic lines developed from rice cv. 'Sasanishiki' and 'Nipponbare' on blast development, *Res. Bull. Aichi Agric. Res. Ctr.* 26, 87-97 (in Japanese, English summary).

Koizumi, S, Tani, T., and Fuji, S. (1996) Control of rice blast disease by multilines, *Nogyo Gijyutu* 51 (2), 89-93 (in Japanese).

MacKenzie, D. R. (1979) The multiline approach to the control of some cereal diseases. in *Proceedings of The Rice Blast Workshop*, International Rice Research Institute, Los Banos, Philippines, pp. 199-216.

Matsunaga, K. (1996) Breeding of a multiline rice cultivar 'Sasanishiki BL' and it's use for control of blast disease in Miyagi Prefecture, *Nogyo Gijyutu* 51 (4), 173-176 (in Japanese).

Mundt, C. C. (1994) Use of host genetic diversity to control cereal diseases: Implications for rice blast, in R. S. Zeigler, S. A. Leong and P. S. Teng (eds.), *Rice Blast Disease*, Cab International and IRRI, Cambrige, pp. 293-308.

Mundt, C. C., and Leonard, K. J. (1985) Effect of host genotype unit area on epidemic development of crown rust following focal and general inoculations of mixtures of immune and susceptible oat plants. *Phytopathology* 75, 1141-1145.

Nakajima, T., Sonoda, R., and Yaegashi, H. (1996a) Effect of a multiline of rice cultivar Sasanishiki and its isogenic lines on suppressing rice blast disease, *Ann. Phytopathol. Soc. Japan* 62, 227-233.

Nakajima, T., Sonoda, R., Yaegashi, H., and Saito, H. (1996b) Factors related to suppression of leaf blast disease with a multiline of rice cultivar Sasanishiki and its isogenic lines, *Ann. Phytopathol. Soc. Japan* 62, 360-364.

Rossman, A. Y., Howard, R. J., and Valent, B. (1990) *Pyricularia grisea* the correct name for the rice blast disease fungus, *Mycologia* 82 (4), 509-512.

Shindo, K., and Horino, O. (1989) Control of rice blast disease by mixed plantings of isogenic lines as multiline cultivars. *Bull. Tohoku Natl. Agric. Exp. Stn.* 39, 55-96 (in Japanese, English summary).

Wolfe, M. S. (1985) The current status and prospects of multiline cultivars and variety mixtures for disease resistance, *Annu. Rev. Phytopathol.* 23, 251-273.

Yamada, M., Kiyosawa, S., Yamaguchi, T., Hirano, T., Kobayashi, T., Kushibuchi, K., and Watanabe, S. (1976) Proposal of a new method for differentiating races of *Pyricularia oryzae* Cavara in Japan, *Ann. Phytopathol. Soc. Japan* 42, 216-219.

Yamanaka, S., and Yamaguchi, T. (1987) *Rice Blast Disease*, Yokendo, Tokyo (in Japanese).

USE OF LINEAGE EXCLUSION IN A MULTI-OBJECTIVE RICE BREEDING PROGRAM

J. W. GIBBONS, D. GONZÁLEZ AND D. DELGADO
FLAR, CIAT, A.A. 67-13, Cali, Colombia.

Introduction

Rice blast disease caused by *Pyricularia grisea* is the single most important disease constraint to rice production in Latin America and the Caribbean (LAC). The Latin American Fund for Irrigated Rice (FLAR), created in 1995, determined that breeding for blast disease resistance coupled with grain quality and high yield potential should be a priority for the entire region. Region specific objectives include iron toxicity and cold tolerance primarily for the Southern Cone of Brazil, Uruguay and Paraguay; and Hoja Blanca virus and its insect vector *Tagosodes orizicolus* tolerance for the Tropical Zone of Colombia, Venezuela, Panama, Costa Rica, Guatemala, and the Caribbean. The goal of the breeding program of FLAR is to efficiently combine these characteristics into cultivars acceptable to producers, millers and consumers.

The International Center for Tropical Agriculture (CIAT) has had considerable success in developing improved germplasm adapted to rice agroecosystems of LAC. Since 1989 sixty new cultivars originating from the CIAT breeding program (L. Berrío, *personal communication*) have been released by National Agriculture Research Systems (NARS). Notable among those are the durable blast resistant cultivar Oryzica Llanos 5 (Leal, *et al*, 1989), and nine cultivars from the cross CT8008 (CT3050/Oryzica 1//IR21015) which are grown in a wide range of agroecosystems from Southern Brazil to Guatemala. Early generation selections which gave rise to these cultivars were made under the high disease pressure "Hot Spot" of the Santa Rosa Experiment Station in the Llanos of Colombia (described in Correa-Victoria and Zeigler, 1995). Although CIAT no longer funds the traditional irrigated rice breeding program, FLAR continues to utilize Santa Rosa as it's main breeding site.

Selection of parents to be used in a multi-objective rice breeding program is based on several factors including plant type, grain quality, reaction to biotic and abiotic stresses, and ancestry. Crosses are made with what are considered to be complementary genotypes and the resulting progeny are tested against the *a priori* objectives. Plants and/or lines which do not conform to the objectives are eliminated. The ability to select parents which combine reasonably well and to differentiate between progeny which do or do not possess the desired combinations determine the efficiency of a breeding program. A well characterized germplasm collection and adequate screening tests are minimum requirements for a successful rice breeding program.

Lineage exclusion has been proposed as an aid to rice breeding programs which consider blast resistance as one of the principle objectives (Zeigler *et al.*, 1994). We have characterized our working germplasm collection (WC) for the prevalent lineages of Santa Rosa (Levy *et al.*, 1993), and are making cross combinations which exclude

D. Tharreau et al. (eds.), Advances in Rice Blast Research, 146–153.

lineage compatibility. Focusing on resistance sources to lineages of the pathogen, rather than on genes resistant to specific pathotypes (Correa-Victoria *et al.*, 1994) or field reaction alone, has aided in our ability to select parents which when combined may confer resistance to all lineages.

Field Evaluation

Field evaluations are done at Santa Rosa under direct seeded, favored upland conditions. Spreader rows (Pulver and Bruzzone, 1985, Correa-Victoria and Zeigler, 1995) are planted perpendicular to prevailing wind direction. The rows are inoculated with fresh dried leaf tissue of spreader component cultivars grown in pure plots. When the rows are heavily infected and actively sporulating (two to three weeks after seeding) the experimental material mixed with the highly susceptible "Fanny" (50% material + 50% Fanny) is drill seeded at a density of 0.3 grams per linear meter perpendicular to the spreader. Leaf blast scores based on the 0 to 9 scale of the standard evaluation system(SES) (IRRI,1996) are taken at about 32 and 45 days after seeding. Neck blast also is evaluated on a 0-9 SES scale at about 25 days after flowering. Because of the high level of segregation in F2 and F3 populations, leaf blast scores of these generations are based on two ratings: maximum lesion type (1 to 4) and the incidence or general number of plants with that lesion type (1 to 3 scale with 1 being few infected plants and 3 if most plants are diseased) observed in the population. Entire populations which have type 4 lesions and incidence of 2 or 3 are discarded. Depending on the generation or uniformity of the material, individual plant or bulk selections are made *in situ.*

Lineage Evaluation

Evaluation of advanced breeding lines and germplasm accessions for lineage susceptibility is carried out under greenhouse conditions. At least two highly virulent isolates from each of the six Santa Rosa lineages (SRL) are used (Table 1). For isolate, twenty rice seedlings of each line are grown in sterile soil in individual pots for 19 days before being transferred to 55cm. long by 35cm. wide by 40 cm. tall aluminum framed, clear plastic covered mist chambers. We are able to evaluate about 100 lines at a time. Twelve lines known to be susceptible to different lineages are included as checks (Table 2). The plants are inoculated by aspersion (Levy *et al.*, 1993) and covered to maintain high humidity conditions in the chamber. Evaluations of lesion type (LT) and % leaf area affected (%LAA) (IRRI, 1996) are made 15 days after inoculation. A final rating scale from highly resistant (HR) to highly susceptible (HS) is used to assign reaction to test entries (Table 3).

Table 1. *Pyricularia grisea* isolates used for lineage evaluation.

ISOLATES	GENETIC LINEAGE
Fanny 54	SRL6
Selecta 3-20(1)	SRL6
Fanny 47-1	SRL5
Isol 6-7-1	SRL5
Oryzica Caribe 8 (31-2)	SRL4
Oryzica Caribe 8 (33-1)	SRL4
Metica 1 (33-18)	SRL3
Metica 1 (33-20)	SRL3
Oryzica Llanos 5 (237-2)	SRL2
Cica 9 (151-1)	SRL2
Cica 9 (15)	SRL1
Cica 9 (52-1)	SRL1

Table 2. Check lines susceptible to different SRLs.

Line	Susceptibility
FANNY	SRL6,5,4,3,2,1
O. LLANOS 5	Nil
O. CARIBE 8	SRL4
CICA 9	SRL6,2,1
CICA 8	SLR5
METICA 1	SLR6,4,3
O. YACÚ 9	SRL6,2
ISOL 1	SRL6,4,2,1
ISOL 6	SRL5
ISOL 8	SRL6,5,4
ISOL 10	SRL6,5,4
ISOL 21	SRL6,5,4

Table 3. Scale used to rate germplasm for reaction to SRLs in the greenhouse.

REACTION	LT[1] (%LAA)[2]
HR (Highly resistant)	1(>0.1%); 2 (<10%)
R (Resistant)	2(>10%); 2,3(<10%)
I (Intermediate)	3(< 8%); 2,3(>10%)
S (Susceptible)	3(> 8%); 4 (< 6%)
HS (Highly susceptible)	4(> 5%)

[1] Lesion type.
[2] % Leaf area affected.

Working Collection (WC)

We have evaluated 382 WC accessions for reaction to the six SRLs. The WC contains entries from CIAT, IRRI, IITA,USA and NARS which are actively used in the breeding program. Over 92 % (355) were resistant (HR or R) to four or more SRLs (Table 4).

Table 4. Reaction of 382 WC accessions to the 6 SRLs in greenhouse and under field conditions at Santa Rosa.

	No. of lineages						
	6	5	4	3	2	1	0
Resistant[1]	121	147	87	19	8	0	0
% of total WC	31.7	38.5	22.7	52.1	0	0	
Resistant in field[2]	105	85	44	7	2	0	0
% of resistant	86.8	57.8	50.6	36.8	25	0	0

[1] HR or R
[2] Leaf and neck blast score <5 on scale of 0 to 9 where 0 = no incidence and 9 = more than 75% LAA or more than 50% severely infected panicles. SES (IRRI, 1996).

Lines included in the WC are selected for some outstanding feature including blast reaction at Santa Rosa. It is not surprising that a large number would show a good spectrum of resistance to the SRLs. About 32% (121) were resistance to all six SRLs. Of these, only 16 were rated susceptible (>4) in the field; 8 only for neck blast (data not shown). Lineage evaluation is done in the leaf stage. There is some evidence that leaf infection does not always predict neck blast (Weeraratne, 1981) so neck susceptibility could have escaped detection in the greenhouse. Although the whole virulence spectrum of a particular lineage may be present in a single isolate (Correa-Victoria *et al.*, 1994), the difficulty of selecting single isolates which show virulence to a large number of lines (Roumen, 1994) could result in some greenhouse escapes. Lineage evaluation is a dynamic process: new isolates continuously are collected and tested for virulence spectrum in order to capture the highest virulence in the population (F. Correa, *personal communication*). Of the 27 WC lines which were susceptible to 3 or more SRLs, 9 were rated as resistant in the field. It is possible that partial resistance harbored in some of these lines could be sufficient to protect them at Santa Rosa. Of the WC lines resistant to 4 or 5 SRLs 55% were resistant in the field. This indicates the importance of developing breeding lines which have resistance to all lineages prevalent in the target rice production area.

Susceptibility to SRL 6 was most frequent in the WC with over 54% of the lines susceptible (Table 5). SRL 4 and SRL 5 followed with 18 and 17 %, respectively. SRL 1, 2 and 3 combined were compatible with 18% of the accessions. SLR 4, 5 and 6 have a much wider virulence spectrum compared to SRL 1, 2 and 3, and the latter are characterized by greater incompatibility than compatibility with known resistance genes (Correa-Victoria *et al.*, 1994). It is important to note that the WC has a high level of diversity of reaction to the SRLs, thus allowing for a very high number of possible combinations which exclude susceptibility to all SRLs.

Table 5.   Frequency of susceptibility of WC accessions to the 6 SRLs.

Lineage						
	6	5	4	3	2	1
No. of Susceptible Accessions [1]	208	71	64	3	31	33
% of total WC	54.4	18.6	16.8	0.8	8.1	8.6

[1] Susceptible = I, S or HS.

F4 Breeding Lines

Reaction to SRLs and Santa Rosa blast conditions was evaluated on 291 F4 lines originating from 4 triple crosses (Table 6). The F4 lines resulted from pedigree selection in the F1, F2 and F3 generations. Selection in the F1 and F3, based on grain and plant type, was made at Palmira, Colombia, a non-stress environment. F2 plant selections were made at Santa Rosa. Quality data from F3 grain were used to aid in selection of F3 families.

Table 6. Designation and cross combinations of the F4 lines.

Designation	Combination
FL001	WC379 / WC351 // WC284
FL003	WC379 / WC351 // WC295
FL006	WC381 / WC351 // WC292
FL007	ACC1302 / WC351 // WC365

Table 7. Rice blast reaction of 8 parental lines to isolates representative of the SRLs and at Santa Rosa.

	PROGENITORS							
	WC284	WC292	WC295	WC351	WC365	WC379	WC381	ACC1302
	CT8198	CT8240	CT8249	CT10865	CT8008	IRGA 411	IRGA 370	IRGA 416
LINEAGE[1]								
SRL1	HR	HR	HR	HR	HR	HR	HR	HS
SRL2	HR	HR	HR	HR	HR	HR	HR	HS
SRL3	HR	_	HR	HR	HR	HR	HR	I
SRL4	HR	HR	I	I	HS	HR	I	HR
SRL5	HR	HR	HR	HR	HR	HR	HR	HS
SRL6	R	HS	HS	HS	HR	HS	HS	S
SANTA ROSA[2]								
Leaf, neck	4.4	1.3	3.4	4.1	6.3	4.5	5.7	7.7

[1] HS, highly susceptible; S, susceptible; I, intermediate; R, resistant; HR, highly resistant.
[2] Scale of 0 to 9 where 0 = no incidence and 9 = more than 75% LAA or more than 50% severely infected panicles. SES (IRRI, 1996).

The parents of these crosses varied from resistant to all SRLs (WC284) to susceptible to 5 SRLs (ACC1302) (Table 7). Field reaction also varied from high to low susceptibility in both leaf and neck. Three parental lines (WC295, WC351 and WC381) which share the same SRL resistance spectrum varied in field reaction. In contrast to WC381, it is possible that WC295 and WC351 contain minor genes which confer field resistance in Santa Rosa.

The parents used in the crosses FL001 and FL007 combined to exclude susceptibility to all SRLs (Table 8). Resistance to all six SRLs was found in 59 lines (31%) from these two crosses. Field leaf blast scores of these lines were significantly lower than others within the same cross. Crosses FL003 and FL006, however, did not exclude SRL 6. Over 90 lines (92%) from these two crosses were susceptible to SRL 6 and only 1 was resistant to all SRLs. It is clear for the crosses studied here that with only one parent of a triple cross excluding a lineage, progeny resistant to that lineage can be recovered. It could be expected that a cross which included two or three excluding parents to a particular lineage would result in a higher percentage of resistant progeny. In contrast, particular lineage would result in a higher percentage of resistant progeny. In contrast, when all parents are susceptible to the same lineage, there is a very high probability that the resulting progeny will be susceptible to that lineage. Combining parents which are susceptible to a lineage known to occur in a target production area would result in susceptible progeny, and a potentially dangerous situation. A surprising number of lines from FL001 and FL006 were susceptible to excluded SRLs 1, 2, 3 and 5. Perhaps isolate virulence was not sufficiently strong to detect susceptibility in the parents. Recombination events or recessive compatible virulence alleles may be expressed in the progeny of this cross. The ability to select only those lines which have resistance to a particular lineage or set of lineages is hampered by the impracticality of greenhouse screening of large numbers of segregating populations. Therefore, FLAR recommends "hot spot" screening to eliminate all susceptible genotypes followed by advanced line testing for lineage compatibility.

Table 8. Summary of the blast reaction of F4 lines to 6 SRLs and at Santa Rosa.

Cross designation	GREENHOUSE		FIELD
(Compatible lineage)	Suscept. to Lineage	No. lines (percent)	average[1]
FLOO1	NIL	32 (20.9 %)	3.28a
(6/ 6,1-4// 0)	SRL4	29 (18.9 %)	3.69ab
	SRL6	9 (5.8 %)	3.89 b
	SRL6-4	21 (13.7 %)	3.81 b
	SRL1-2-3-4-5 or 6	62 (40.5 %)	4.17 b
		n=153	
FLOO3	NIL	1 (2.9%)	4.00 NS
(6/ 6,1-4// 6, 1-4)	SRL6	10 (29.4%)	3.60
	SRL4	8 (23.5%)	3.75
	SRL6-4	15 (44.1%)	4.13
		n= 34	
FLOO6	SRL6	62 (92.5%)	3.50 NS
(6, 1-4/ 6,1-4// 6)	SRL6-5	4 (5.9%)	3.75
	SRL4-5-6	1 (1.4%)	4.00
		n= 67	
FLOO7	NIL	27 (72.9%)	2.85a
(1,2,5,6, 1-3/ 6, 1-4// 4)	SRL4	2 (5.4%)	4.00 b
	SRL6 - 5 – 4	8 (21.6%)	3.50ab
		n= 37	

[1] Means with the same letter are not significantly different. NS = Not significant at 0.05 level. Completely randomized design ANOVA, Duncan multiple range test (SAS, 1996).

Conclusions

 Lineage exclusion is a valuable tool in the FLAR breeding program. The combination of "hot spot" screening for massive line selection and greenhouse lineage evaluation of parental lines and advanced generations is resulting in an accumulation of genotypes which combine resistance to all the SRLs. Internationally, progress is being made. On the national level, however, much work is needed. Too few programs in LAC screen under "hot spot" conditions; lineage characterization in rice production areas is limited; and, germplasm banks have yet to be properly evaluated for lineage compatibility.

National breeding plans prepared by crop improvement teams at the local level, are an important component in the flow of improved technology within FLAR. These plans are basically an analysis of cultivar needs and development, and most often result in alternatives for more efficient screening, selection and release. Breeding plans also clarify the role of international and national organizations in cultivar development. Combining lineage exclusion with the other major breeding objectives in LAC requires much cooperative effort. FLAR is confident that the mechanisms are in place to continue the recent progress in blast resistance breeding made in LAC.

Acknowledgments

We wish to thank James Silva for help with data handling.

References

Correa-Victoria, F.J., and Zeigler, R.S. 1995. Stability of partial and complete resistance in rice to *Pyricularia grisea* under rainfed upland conditions in Eastern Colombia. Phytopathology 85:977-982.
Correa-Victoria, F. J., Zeigler, R. S., and Levy, M. 1994. Virulence characteristics of genetic families of *Pyricularia grisea* in Colombia. Pages 211-229 in: Rice Blast Disease. R.S. Zeigler, P. S. Teng, and S. A. Leong, eds. Commonwealth Agricultural Bureaux International, Wilmington, England.
Leal, D., Davilos, A., Delgado, H., and Uruena, E. 1989. Dos nuevas variedades de arroz para el piedemonte llanero Oryzica Llanos 4 y Oryzica Llanos 5. Arroz 38:11-21.
Levy, M., Correa-Victoria, F. J., Zeigler, R. S., Xu, S., and Hamer, J. E. 1993. Genetic diversity of the rice blast fungus in a disease nursery in Colombia. Phytopathology 83:1427-1433.
IRRI. 1996. Standard Evaluation System for Rice. P.O. box 933, Manila, Philippines, 52pp.
Pulver, E. L., and Bruzzone, C. 1985. Presion alta y uniforme de piricularia con fines de seleccion. Arroz en las Americas 6:1-12.
SAS/STAT user's guide release 6.12. 1996. SAS Institute Inc., Gary, NC, USA.
Roumen, E. C. 1994. A strategy for accumulating genes for partial resistance to blast disease in rice within a conventional breeding program. Pages 245-265 in: Rice Blast Disease. R.S. Zeigler, P. S. Teng, and S. A. Leong, eds. Commonwealth Agricultural Bureaux International, Wilmington, England.
Weeraratne, H., Martinez, C., and Jennings, P. R. 1981. Genetic strategies in breeding for resistance to rice blast *Pyricularia oryzae* in Colombia. Pages 305-311 in: Proc. Symp. Rice Resistance Blast. Centro Internacional de Agricultura Tropical, Cali, Colombia.
Zeigler, R. S., Tohme, J., Nelson, R., Levy, M. and Correa-Victoria, F. 1994. Lineage exclusion: A proposal for linking blast population analysis to resistance breeding. Pages 267-291 in: Rice Blast Disease. R.S. Zeigler, P. S. Teng, and S. A. Leong, eds. Commonwealth Agricultural Bureaux International, Wilmington, England.

LINEAGE-EXCLUSION TESTS FOR BLAST RESISTANCE IN SOUTHERN INDIA

R. Sivaraj[1], S.S. Gnanamanickam[2], and M. Levy[1]

[1] Department of Biological Sciences, Purdue University, West Lafayette, IN USA 47907-1392.

[2] Centre for Advanced studies in Botany, University of Madras, Madras 6000 25, India.

Abstract

We have identified 25 MGR-fingerprint lineages of the rice blast fungus in southern India in samplings from 1991-95; some lineages appear to be geographically limited to certain states. In artificial inoculation assays, each lineage has a definable virulence spectrum on rice cultivars that is limited by avirulence shared by all lineage members. We are testing a lineage-exclusion resistance breeding strategy, i.e., combining resistances that, in complement, exclude all lineages in the pathogen population, in contrast to the conventional strategy that combines resistances to exclude detected pathotypes. In two-semester trials in a blast nursery at Pattambi, Kerala we evaluated field performance of rice lines constructed to pyramid two dominant blast resistance genes from the near-isogenic lines C101A51 (bearing Pi-2(t)) and C101LAC (bearing Pi-1(t)); previous inoculation assays indicated that no lineage in Kerala or adjoining states was compatible with Pi-2(t) and that Pi-1(t) was effective against most lineages. All pyramids remained uninfected during trials while many cultivars, including C101LAC but not C101A51, were heavily infected. Our pathogenicity survey results indicate, however, that the durability of these pyramids is better tested in other states (Tamil Nadu, Maharastra and Orissa), where we have detected isolates differentially compatible with both resistance genes. Additional pyramids of C101A51 and C103TTP (allegedly bearing a version of the Pi-1 (t) gene) were constructed. Using isolates from both India and Colombia, we demonstrated that the gene from C103TTP (renamed as Pi-1b(t)) was expressed as a recessive phenotype. We are currently working on introgressing commercial cultivars used in various regions of India with available lineage-excluding resistances.

1. Introduction

Rice blast disease, caused by *Pyricularia grisea* (*Magnaporthe*) Sacc., is an endemic constraint to rice production throughout India (Padmanabhan, 1965) and in many rice growing regions worldwide. Breeding for durable host resistance using single major genes has been frustrated by the rapid adaptation of the pathogen population (Kiyosawa, 1982; Ou, 1965). Conventionally, useful resistant genes/genotypes are selected based

D. Tharreau et al. (eds.), Advances in Rice Blast Research, 154–161.

on their incompatibility with, or exclusion of, pathotypes (virulence variants or physiologic races) of the fungus that are detected during the breeding process (Padmanabhan *et al.*, 1970). Resistance breakdown in the field is associated with the appearance of non-excluded, compatible pathotypes. The origins of such newly detected pathotypes, whether by increased frequency or migration of already existing genotypes or by evolution of novel genotypes via mutation or recombination, remain unclear. However, it is clear that resistance combinations designed to exclude detected pathotype diversity alone do not produce long lasting pathogen management.

Recently, Zeigler *et al.* (1994) proposed an alternative strategy for using major genes to achieve durable resistance. This strategy is called lineage-exclusion and is based on the experimental observations that, typically: 1) rice blast fungus populations are composed of a limited number of distinct genetic families, called lineages, as defined by MGR-DNA fingerprint analysis, and 2) each MGR-defined lineage has a virulence spectrum that is limited by at least one avirulence (incompatibility with a single cultivar or resistance gene) that is shared by all isolates in the lineage (Levy *et al.*,1991; Levy *et al.*,1993; Correa-Victoria *et al.*, 1994; Zeigler *et al.*,1995; Roumen *et al.*, 1997). The most important assumptions of the lineage-exclusion strategy are that there is a close relationship between the pathogen's phylogenetic history and its immediate virulence potential, and that this relationship is constrained by natural selection (positive fitness) on the pathogen that favors the conservation of lineage-specific avirulence genotypes. Operationally, lineage-exclusion breeding can use even already defeated major genes, so long as each gene is overcome differentially by only some lineages in a population and at least two such genes complementarily exclude all of the lineages present.

This report summarizes our ongoing studies for testing lineage-exclusion resistance breeding for rice blast disease management in southern India. We describe first the genetic structure of sampled rice blast diversity in the region, in terms of MGR-lineages, and then the virulence spectrum of these lineages on particular near-isogenic lines (NILs) reported to contain different single resistance genes. Finally, we report on the construction and field performance of resistance gene pyramids obtained from the NILs and designed by lineage-exclusion considerations from our population analyses.

2. Rice Blast Population Structure in Southern India

2.1 MGR-LINEAGE DIVERSITY

We have used MGR-DNA fingerprint diversity (as in Levy *et al.*, 1993) to evaluate the genealogical relationships among 350 field isolates of the rice blast fungus collected during 1991-95 from a total of 80 cultivars and 11 locations in five different states of southern India (*Figure 1*). The isolates expressed fingerprint profiles containing ca. 45-55 *Eco*RI RFLPs in a fragment size range of 2.0 - 18.0 Kb. Using UPGMA clustering analysis we have identified 25 discrete fingerprint lineages. Isolates within the same lineage express 88% or greater fingerprint similarity while different lineages express an average of less than 75% similarity.

Six lineages (A, G, H, I, J and L) were detected in multiple years and/or at multiple sites and 19 lineages were detected only in a single sampling, at one site in one state (*Figure 1*). Generally, the frequently detected lineages were collected from several cultivars while the infrequent lineages were collected from one or two cultivars. Lineage I was the most widespread in our sampling and the predominant lineage attacking the regionally important cultivar IR50. The same lineage also has been detected on IR50 grown in Bangladesh (Shahjahan, Rojas and Levy, unpublished data). The apparent geographic variation and cultivar-specificity suggests that there may be substantial lineage diversity remaining to be sampled. However, repeated sampling in a blast nursery in Kerala reveals a relatively stable lineage composition (Gnanamanickam *et al.*, these proceedings).

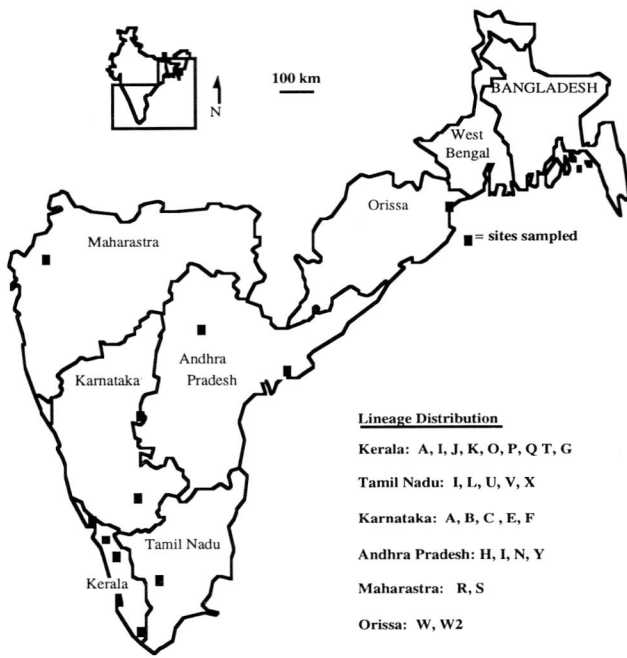

Figure 1. Sampling sites and MGR-lineage distribution in southern India.

2.2 PATHOTYPE DIVERSITY, VIRULENCE SPECTRUM AND LINEAGE-EXCLUSION

We determined pathotype diversity for representative isolates from each lineage by inoculation assay on NILs produced by Mackill and Bonman (1992) with the generally susceptible *indica* cultivar, CO39, as the recurrent parent. We used the NILs C101LAC, C101A51, C104PKT, C101PKT, C105TTP-4L23 and C103TTP that, respectively, are reported to contain the resistance genes, Pi-1(t), Pi-2(t), Pi-3(t), Pi-4a(t), Pi-4b(t) and a version of Pi-1 (t) that we have tentatively called Pi-1b(t); Inukai *et al.*, (1994) indicate that resistance genotype of C105TTP-4L23 may be Pi-4a(t) plus an unknown additional gene. We assayed infection responses for two-week old seedlings (3-4 leaf stage) under controlled conditions favorable for infection of the CO39 parent, as a susceptible check, i.e., sufficient to produce at least 20% diseased leaf area with predominant, fully sporulating lesions (type 4) with a 15 ml aqueous conidial suspension (ca. 100,000 conidia per ml).

We summarize the pathogenic variation detected in TABLE 1. Collectively, the 117 isolates characterized infected all of the resistance genes tested but with great differences in the relative frequency of compatibility. Very few isolates overcame Pi-2(t), and Pi-1(t) and Pi-1b(t) were effective against more than 80% and 60% of the isolates, respectively. The three other resistances were overcome by a large majority of isolates. We found a broad array of pathotypes representing nearly all possible permutations of compatibilities on the commonly infected resistances. The most notably absent pathotypes were those that were dually compatible with Pi-2(t) and either Pi-1(t) or Pi-1b(t).

TABLE 1. Lineage-specific resistance spectra (R = all isolates incompatible; S = some isolates compatiblle of rice blast resistance genes against rice blast isolates from southern India.

	n[b]	R-genes[a]					
		Pi-1	Pi-2	Pi-3	Pi-4a	Pi-4b	Pi-1b
isolates compatible	117	21	6	86	91	69	44
MGR-Lineages:							
A,G,H,I,O,P,S,Z	86	S	R	S	S	S	S
R,V,W,W2,X	11	R	S	S	S	S	R
B,C,E,F,L,N,Q,U,Y	15	R	R	S	S	S	R
K	1	R	R	S	R	R	R
J	3	R	R	R	R	R	S
T	1	R	R	R	R	R	R

[a] As contained in the near-isogenic lines: C101LAC, C101A51, C104PKT, C101PKT, C105TTP-4L23 and C103TTP, respectively (Mackill and Bonman, 1992); [b] Number of isolates tested.

The distribution of compatibility with the more effective resistance genes indicated a strong and differential correspondence between lineage and virulence spectrum, although several lineages shared the same spectrum on the few genes characterized (TABLE 1). Pi-2(t) compatibilty was detected in only five lineages, Pi-1(t) in eight lineages and Pi-1b(t) in 15 lineages; lineages compatible with Pi-2(t) were detected only in Tamil Nadu, Maharastra and Orissa while lineages compatible with Pi-1(t) and Pi-1b(t) were widespread. As will be discussed later, importantly, no lineage expressed compatibility with both Pi-2(t) and either of the Pi-1(t) genes, either dually in one isolate or differentially among separate isolates in the lineage. Consequently, in terms of lineage-specific resistance, Pi-2(t) excluded 20 lineages, Pi-1(t) excluded 17 lineages and Pi-1b(t) excluded 11 lineages. In this sampling, hypothetically, the complement of Pi-2(t) with either of the Pi-1(t) genes would exclude all lineages (and all isolates) tested.

3. Lineage-excluding Resistance Gene Pyramids

Based upon the above studies we constructed two different sets of resistance gene pyramids, the first combining Pi-2(t) with Pi-1(t) and the second combining Pi-2(t) with Pi-1b(t). For both constructions we crossed the appropriate NIL parents and, ultimately, selected for F3 lines expressing non-segregating resistance to a mixed inoculum of isolates that were differentially compatible with both component resistance genes. In the course of constructing the latter set, the segregation analyses of F2 population samples indicated that resistance due to Pi-1b(t) was inherited as a single locus recessive (TABLE 2). These results were confirmed twice using separate groups of isolates, from India and from Colombia, that each produced the appropriate mixture of virulences. Our findings are consistent with the excessive susceptibility previously observed among advanced generation progeny from crosses involving C103TTP (Mackill and Bonman, 1992; Inukai et al., 1994).

TABLE 2. Inoculation reactions of Pi-1b(t) and Pi-2(t) parental NILs and an F2 population of their hybrids.

Cultivar	Virulence of Inoculation Set		
	Pi-1b$^+$	Pi-2$^+$	Pi-1b$^+$ + Pi-2$^+$
CO39	S	S	S
C101A51 (Pi-2 (t))	R	S	S
C103TTP (Pi-1b (t))	S	R	S
C101A51 x			
C103TTP (F2)	60 R : 20 S	17 R : 63 S	33 R : 12 S
(accepted R:S ratio)	3 : 1 [a]	1 : 3 [b]	3 : 13 [c]

[a] Indicates that Pi-1b(t) is a single gene recessive; [b] Indicates that Pi-2(t) is a single gene dominant;
[c] Indicates an interaction between one recessive and one dominant gene.

We tested the performance of the pyramids containing Pi-1(t) and Pi-2(t) in "hot spot" nurseries located at Ponampet, Karnataka in 1996 and at Pattambi, Kerala in 1996 and 1997. In each test, all seven pyramid lines remained essentially uninfected (sparse hypersensitive flecks at most) while many cultivars were heavily infected (TABLE 3).

TABLE 3. Field reactions of parental NILs and derivative Pi-1(t) + Pi-2(t) resistance pyramids in blast "hotspot" nurseries in southern India.

Cultivar (resistance gene)	Field site-year		
	Pattambi, Kerala 1996	Ponampet, Krnt. 1997	Pattambi, Kerala 1997
CO39	S	S	S
C101LAC (Pi-1 (t))	R	R	S
C101A51 (Pi-2 (t))	R	R	R
seven pyramid lines (Pi-1 (t) + Pi-2(t))	R [a]	R [a]	R [a]

[a] Effective resistance is established but without confirmation of complementary lineage-exclusion because one or both parental NILs were uninfected.

However, the parental NILs were not simultaneously infected in any test; only C101LAC, the Pi-1(t) parent, was severely infected in the 1997 Kerala test. Consequently, while the pyramids are clearly protected, the results do not resolve whether complementary lineage-exclusion or simply the absence of isolates compatible with Pi-2(t) is enforcing the protection.

4. Discussion

Figure 2 illustrates two related, but contrasting, strategies of blast resistance breeding, each using two major genes (Pi-1 and Pi-2), against a pathogen population composed of isolates that are either incompatible or differentially compatible with these resistance genes, i.e., there is no observed pathotype that is compatible with the pyramided genotype, Pi-1+Pi-2. Pyramiding resistances against pathotypes (races) detected in blast nurseries or farmer's fields is the conventional way rice blast resistance breeding has proceeded for decades. This strategy, which could be called "pathotype-exclusion", frequently breaks down when appropriate pathotypes appear within 1-2 years after such resistance is deployed in large areas. Lineage-exclusion modifies the conventional strategy by additionally considering the distribution of pathotypes (or more strictly, the component virulences and avirulences) among genetic (MGR-fingerprint based) lineages of the pathogen. In this sense, lineage-exclusion becomes a form of "phylogenetic pathotype-exclusion". Lineage-exclusion presumes that lineage-specific avirulences represent an evolutionary genetic barrier to pathotype diversification within the lineage.

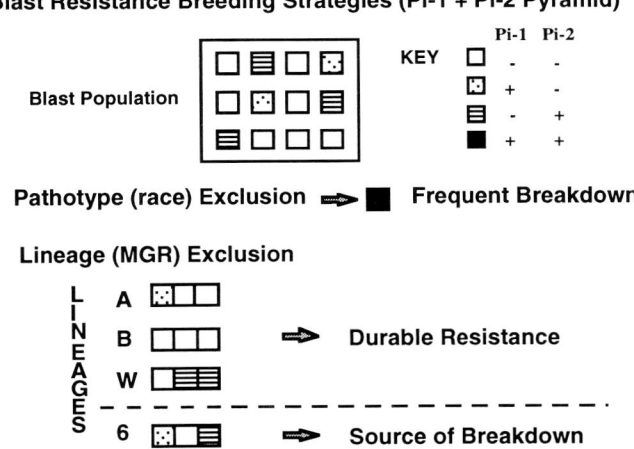

Figure 2. Blast resistance breeding strategies using a two resistance gene, Pi-1 + Pi-2, pyramid against a rice blast population with isolates that are differentially compatible with each resistance gene. Conventional pathotype (race) exclusion, directed against pathotypes not yet detected, leads to frequent breakdown. Lineage-exclusion, directed against the distribution of virulence and avirulence among genetic lineages, predicts durable resistance when each lineage is avirulent with, at least, one component resistance.

The lineage-exclusion model proposes explanations for both durable resistance and for resistance breakdown. When compatibility with the component resistance genes is distributed among lineages (as in lineages A, B and W of southern India, *Figure 2*), then the pyramided resistance is hypothesized to be durable. Durability is expected because, in the absence of interlineage recombination, no single lineage exhibits the genetic capacity to overcome both resistances. Alternatively, an MGR-lineage with isolates that differentially overcome each resistance gene already exhibits this capacity, even when the mutually compatible pathotype is not yet present. Such a lineage is hypothesized as a likely source of resistance breakdown. Rice blast lineage SRL-6 from Colombia is an example of the latter type of lineage, which Tapiero and Levy (these proceedings) observed as the source of novel breakdown of the Pi-1(t) + Pi-2(t) pyramids in hot-spot nursery experiments in Santa Rosa, Colombia.

The successful field trials of our first resistance gene pyramids designed by lineage-exclusion offer support for the strategy, in addition to establishing the value of Pi-1(t) and Pi-2(t) for rice blast resistance breeding for southern India. However, because the field trials were conducted in Kerala and Karnataka, where no Pi-2(t)-compatible pathotypes have been detected by either field test or inoculation assay, it was not possible to resolve lineage-exclusion effects from a successful pathotype-exclusion process. We also recognize that much further sampling will be necessary to confirm the apparent lineage-specific distribution of avirulences (especially for Pi-2(t)) as well as the overall diversity of the rice blast population in the region.

Nonetheless, the prospects for further direct evaluation of this strategy are promising. Our population analysis revealed that lineages differentially compatible with Pi-1(t) and Pi-2(t) co-occur at sites in both Tamil Nadu and Maharastra. The blast nursery at Lonavala, Maharastra, shown to support severe Pi-2(t) field infection (Chen *et al.*, 1996), should be an especially informative site for evaluating the durability of the pyramids previously tested. The economic impact of lineage-exclusion resistance breeding also will be assessable in the near future, after completion of Pi-2(t) introgression into several broadly susceptible, commercial cultivars of southern India, e.g., IR50 and Jothi, where our analyses have indicated that all of the several lineages infecting these cultuivars are excluded by Pi-2(t).

5.   References

Chen, D.H., Zeigler, R.S., Ahn, S.W., and Nelson, R.J. (1996) Phenotypic characterization of the rice blast resistance gene *Pi-2*(t), *Plant Dis.* **80**, 52-56.
Correa-Victoria, F., Levy, M., and Zeigler, R.S. (1994) Virulence characteristics of genetic families of *Pyricularia grisea* in Colombia, in R. S. Zeigler, S. Leong and P. Teng (eds.),. *Rice Blast Disease*, CAB International, UK, and IRRI, The Philippines, pp. 211-229.
Inukai, T., Nelson, R.J., Zeigler, R.S., Sarkarung, S., Mackill, D.J., Bonman, J.M., Takamure, I., and Kinoshita, T. (1994) Allelism of blast resistance genes in near-isogenic lines of rice, *Phytopathology* **84**, 1278-1283.
Kiyosawa, S., (1982) Genetic and epidemiological modelling of breakdown of plant disease resistance, *Annu.Rev.Phytopathol.* **20**, 93-117.
Levy, M., Romao, J., Marchetti, M.A., and Hamer, J.E. (1991) DNA fingerprinting with a dispersed repeated sequences resolves pathotype diversity in the rice blast fungus, *The Plant Cell* **3**, 95-102.
Levy, M., Correa-Victoria, F.J., Zeigler, R.S., Xu S., and Hamer, J.E. (1993) Genetic diversity of the rice blast fungus in a disease nursery in Colombia, *Phytopathology* **83**, 1427-1433.
Mackill, D.J. and Bonman, M.J. (1992) Inheritance of blast resistance in near-isogenic lines of rice, *Phytopathology* **82**, 746-49

Ou, S.H. (1985) *Rice Diseases,* 2nd. edn., Commonwealth Mycological Institute, Kew, UK.

Padmanabhan, S.Y. (1965) Estimating losses from rice blast in India, in *The Rice Blast Disease*, John Hopkins Press, Baltimore, MD, pp. 203-221.

Padmanabhan, S.Y., Chakrabarti, N.K., Mathur, S.C., and Veeraraghavan, J. (1970) Identification of pathogenic races of *Pyricularia oryzae* in India. *Curr. Sci.* **44**, 1574-1577.

Roumen, E., Levy, M., and Notteghem,J.L. (1997) Characterization of the European pathogen population of *Magnaporthe grisea* by DNA fingerprinting and pathotype analysis. *European Journal of Plant Pathol.* **103**, 363-371.

Shen, Y., Zhu, P., Yuan, X., Zhao, X., Levy, M. (1998) Genetic diversity and geographic distribution of *Magnaporthe grisea* in China. *J. Zhejiang Agricultural Univ.* **24**, 493-501.

Zeigler, R. S., Tohme, J., Nelson, R., Levy, M. and Correa-Victoria, F. (1994) Lineage-exclusion: A proposal for linking blast population analysis to resistance breeding, in R. S. Zeigler, S. Leong and P. Teng (eds.),. *Rice Blast Disease*, CAB International, UK, and IRRI, The Philippines, pp. 267-292.

Zeigler, R.S., Cuoc, L.X., Scott, R. P., Bernardo, M.A., Chen, D.H., Valent, B., and Nelson, R.J. (1995) The relationship between lineage and virulence in Pyricularia grisea in the Philippines. *Phytopathology* **85**, 433-451.

SPATIAL AND TEMPORAL STABILITY OF GENETIC RESISTANCE TO RICE BLAST

Searching for durable blast resistance

SANG-WON AHN
International Network for Genetic Evaluation of Rice, Genetic Resources Center, International Rice Research Institute, PO Box 933, 1099 Manila, Philippines

1.   Summary

Knowledge of population dynamics of the blast pathogen over space and time in interaction with rice cultivars under diverse environments is essential for the strategic and effective use of genetic resistance. Spatial and temporal stability of genetic resistance to rice blast is closely associated with the stability of blast population in an agroecoystem. Recent progress in molecular techniques has accelerated the development of rice cultivars with various blast resistance gene combinations. The major challenge to researchers is the selection of appropriate resistance genes and the prediction of spatial and temporal stability in rice cultivars with such gene combinations under different ecosystems. The conventional evaluation scheme has mainly focused on the static and short-term interaction of test entries with existing sets of major race groups, phylogenic groups (lineages), or natural inoculum present in selected test sites. A multisite evaluation of the International Rice Blast Nursery (IRBN) has been an effective means to determine the stability of genetic resistance over a wide range of geographical locations. The conventional varietal evaluation scheme is based on the assumption that stability of resistance over existing diverse blast groups is highly correlated with temporal stability in a given rice cultivar. However, no close correlation between stability over space and stability over time was observed in nondurable rice cultivars. Most known durably resistant rice cultivars exhibited stability both over time and space. Thus, stability over space may not be a sufficient evaluation parameter for durably resistant rice cultivars. To evaluate the temporal stability of partial resistance, the sequential evaluation scheme was developed. Using this field technique for a biological endurance test, two types of pathogenic variations were observed –1) development of new pathogenic races within a lineage and 2) gradual change in aggressiveness within a race-lineage combination. In the future, emphasis should be given to the study of blast population dynamics (genetic as well as pathogenic) over space and time in close association with rice genotypes. Knowledge on this dynamic interaction may provide insights into the nature of genetic durability of host resistance and efficient and effective ways to use host resistance over space and time.

D. Tharreau et al. (eds.), Advances in Rice Blast Research, 162–171.
© 2000 *Kluwer Academic Publishers. Printed in the Netherlands.*

2. Introduction

Rice cultivars with durable blast resistance is the most preferred option for blast management in many blast-prone rice-growing areas. Demand for such rice cultivars with other commercial values is rapidly growing among rice farmers and consumers due to its economic environmental health-conscious society.

The inherent ability of the blast fungus to generate genetic variability (Ou, 1985; Zeigler et al 1997) and its complex interaction with the rice genotype over space and time (Ahn, 1994) are well known. Various breeding strategies to develop rice cultivars with stable genetic resistance over space and time have been formulated. Also, different assessment techniques have been applied to identify suitable breeding lines for large scale cultivation in blast -prone areas. Although some rice cultivars with durable blast resistance have been successfully developed in different countries, our ability to predict the future performance of different breeding lines after large scale cultivation in farmer's plot is quite limited at the moment. Often short-lived resistance of many rice cultivars is due to inability to predict the spatial and temporal dynamics of the blast pathogen as well as a narrow genetic base of blast resistance in rice cultivars. Accordingly, our effort is focused more to expanded knowledge about spatial and temporal interaction of rice blast with rice genotype. This will provide a base to improve the evaluation scheme for blast resistance.

This paper describes the magnitude of variation of the rice blast fungus over space and time, and its implication on resistance evaluation strategy.

3. Structure and dynamics of rice blast fungus over space and time

The blast fungus, *Magnaporthe griesea* (anamorph *Pyricularia grisea* (Cooke) Sacc. Syn. *P.oryzae* Cav.) has been well known for its genetic as well as pathogenic variation (Ou 1980, 1985). Recent study on blast pathogen population using DNA fingerprint analysis with the middle repetitive element MGR586 (Hamer et al.,1989) showed that several distinct groups of pathogen populations, presumably representing clonally propagated lineages exist in different rice-growing areas (Chen et al., 1995; Hamer et al., 1989; Han et al. 1993; Shen et al., 1996). It is believed that all isolates belonging to each lineage were derived from a common ancestor. Some lineage consists of various pathogenic races, while others apparently consist of a single race. The relationship between lineage and pathogenic race needs further study. There was no one to one relationship between lineage and pathotype except few lineages in the Philippines (Chen, et al., 1995). Each lineage was associated with a specific group of pathotypes in Colombia (Levy et al., 1991). However, Chen et al., (1995) found that the probability of correctly predicting pathotype from lineage was only 33% and that of predicting lineage from pathotypes was 65%. Thus, the combined use of pathogenic grouping (race) and lineage/haplotype grouping of genetic diversity would facilitate more meaningful analysis and application.

High haplotypic diversities within the lineage are commonly observed, indicating a high rate of mutation or parasexual exchange of DNA (Zeigler et al., 1997). In some cases, genetic instability in the fungus appears to be a consequence of the chromosomal location of the gene (Valent and Chumley, 1994). Transposable elements also play a role in the variability of the blast fungus (Valent and Chumley, 1991). These mechanisms together with migration may be major forces that generate variation as well as to eliminate any disadvantageous genetic elements accumulated in clonal lineages. Zeigler et al., (1994) presumed that individual lineages may have their specific virulence spectrum or members of a lineage may share a common limit to virulence. They observed that resistance in certain cultivars can not be overcome by individual isolates of a particular lineage. In other cultivars, compatibility is frequent in the lineage. This observation allowed the formulation of the idea of "lineage exclusion" in which resistance genes conditioning resistance to the full spectrum of pathogen lineage can be selected and combined in a rice genotype. However, the concept may have the same consequence of race exclusion, if significant amounts of recombination occur such that limits to virulence are readily overcome. The question is how stable the virulence spectrum of isolates belonging to same lineage group is and what the adequate sample size is to determine the incompatibility of member isolates of a lineage to certain rice cultivars.

In the blast nursery at IRRI, Philippines, no single isolate belonging to lineage 7 showed virulence to Pi-2. Combination of Pi-1 and Pi-2 was expected to exclude all lineages present in the blast nursery. However, soon after a continuous exposure of two test lines having a combination of two major genes, compatible isolates were found and two lines were severely damaged in a sequential planting plot (unpublished). All these compatible isolates were classified as lineage 7 (H. Leung, personal communication) and possess virulence to Pi-2 (Table 2). Some may think this is a matter of sampling. However, these unique isolates may not persist in the blast population, unless there is a strong selection pressure favoring such isolates. No simple rules are available at the moment to select representative isolates for a lineage except to include all range of races using conventional method.

Race composition of the rice blast pathogen varies among different geographical areas(Ou,1985). Due to this spatial variation in race composition, different rice cultivars show a remarkable degree of differential reaction to rice blast (Table 1). Recent analysis of genotype (G) x environment (E) interaction in multilocational trial of rice blast has indicated that a significant proportion of variation due to G x E interaction, which is often higher than that due to genotype or environmental effect alone, can be attributed to spatial variation of virulence frequency in test sites (Ahn and McLaren, 1995). This spatial variation is closely associated with varietal type (indica or japonica) of commonly cultivated rice in the area (Ahn et al., 1997). The blast population between two test sites in the Philippines , 40 km apart, was quite distinct(Chen et al., 1995). Apparently, predominant varietal type of test materials in both test sites is associated with the type of blast lineage. However, the spatial differences would not remain constant if the type and acreage of cultivated rice continue to change over time.

TABLE 1. Differential reactions of some rice cultivars to blast in the 1995 IRBN test sites

Rice cultivars	Suweon, Korea	Sakha, Egypt	Los Baños, Philippines	Cavinte, Philippines	Ibadan, Nigeria	Sta. Rosa, Colombia
Oryzica Llanos 5	5	2	1	2	5	4
Taikeng Yu 1420	9	2	1	1	0	1
IR48088-PM-1-8-1-4-1	6	7	5	1	4	1
CO 25	3	5	5	5	0	8
IR50	1	4	8	8	8	6
IR64	2	2	5	6	3	3

TABLE 2. Virulence change of lineage 7 before and after the exposure of BL lines having Pi-1 and 2.

R-gene	Before BL12				After BL12
	1	7	17	44	7
Pi-1	R	S	S	R	S
Pi-2 ·	R	R	R	S	S
Pi-1 + Pi-2	R	R	R	R	S

Population structure of the pathogen would sensitively respond over time to any changes in biotic and abiotic elements of the environment. The frequency, composition and predominant haplotypes in blast lineages greatly varied with the season (Chen et al.,1995). Also predominant race may change between leaf blast and panicle blast phases in the same cultivars within the same plot. Using a combination of haplotype and race group as a criterion for an identical group of isolates, Han et al (1997) estimated that approximately 30 to 50% of total isolates present during panicle blast phase was believed to originate from the population present during the leaf blast phase in the same field of a rice cultivar. Origin of other isolates obtained from the panicle blast phase is not known. This temporal variation within a crop season may also partly explain differential reaction of some cultivars to leaf blast and panicle blast infection.

Early studies have indicated that an increase of a certain pathotype as well as highly adapted isolates of blast pathogen in farmers' fields was closely correlated with the planting area of rice cultivars of particular genotypes (Kiyosawa and Shiyomi, 1976). Extensive and intensive use of a certain resistance gene in one or several rice cultivars would certainly exert a strong selection pressure on blast population favoring a rapid increase of a particular pathotype with corresponding virulence gene as well as certain lineage group. This was clearly observed in several countries.

Tong-il type rice cultivars, high-yielding semidwarf Korean rice cultivars derived from the cross between japonica and indica parents, were favorably promoted by the government since 1972 due to their remarkably high yield potential and their resistance to stripe virus disease and rice blast. Self-sufficiency in rice in 1977 was achieved through wide and intensive cultivation of these cultivars, and the industrial process of Korean economy in mid-1970s was able to take off without any food problem and associated social problems. Occurrence of blast on Tong-il type cultivars in farmers' fields was first observed in 1976 in one small area. However, a severe epidemic on two Tong-il types occurred in 1978. Long term monitoring of blast race composition during this period clearly indicated that the frequency of race specific to japonica and indica cultivars closely followed the change of the proportion of planting area of these two varietal groups (Ryu et al., 1987).

In the past, the average span of durability of blast resistance in rice cultivars in Meta region of Colombia was 1-2 years (Ahn, 1982). Oryzica Llanos 5 was developed using various resistance sources as parents, and released in blast-prone areas in Colombia in 1989. This variety is being planted in more than 50,000 ha, and no typical susceptible blast lesions were observed until 1995. Some susceptible lesions were observed in evaluation plots in 1996 and in farmers' fields in 1997, although severity was quite low (Correa-Victoria, personal communication). All compatible isolates belong to one lineage SRL-4. These occurred in a low frequency in the past and was rarely observed in evaluation plot. It is suspected that this increase in frequency is due to the large planting area of Oryzica Caribe 8, a highly susceptible cultivar to SRL-4 , and high selection pressure exerted by Oryzica Llanos 5 in commercial fields.

The damage level of rice cultivars could change over time not only due to the change of race but also due to a gradual change in aggressiveness of the blast isolate belonging to same race or lineage(Ahn et al., 1996). Several rice cultivars and corresponding compatible isolates on each cultivars have been evaluated using the sequential evaluation technique (Fig. 1). Sequential evaluation was made by continuously monitoring disease severity in a series of staggered plantings of the same host cultivars inoculated with all available compatible isolates (Ahn and Ou, 1982, Kim and Ahn, 1991). Blast isolates from the first and seventh crop cycles were examined for their lineage group, virulence group, and aggressiveness on their own hosts of isolation. A significant gradual increase in aggressiveness of blast isolates was observed in the seventh crop cycle of C101A51 having a Pi-2 gene. On the other hand, aggressiveness of isolates from known durably resistant IR64 and IR36, was much lower than C101A51 (Table 3).

TABLE 3. Diseased leaf area (%) of 4 rice cultivars inoculated with *Pyricularia grisea* isolates from the blast population maintained on the same hosts in 7 sequential plantings

Test cultivar	Origin of Host							
	CO 39		C101A51		IR36		IR64	
	1[a]	7	1	7	1	7	1	7
CO 39	35.8	46.1	3.8	38.1**[b]	37.1	31.3	45.8	36.3
C101A51	0.2	0.5	13.3	44.6**	0.2	0.2	0.3	0.3
IR36	7.5	3.8	0	0	6.2	8.3	1.8	11.3*
IR64	0.2	0.4	0	0	0.1	0.1	0.8	3.0*

[a]Sequential planting number
[b]*,** = significantly different at P = 0.05 and 0.01, respectively by t-test.

4. Resistance breeding and evaluation

A varietal improvement program could employ single approach or a combination of various approaches considering available resources and expertise as well as situation of the target environment. Various approaches can be classified into several strategies (Table 4). It is extremely difficult to determine which approach will be highly effective and successful for each circumstance without a careful analysis of the pathogen population, and characterization of resistance genes effective in the region. With the recent progress in molecular maps of the rice genome, mapping of rice blast resistance genes is making a rapid advance (McCouch et al., 1988). Marker aided selection is routinely applied in selecting specific combination of major genes as well as QTLs in breeding program. Many breeding lines containing specific genetic combination will soon be readily available for evaluation.

TABLE 4. Classification of the utilization of genetic resistance in blast management

Strategy	Tactic	Method
Cyclic disturbance of coevolutionary process	Recycling of R-genes or alternation with non-host (diversity over time)	• Rotation or sequential release of varieties with distinct R-genes • Crop rotation
Continuous disturbance of coevolutionary process	Simultaneous use of R-genes in an agroecosystem (diversity over space)	• Varietal mixture • Multi-line • Intra-regional deployment of varieties with different R-genes
Delay or slow down of coevolution	Combination of different resistance mechanism	• Combining genes for qualitative resistance (pyramiding) • Use of partial resistance alone or combination with qualitative genes
Blocking co-evolutionary process	Introducing immune system	• Transfer of alien genes for immune system

Blast evaluation is a continuous step-wise process, which should be well integrated into every phase of varietal improvement. The most widely used method at the initial phase of blast evaluation in many rice programs may be a uniform blast nursery test (Ou, 1965). It aims to evaluate the leaf blast reaction of test entries at the early vegetative stage of the rice plant, a vulnerable period for blast infection. This method can accommodate many test entries requiring small seed quantity. Breeding lines showing promising reactions in the nursery are further observed for panicle blast reactions under natural conditions.

In advanced generations, efficacy and stability of partial resistance or the spectrum of qualitative resistance against major race groups or in different test sites are evaluated. Because of the global importance and complexity of this disease, international collaboration on breeding for rice blast resistance has been carried out since the 1960s. The International Network for Genetic Evaluation of Rice (INGER, formerly the International Rice Testing Program) has been coordinating the International Rice Blast Nursery (IRBN) since 1975. This trial is specifically designed to provide rice scientists working on rice blast easy access to diverse resistance donors and improved breeding lines as well as information on spatial stability over a wide range of test environments (Ahn, 1994).

One common aspect of the durably resistant cultivars in the IRBN test was low disease severity (DSI) values, mostly near 5.0, whereas the DSIs of all the known susceptible cultivars are higher than 6.0. Also the frequency of incompatible reactions of durably resistant cultivars is higher than 60%, whereas the frequency of extremely high reaction scores, 8-9, is much lower than those of nondurable resistant cultivars. Accordingly, these parameters have been utilized in identifying promising entries from IRBN trials. Cultivars such as IR64 from the Philippines, Oryzica Llanos 5 from Colombia , and Samgangbyeo from Korea showed a broad spectrum of spatial resistance in the IRBN and highly stable temporal resistance in farmers' fields and/or sequential evaluation plots (Fig. 2). However, we found out that some cultivars exhibiting spatial stability were not necessarily stable over time under intensive cultivation over wide planting area or sequential planting trial at IRRI (Ahn and Toledo, 1998). Tetep, Carreon, CICA 8, CICA 9, and C101A51 did not show stability over time in respective areas of cultivation or in sequential evaluations, even though these cultivars exhibited a broad spectrum of spatial resistance in IRBN. Todorokiwase, a japonica rice, exhibiting an intermediate degree of spatial stability is known as durably resistant in Japan, but not in Taiwan, where it was moderately susceptible to leaf blast and susceptible to panicle blast (Council of Agriculture, 1987). This analysis indicated that the evaluation of spatial stability alone is not sufficient to identify potentially durable resistance over time. Accordingly, the assessment of spatial stability using multilocation evaluation should be followed by the assessment of temporal stability of blast resistance.

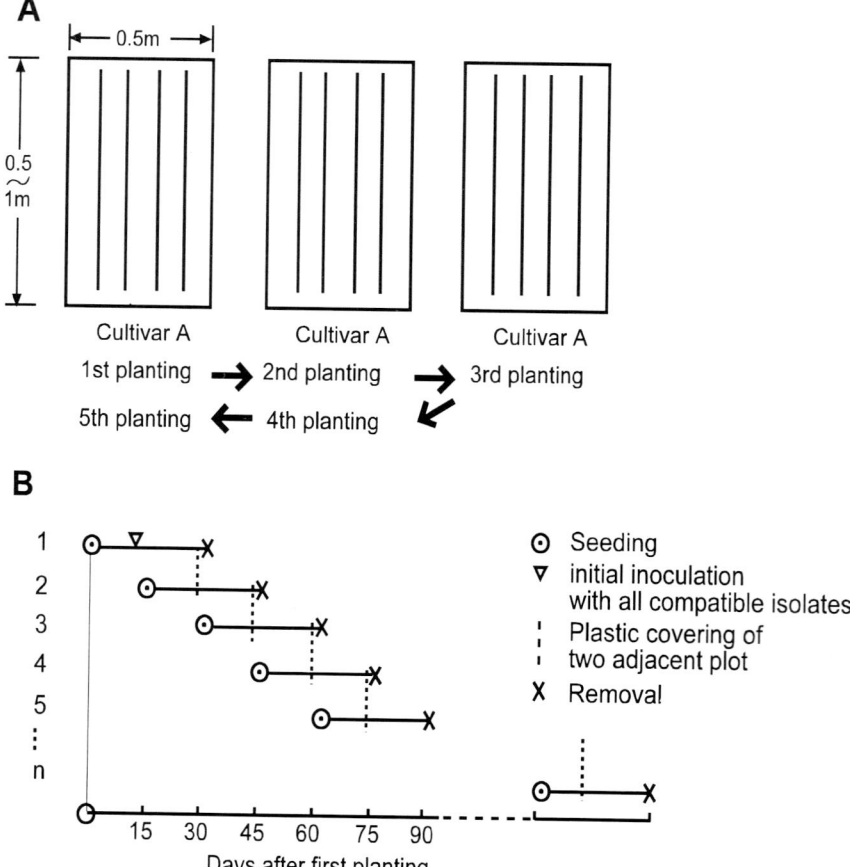

Fig. 1. **A**, Layout of sequential planting. **B**, planting schedule and activities.

No simple method to determine the temporal stability is available yet. Yearly testing of blast resistance for short period can not reliably assess temporal stability, because it does not allow a prolonged interaction between rice cultivars and potentially compatible pathogen population. The sequential evaluation technique appears suitable for determining the stability of partial resistance over time. Using this field evaluation technique, two types of pathogenic variation are observed: change in aggressiveness within a race-lineage combination and the development of a new race within a lineage. These kinds of dynamic changes over time could not readily be detected in conventional methods of static blast evaluation. Blast reaction scores of nine rice cultivars in sequential evaluation plot and farmers' fields are highly correlated. This field technique can be modified for greenhouse or laboratory assessment depending on evaluation objectives and conditions.

Our understanding of long-term dynamic evolution of blast · pathogen population in relation to rice genotypes under different rice ecosystems, is quite limited. This may generally be true with other microorganisms. In his recent review, Mundt (1995) concluded that fitness of a microorganism in nature is difficult to measure or predict with

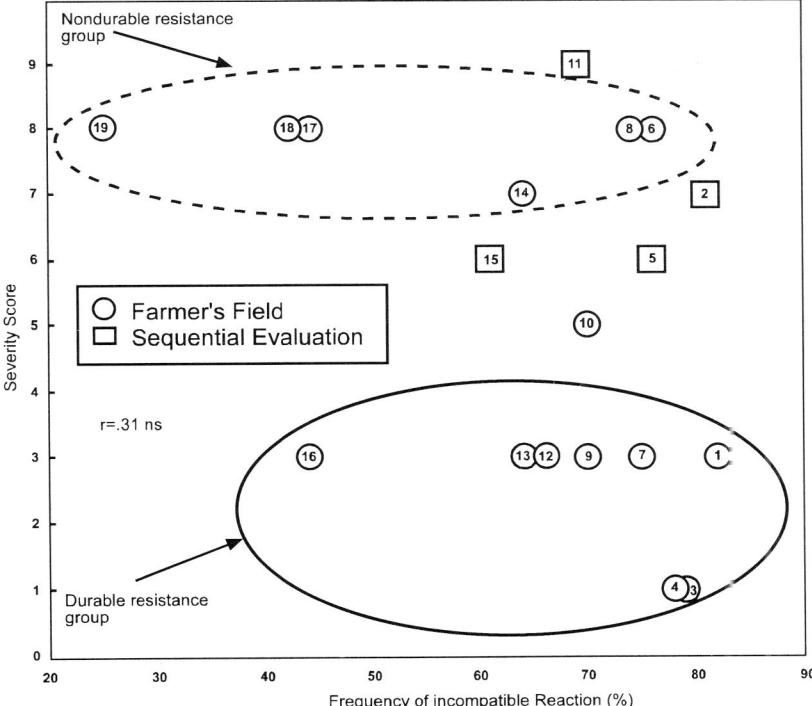

Fig. 2. Correlation between frequency of incompatible scores (0-3) in the International Rice Blast Nursery (1975-1995) and severity scores in farmers' fields or in the evaluation by sequential planting. 1=IRAT 13, 2=Tetep, 3= San Huang Zhan No. 2, 4=Oryzica Llanos 5, 5=Carreon, 6=CICA 8, 7= Morobekan, 8=CICA 9, 9=IR64, 10=OS6, 11=C101A51, 12= Samgangbyeo, 13=IR36, 14=ITA212, 15=Taebaeg, 16=Todorokiwase, 17=CICA4, 18=IR50, 19=Yushin.

much accuracy, and is unlikely to be constant with respect to either environmental conditions or time. Nevertheless, there is a strong need for broadening our knowledge on long-term genetic interaction between blast population and host resistance under diverse conditions to develop a more reliable strategy on genetic deployment to blast resistance. Recent progress in the rapid population analysis would accelerate the advancement of this study (George et al., 1998).

5.   References

Ahn, S.W. (1982) The slow blasting resistance to rice blast. In: IRAT/GERDAT (ed.), *Proceedings of the Symposium on Rice resistance to Blast.* 18-21 March 1981, IRAT/GERDAT, Montpellier, France, pp. 343-370.

Ahn, S.W. (1994) International collaboration on breeding for resistance to rice blast, in: R.S. Zeigler, S.A. Leong, and P.S. Teng (eds.), *Rice Blast Disease,* CAB International and International Rice Research Institute, Wallingford, UK, pp. 137-153.

Ahn, S.W. and Ou, S.H. (1982) Epidemiological implications of the spectrum of resistance to rice blast. *Phytopathology* **72**, 282-284.

Ahn, S.W., Makihara, D., Imbe, T., Chen, D.H., and Barrios H.C. (1996) Population dynamics of *Pyricularia grisea* in rice cultivars with partial resistance. *Phytopathology* **8(11), 646** (Abs).

Ahn, S.W., and McLaren, C.G. (1995) Multilocation evaluation of quantitative resistance to rice leaf blast. *European J. Plant Path.* In Abstract of 13th International Plant Protection Congress, Kluwer Academic Publishers, p. 469.

Ahn, S.W. and Toledo, M.C. (1998) Evaluation of spatial and temporal resistance to rice blast, Abstracts of the 7th International Congress of Plant Pathology, 9-16 August 1998, Edinburgh, Scotland 3, 3.4.30.

Ahn, S.W., Ynalvez, A.H., McLaren, C.G., Lopez, V.C., and Barrios, H.A. (1997) Rice genotype by environment (G x E) interaction in multilocation evaluation of rice blast. *Phytopathology* **87** (Supplement), S3 (Abstr.)

Chen, D.H., Zeigler, R.S., Leung, H., and Nelson, R.J. (1995) Population structure of *Pyricularia grisea* at two screening sites in the Philippines. *Phytopathology* **85**, 1011-1020.

Council of Agriculture (1987) Rice varieties in Taiwan, 1930-1987. Report Agric. & Forest., Taiwan Provincial Government.

George, M.L.C., Nelson, R.J., Zeigler, R.S., and Leung, H. (1998) Rapid population analysis of *Magnaporthe grisea* by using rep-PCR and endogenous repetitive DNA sequence. *Phytopathalogy* **88**, 223-229.

Hamer, J., Farrall, L., Orbach, M.J., Valent, B., and Chumley, F.G. (1989) Host species-specific conservation of a family of repeated DNA sequences in the genome of a fungal plant pathogen. *Proc. Natl. Acad. Sci.* USA 86-9981-9985.

Han, S.S., Ra, D.S., Choi, S.H., and Kim, C.K. (1997) Population dynamics of *Pyricularia grisea* during leaf and panicle blast stages in the same field. *Korean J. Plant Pathol.* **13(6), 408-415.**

Han, S.S., Ra, D.S. and Nelson, R.J. (1993) Comparison of RFLP-based phylogentic trees and pathotypes of *Pyricularia oryzae* in Korea, *RDA J. Agri. Sci.* **35(1),** 315-323.

Kim, C.H., and Ahn, S.W. (1991) Sequential evaluation for field assessment of partial resistance to rice blast. *Korean J. Breed.* **23(2),** 103-110.

Kiyosawa, S. and Shiyomi, M. (1976) Simulation of the process of breakdown of disease resistant varieties. *Japan. J. Breeding* **26,** 339-352.

Levy, M., Correa-Victoria, F.J., Zeigler, R.S., Yu, S. and Hamer, J.E. (1991) Organization of genetic and pathotype variation in the rice blast fungus at a Colombia "hot spot." *Phytopathology* **81**, 1236-1237.

McCouch, S.R., Kochert, G., Yu, Z.H., Wang, Z.Y., Khush, G.S., Coffman, W.R. and Tanksley, S.D. (1988) Molecular mapping of rice chromosomes. *Theor. Appl. Genet.* **76**, 815-829.

Mundt, C.C. (1995) Models from plant pathology on the movement and fate of new genotype of microorganisms in the environment. *Annu. Rev. Phytopathol.* **33,** 467-488.

Ou, S.H. (1965) A proposal for an international program of research on the rice blast disease. In *The Rice Blast Disease,* The Johns Hopkins Press, Baltimore, Maryland, pp. 441-446.

Ou, S.H. (1980) Pathogenic variability and host resistance in rice blast disease. *Annu. Rev. Phytopathol.* **18,** 167-187.

Ou, S.H. (1985) *Rice Diseases.* 2nd edn. Commonwealth Mycological Institute, Kew Surrey, UK.

Ryu, J.D., Yeh, W.H., Han, S.S., Lee, Y.H., and Lee, E.J. (1987) Regional and annual fluctuation of races of *Pyricularia oryzae* during 1977-1985 in Korea. *Korean J. Plant Pathol.* **3**, 174-179.

Shen, Y., Zhu, P., Yuan, X., Zhao, X., Manry, J., Rojas, C., Shahjahan, A.K.M., and Levy, M. (1996) The genetic diversity and geographic distribution of *Pyricularia grisea* in China. *Scientia Agricultura Sinica* **29(4)**, 39-46.

Valent, B. and Chumley, F.G. (1991) Molecular genetic analysis of the rice blast fungus, *Magnaporthe grisea*. *Annu. Rev. Phytopathology* **29**, 443-467.

Valent, B., and Chumley, F.G. (1994) Avirulence genes and mechanisms of genetic instability in the rice blast fungus, in R.S. Zeigler, S.A. Leong, and P.S. Teng (eds.) *Rice Blast Disease*. Commonwealth Agricultural Bureaux International, Wallingford, UK, pp. 111-134.

Zeigler, R.S., Scott, R.P., Leung, H., Bordeos, A.A., Kumar, J., and Nelson, R.J. (1997) Evidence of parasexual exchange of DNA in the rice blast fungus challenges its exclusive clonality. *Phytopathology* **87**, 284-294.

Zeigler, R.S., Tohme, J., Nelson, R.J., Levy, M., and Correa-Victoria, F.J. (1994) Lineage exclusion: A proposal for linking blast population analysis to resistance breeding, in R.S. Zeigler, S.A. Leong, and P.S. Teng (eds.) *Rice Blast Disease*. Commonwealth Agricultural Bureaux International, Wallingford, UK, pp. 267-292.

LINEAGE-EXCLUSION RESISTANCE BREEDING: PYRAMIDING OF BLAST RESISTANCE GENES FOR MANAGEMENT OF RICE BLAST IN INDIA

S. S. GNANAMANICKAM[1], LAVANYA BABUJEE[1], V. BRINDHA PRIYADARISINI[1], B. V. DAYAKAR[1], D. LEENAKUMARI[2], R. SIVARAJ[3], M. LEVY[3], and S. A. LEONG[4]

[1]*Centre for Advanced Studies in Botany, University of Madras, Madras 600025, India;* [2]*Regional Agricultural Research Station, Kerala Agricultural University, Pattambi 679306, India;* [3]*Dept. of Biological Sciences, Purdue University, W.Lafayette, IN47907, USA, and* [4]*USDA-ARS, Dept of Plant Pathology, University of Wisconsin, Madison, WI53706, USA.*

1. Abstract

By MGR-DNA fingerprinting analyses, we identified 29 lineages of *Magnaporthe grisea* among rice blast pathogen population in Southern India. The population structure of *M. grisea* in different states of India was also characterized by a *Pot2*-based rep-PCR method. The latter method identified 1 to 22 bands in the fingerprints of rice-infecting and less than 10 bands in the fingerprints of non rice-infecting strains of *M. grisea* which were in the size range of 0.4 to 23 kb. Lineage-exclusion assays showed that a combination of *Pi-1 and Pi-2* genes for blast-resistance can exclude all the lineages identified. This virulence characteristic of *M. grisea* remained stable while its lineage-composition showed drastic changes over the period 1993-97 in rice-growing areas of Kerala where annual blast epidemics is common. Therefore, gene pyramids of *Pi-1/Pi-1b+Pi-2* were constructed in the CO39 (blast-susceptible) genomic background and such pyramids were resistant to blast in hot-spot locations. The CO39 pyramid now serves as the R-donor for introgressing the *Pi-1+Pi-2* genes for blast resistance into cultivars such as Jyothi and IR50 which are high-yielding but blast-prone to make them durably resistant.

2. Introduction

Of the most devastating fungal diseases that constrain rice production world-wide, blast caused by *Magnaporthe grisea* (anamorph:*Pyricularia grisea*) ranks first because of its severity under conducive conditions. The use of host plant resistance as a management strategy to overcome rice yield losses due to blast has been constantly challenged by the variability of the pathogen population, especially in areas under severe disease pressure.

Advances in molecular techniques have greatly helped in our understanding of the structure and dynamics of the blast pathogen population. The middle repetitive DNA

D. Tharreau et al. (eds.), Advances in Rice Blast Research, 172–179.
© 2000 *Kluwer Academic Publishers. Printed in the Netherlands.*

sequence MGR586 (Hamer *et al*,1989) has been widely used as probe to delineate *M. grisea* pathogen population by RFLP analyses. Information generated from these analyses is used to devise novel and efficient strategies for the management of rice blast in India (Zeigler et al, 1994).

By RFLP analyses we delineated 29 families of *M. grisea* in Southern India. From plant inoculation assays conducted in the greenhouse as well as in the field we observed that the major resistance genes *Pi-1+Pi-2* excluded the entire population of *M. grisea* (Sivaraj et al, 1995; Babujee, 1997). On the basis of the lineage exclusion data, we are now mobilizing the gene pyramid *Pi-1+Pi-2* into high yielding but blast-susceptible cultivars such as Jyothi and IR50 to make them durably resistant to blast.

Research Objectives

Our research objective is to develop durably blast resistant rice cultivars which would be suitable for cultivation by rice farmers of India. The following are our specific objectives:
* To link the information that emerges from blast population analyses in India to resistance breeding
* To develop a lineage-exclusion resistance breeding strategy that will combine major genes for blast resistance to exclude the known lineages of *Magnaporthe grisea* in target regions of India
• To pyramid blast resistance genes into agronomically useful, high-yielding but blast-prone rice cultivars of Southern India.

3. Materials and Methods

A total of more than 650 monoconidial isolates of the rice blast fungus in Southern India were collected from the states of Andhra Pradesh, Karnataka, Kerala, Tamil Nadu, Maharashtra and Orissa (Sivaraj, 1995). The pathogen population in Kerala was repeat sampled in 1997 (Babujee, 1997). Genomic DNA from the monoconidial isolates was subjected to RFLP analysis. The conserved middle repetitive DNA sequence MGR586 was used as the probe to resolve the population structure and genetic organization of the blast fungus in Southern India.

In addition, genomic DNA samples of 145 *M. grisea* strains collected from different states (Andaman and Nicobarislands, Himachal Pradesh, Karnataka, Kerala, Meghalaya and Punjab) of India were finger-printed by the *Pot*2 repetitive element-based polymerase chain reaction (rep-PCR with two primer sequences of the *Pot*2 element) (Kachroo et al, 1994). The primer sequences, *Pot*2-1 (5'-CGG AAG CCC TAA AGC TGT TT-3') and *Pot*2-2 (5-CCC TCA GTC ACA CGT TC-3') amplify sequences between randomly dispersed copies of the element in a genome of *M. grisea*. PCR (Long PCR conditions including a higher pH, 9.2) and electrophoresis protocols used were those specified by George et al., (1998). Fingerprint patterns were visualized by electrophoresing 10 μl of the PCR products on a gel (0.5% agarose and 0.75% synergel) at 120 volts followed by EtBr staining. The correspondence between the MGR-RFLP lineage groups and the *Pot2*-PCR fingerprint groups was examined for the Kerala strains of *M. grisea*. The fingerprint patterns generated were analyzed by GelCompar and a dendrogram consisting of clusters was derived.

Conidial suspension from representative isolates of each lineage was sprayed on to a set of near-isogenic rice lines (NILs) (Mackill and Bonman, 1992) to determine their virulence spectrum. Each of the NILs carries a single major gene for blast resistance in a susceptible CO39 genomic background

Gene(s) which excluded the different lineages prevalent in S. India were identified. Pyramid rice lines carrying the blast resistant genes *Pi-1+Pi-2* were generated and screened at two hot-spot locations in Southern India

This CO39 gene pyramid (*Pi-1+Pi-2*), is now being introgressed into high-yielding but blast-susceptible cultivars, Jyothi and IR50 by conventional breeding and marker assisted selection as shown below:

Backcrossing scheme to transfer dominant genes for blast resistance (Pi-1+Pi-2) from a pyramid rice line to high-yielding but blast-prone rice cultivars (Jyothi, IR50 etc.)

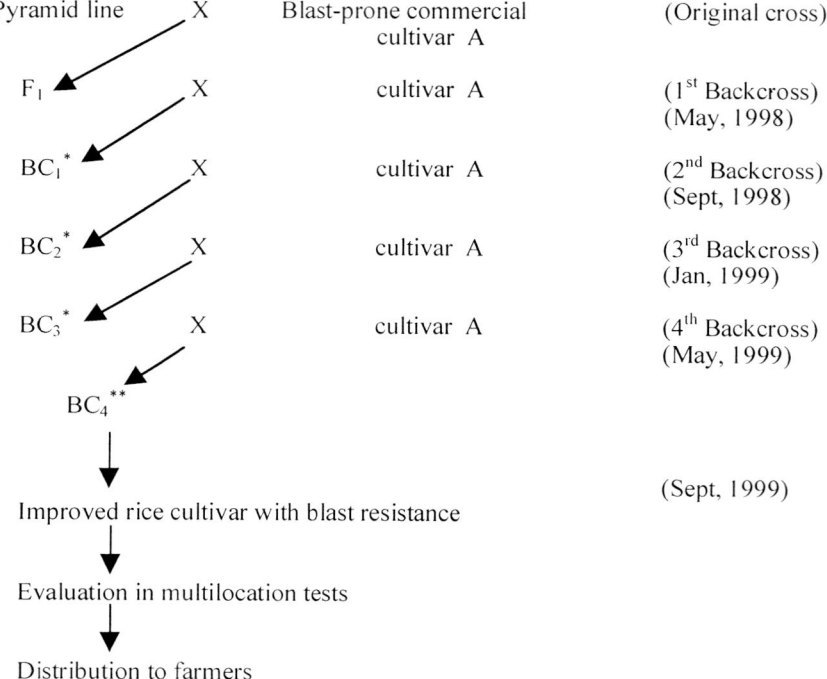

Pyramid line   X   Blast-prone commercial   (Original cross)
                                       cultivar A

F_1 X   cultivar A   (1st Backcross)
                                                    (May, 1998)

BC_1^* X   cultivar A   (2nd Backcross)
                                                    (Sept, 1998)

BC_2^* X   cultivar A   (3rd Backcross)
                                                    (Jan, 1999)

BC_3^* X   cultivar A   (4th Backcross)
                                                    (May, 1999)

BC_4^{**}

                                                    (Sept, 1999)

Improved rice cultivar with blast resistance

Evaluation in multilocation tests

Distribution to farmers

*Blast resistant progeny identified on the basis of phenotypic evaluation by spray inoculation and also by molecular marker assisted selection and used in succeeding generation.
**Blast resistant plants are selfed to obtain homozygosity.

4. Results and Discussion
5.
4.1. Population structure of M. grisea

The S. Indian *M.grisea* isolates were grouped into 29 distinct clonal lineages by DNA fingerprinting uses MGR586 sequence as the probe (Sivaraj *et al* 1996). The lineages are listed in Table 1. Genetic diversity of *M. grisea* was maximum (11 lineages) in Kerala (Fig. 1) during 1991-93. This region was sampled in 1997 and the lineage composition was found to be drastically different (Fig. 2, Table 2). The lineage G, originally isolated from

Fig. 1

Fig. 2

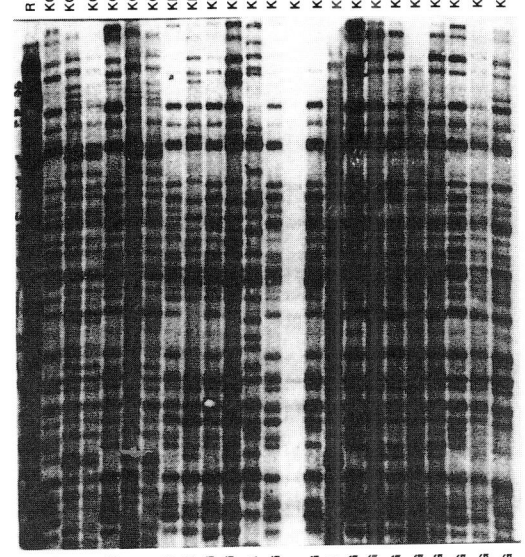

Figure 1. A map of Kerala state of Southern India. The rice-growing Kuttanad area has been darkened. On the basis of population analysis of *Magnaporthe grisea* prevalent in this area and lineage-exclusion assays, introgression of *Pi-1+Pi-2* genes for blast-resistance into rice cv. Jyothi, a high yielding but blast-prone cultivar that is grown in large acreages (ca. 88,000 acres) in Kerala, is in progress.

Figure 2. An autoradiogram showing the pattern of hybridization of *Eco* R1 DNA fragments of *Magnaporthe grisea* from Kerala. Similarity coefficients were used to group them into lineages. Lineage G is dominant and is widely distributed on rice cv. Jyothi (66%) in epidemic areas.

Panicum repens was dominant in this region and had a frequency distribution of 66% on cv. Jyothi (Babujee, 1997). This was followed by lineages 5(AA), J, A and 3 which had smaller frequency distibutions of 17%, 3.5%, 8% and 4.6%, respectively.

TABLE 1. Virulence spectrum of *Magnaporthe grisea* lineages from Southern India on near-isogenic rice cultivars bearing single resistance genes (The number of susceptible interactions is indicated; dark boxes indicate that no member of a lineage is compatible).

Lineage	No. isolates	Pi-1	Pi-2	Pi-3	Pi-4a	Pi-4b	Pi-1b
A	14	1	0	14	14	7	0
B	2	0	0	1	2	2	0
C	1	0	0	1	1	1	0
E	3	0	0	3	2	3	0
F	2	0	0	2	1	1	0
G	6	0	0	2	3	1	1
H	8	1	0	3	8	8	5
I	34	16	0	29	31	29	19
J	2	0	0	0	0	0	1
K	1	0	0	1	0	0	-
L	3	0	0	1	2	2	0
M	1	0	0	0	0	0	1
N	1	0	0	0	0	0	0
O	1	1	0	1	1	0	0
P	1	1	0	1	0	0	.1
Q	1	0	0	0	0	0	1
R	3	0	1	3	3	1	0
S	2	1	0	2	2	1	1
T	1	0	0	0	0	0	0
U	1	0	0	1	1	1	0
V	3	0	2	3	3	3	0
W	3	0	2	3	3	0	0
X	1	0	1	1	1	0	0
Y	1	0	0	1	1	1	0
Z	4	2	0	4	4	4	4

Resistance genes[a]

[a]The near-isogenic lines are C101LAC (Pi-la(t)), C101A51 (Pi-2(t)), C104PKT (Pi-3(t)), C101PKT (Pi-4a(t)), C105TTP-4L23 (Pi-4b(t), and C103TTP (Pi-1b) (Mackill and Bonman (1992).

TABLE 2. Comparison of lineage composition and virulence characteristics of *Magnaporthe grisea* in Kerala state of India during 1993 (Sivaraj, 1995) and 1997 (Babujee, 1997)

M. grisea characteristic	Status in 1993	Status in 1997
MGR lineages	A,G,H,I,J,K,M, O, P,Q,T	A,G,J,3,5
Virulence of lineages to CO39 pyramid (*Pi-1+Pi-2*)	Avirulent	Avirulent

In the *Pot2*-based rep-PCR analyses, the *M. grisea* genome of rice-infecting strains gave rise to 1 to 22 bands that ranged from 0.4 to 23 kb size (Fig.3). In the DNA of non

Fig. 3

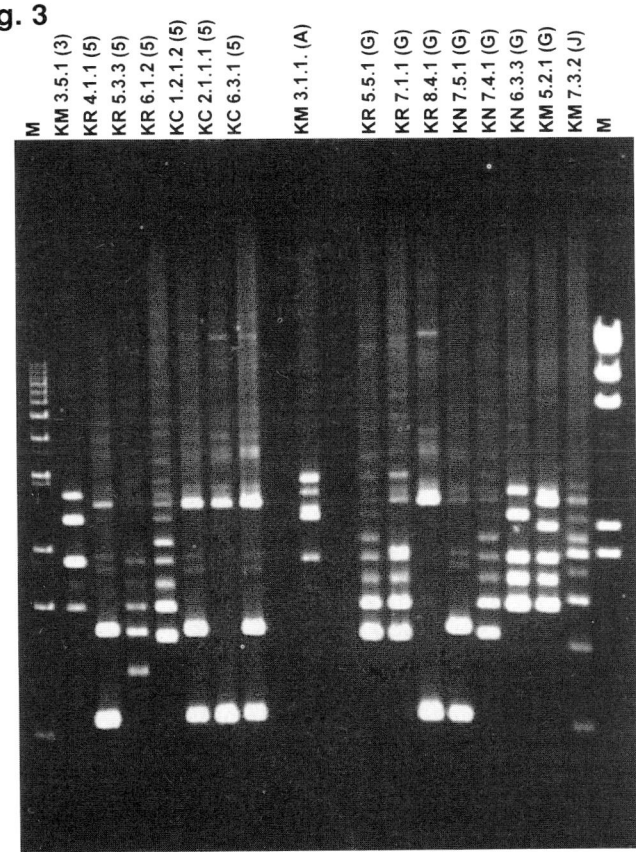

Figure 3. Pot2-based rep-PCR analysis shows fingerprints of a set of *M. grisea* strains sampled from epidemic areas of Kerala state in Southern India. These strains also had MGR lineage (A, G, j, 3 and 5) assignments. GelCompar analysis of the fingerprint data, however, did not group all the strains of major lineages (G and 5) in the same cluster.

rice-infecting strains of the pathogen fewer than ten bands of 0.4 to 23 kb were observed. Among the commercial *Taq* polymerases evaluated for their efficiency, a *Taq* polymerase mix, "Expand Long Template PCR System" marketed by Boehringer Mannheim was efficient in generating the above fingerprint patterns. However, there appears to be a lack of complete correspondence between the MGR lineage groups and the *Pot2*-based rep-PCR fingerprint groups. While several strains that belonged to the new lineages, "5" and "G" of

Kerala had grouped together in the same cluster arrived by *Pot2*-rep PCR analysis, there were other strains in these groups which did not show this consensus (Fig.3).

4.2. Virulence characteristics of M. grisea, lineage-exclusions and pyramiding of blast-resistance genes.

The virulence properties of the lineages were characterized on a set of near-isogenic lines. The resistance genes *Pi-1* and *Pi-2* complemented each other: the pyramid of these two genes excluded the entire S. Indian blast fungus population (Table 1). No changes were observed in virulence properties of the new lineages of *M. grisea* observed in Kerala during 1997 and 1998. None of them were able to overcome the gene pyramid *Pi-1+Pi-2* (Table 2).

This CO39 gene pyramid is now being introgressed into high-yielding but blast-susceptible cultivars such as Jyothi and IR50 by conventional breeding (Fig.4).

Figure 4. A rice panicle that bears the F₁ seeds formed from the cross between cv. CO39 (*Pi-1+Pi-2*) and cv. Jyothi, greenhouse, Pattambi, 1998.

Progeny from each set of crosses will be phenotypically evaluated for blast resistance in the greenhouse. Molecular evaluation of resistant progeny will be made using the specific amplicon polymorphic (SAP) marker RG64 for *Pi-2* and RFLP marker RZ536/RG424 for *Pi-1* (Hittlamani et al, 1995).

The improved rice cultivars with blast resistance will be evaluated unbiasedly by other rice scientists in multilocation tests before distribution to the rice farmers of Southern India.

5. Conclusions

The Southern Indian population of *M. grisea* is composed of 29 distinct families or lineages. A Pot2-based rep-PCR analysis served as a rapid tool to fingerprint the *M. grisea* strains from different states of India. However, there was lack of complete correspondence between the 2 sets of fingerprints among the *M. grisea* populations of the South Indian state, Kerala.

All the S. Indian lineages were excluded by the combination of genes *Pi-1+Pi-2* in greenhouse and field tests

The gene pyramid comprising the blast-resistance genes *Pi-1 +Pi-2* when mobilized into susceptible cultivars such as Jyothi and IR50 would render them durably resistant to blast. These lineage-exclusion resistance breeding efforts will be completed in the year 2000.

Acknowledgement

This research was supported by research grants from the Rockefeller Foundation, New York, U.S.A.

6. References

Babujee, L.(1997). Assessment of the stability of the rice blast fungus *Magnaporthe grisea* in Kerala rice fields by DNA fingerprinting and plant inoculations. M. Phil thesis, University of Madras, 45 p.

Hamer, J.E., Farrall, L., Orbach, M.J., Valent, B., and Chumley, F. (1989). Host species specific conservation of a family of repeated DNA sequence in the genome of a fungal plant pathogen. Proc. Natl. Acad. Sci. USA **86**, 9981-9985.

George, M.L.C., Nelson, R.J., Zeigler, R.S. and H. Leung. (1998). Rapid population analysis of *Magnaporthe grisea* using rep-PCR and endogenous repetitive DNA sequences. Phytopathology **88**, 223-229.

Hittalmani, S., Foolad, M. R., Mew, T. V., Rodriquez, R. L, and Hunag, N. (1995). Development of PCR-based marker to identify rice blast gene, *Pi-2*(t) in a segregating population. Theor. Appl. Genet.**91**: 9-15.

Kachroo, P., Leong, S.A. and Chattoo, B.B. (1994). *Pot2*, an inverted repeat transposon from the rice blast fungus *Magnapor the grisea*. Mol. Gen. Genet. **245**, 339-348.

Mackill, D.J., and Bonman, J.M. (1992). Inheritance of blast resistance in near-isogenic lines of rice. Phytopathology **82**: 746-749.

Sivaraj, R. (1995). Genetic structure and pathotype organization of the rice blast fungus *Pyricularia grisea*Sacc. in Southern India. Ph. D thesis, University of Madras, 121 p.

Sivaraj, R., Gnanamanickam, S.S., and Levy, M. (1996). Studies on the genetic diversity of *Pyricularia grisea*: a molecular approach for management of rice blast, in *Rice Genetics III* (Proc. 3rd Int'l Rice Genetics Symp.), 16-20 October, 1995, Manila, IRRI Publication, pp. 958-962.

Zeigler, R.S., Tohme, J., Nelson, R., Levy, M. and Correa-Victoria, F.J. (1994). Lineage-exclusion: a proposal for linking blast population analysis to resistance breeding, in R.S. Zeigler, S.A. Leong, and P. Teng (eds.), *Rice Blast Disease*, CAB international, Wallingford, U.K, pp. 267-292.

SILICON MANAGEMENT OF BLAST IN UPLAND AND IRRIGATED RICE ECOSYSTEMS

LAWRENCE E. DATNOFF [1], FERNANDO J. VICTORIA-CORREA [2], KENNETH W. SEEBOLD [1] and GEORGE H. SNYDER [1]
[1]University of Florida-IFAS, Everglades Research and Education Center, Belle Glade and [2]CIAT, Cali Colombia

Introduction

Silicon (Si) is one of the most abundant elements in the earth's crust, and most soils contain considerable quantities of the element (Savant et al, 1997a). However, repeated cropping can reduce the levels of plant-available Si to the point that supplemental Si fertilization is required for maximum production, and some soils contain little plant-available Si in their native state (Savant et al, 1997a; Savant et al, 1997b). Low-Si soils are typically highly weathered, leached, acidic and low in base saturation. Thus, highly weathered soils such as Oxisols and Ultisols can be quite low in soluble·Si. Highly-organic Histosols that contain little mineral matter may contain little Si. Interestingly, soils comprised mainly of quartz sand (SiO_2) such as sandy Entisols also may be very low in plant-available Si. These conditions are found in many crop producing areas of Africa, Asia, Latin America and the southeastern USA.

Silicon is considered a plant nutrient "anomaly" because it is presumably not essential for plant growth and development (Epstein, 1994). However, soluble Si has enhanced the growth and development of several plant species including rice *(Oryza sativa L.)*, sugarcane *(Saccharum officinarum L.)*, most other cereals and several dicotyledons. Silicon is absorbed as $Si(OH)_4$ by rice from soil in large amounts that are several fold higher than those of other essential macronutrients. For example, Si accumulation is about 108% greater in comparison to N (Savant et al,1997a). In general, a rice crop producing a total grain yield of 5 Mg (Megagram) ha^{-1} will remove from 0.23 to 0.47 Mg Si ha^{-1} from the soil (Savant et al, 1997a).

In the 1930s and 1940s, pioneering work by Japanese researchers first indicated that Si was effective in controlling rice diseases (Datnoff et al, 1997). These studies demonstrated that applications of 1.5 to 2.0 Mg ha^{-1} of various Si sources to Si-deficient paddy soils dramatically reduced the incidence and severity of blast, caused by *Magnaportha grisea,* and other rice diseases. Since the first reports by the Japanese, many researchers in other countries also have investigated the use of Si for controlling blast.

By using Si with other crop production inputs such as macro- and micro-nutrient fertilizers and fungicides, rice growers can maximize their yields. Research in Florida (Datnoff, 1994; Datnoff et al, 1997) and Colombia (Correa-Victoria et al, 1994) suggests that production inputs can be better managed without compromising yields when Si is used, i. e. minimizing production inputs while still maximizing outputs (yields). This Si management strategy has been demonstrated experimentally in both

D. Tharreau et al. (eds.), Advances in Rice Blast Research, 180–187.
© 2000 *Kluwer Academic Publishers. Printed in the Netherlands.*

irrigated and upland rice. This paper will provide an overview on the application of Si and its interaction with fungicides and phosphorus for managing rice blast.

Disease Suppression with Broadcast Silicon

Trials in commercial rice fields were conducted to evaluate Si rates and their residual effect on suppressing blast development in rice (Datnoff et al, 1997).   Silicon was applied as calcium silicate slag at 0, 5 Mg slag ha^{-1} ($\sim$ 1 Mg Si ha^{-1}) ,10 Mg slag ha^{-1} ($\sim$ 2 Mg Si ha^{-1}) and 15 Mg slag ha^{-1} ($\sim$ 3 Mg Si ha^{-1}) and the trial was duplicated on new plots in 1988. In addition, specific plots receiving Si in 1987 were fertilized again in 1988 with 1 Mg Si ha^{-1} while others received no additional fertilizer.   Thus, evaluations of blast development in 1988 could be made on the plots containing only residual Si from 1987, plots containing residual Si from 1987 plus an additional 1 Mg Si ha^{-1}, and also newly-fertilized 1988 plots.

Neck blast incidence decreased significantly with increasing Si rates (Figure 1) .   Neck blast incidence at the 3 Mg Si ha^{-1} rate was 29% less in the residual plots in comparison to the control. In the amended residual treatment, blast was reduced by 32%, similar to the 27% in the newly-applied treatment. It is clear that the residual Si in the soil can be very effective in blast control of a rice crop planted the following year. This has obvious economic ramifications for growers.   Rice yields increased significantly with increasing Si rates (data not shown), and the greatest yield increases were realized on plots receiving new applications of Si. The Si content in the plant tissue increased with increasing rates of Si, while the Ca did not change (Figure 2). The Si content increased from 1.8% in the control to 4.4% at the highest rate.

Figure 1.   Relationship between neck blast incidence and rate of applied silicon. Residual Si applied only in 1987, Y = 36.0 - 1.7 (x); Fresh Si applied only in 1988, Y = 35.5 - 3.9 (x); and Residual + Fresh Si applied in 1987 and in 1988, Y = 35.1 - 4.5 (x) + 0.2 (x^2).   The mean percent blast incidence for each replication was used for calculating regression equations (n=5).

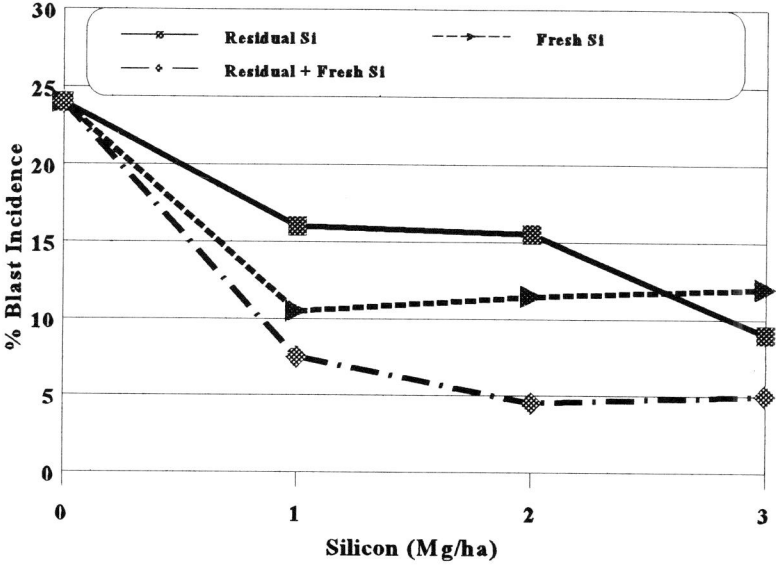

Figure 2. influence of silicon fertility on the cation content of Ca and Si in rice tissue.

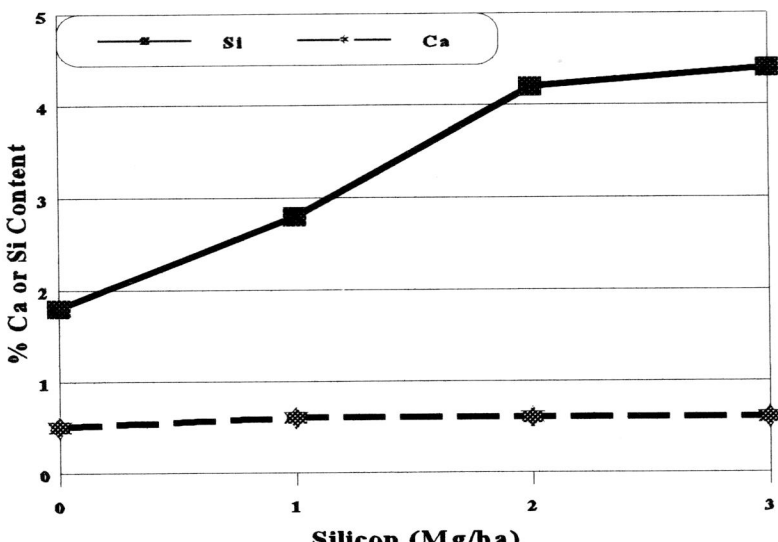

Interaction of Silicon and Fungicides

An evaluation of Si fertilization in combination with benomyl was undertaken for managing blast (Datnoff et al, 1997). A rice crop was treated with Si at 0 and 2 Mg Si ha-1 and benomyl at 0 and 1.68 kg ha-1. Fungicide sprays were applied at 2.1 x 105 Pa with a CO_2 backpack sprayer equipped with three Cone-Jet nozzle tips on a hand-held boom at panicle differentiation, boot, heading and heading + 14 days. During these experiments, environmental conditions (frequent rainfall and high relative humidity) were favorable for rice blast development. Blast incidence was 73% in the non-Si, non-fungicide control plots and 27% in the benomyl treated plots (Figure 3). Where Si was applied blast incidence was 36% in the non-fungicide plots and 13% in the benomyl treated plots. The same degree of blast control was generally obtained when either the benomyl or Si were applied individually. The greatest blast control was obtained by using both treatments together.

Figure 3. Influence of silicon fertilization and benomyl foliar spray on blast incidence. Values with the same letter are not significantly different based on Fisher's LSD (P=0.05).

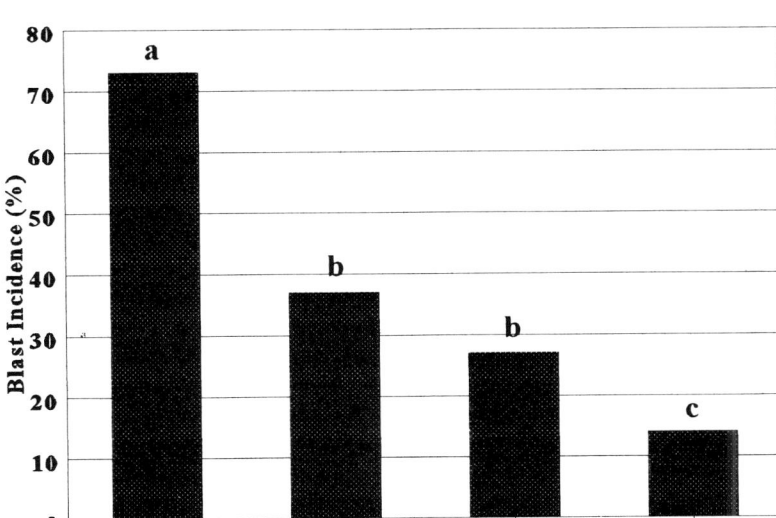

Because Si can control blast to the same general degree as a fungicide, it is possible that Si might help reduce the number of fungicide applications or even the rate of application.   To test these hypotheses, several field experiments were conducted in savannahs of Colombia (Seebold et al, 1995; Seebold et al, 1997; Seebold et al, 1998). Silicon was applied as wollastonite, 2 Mg ha-1, and Oryzica Sabana 6 was seeded at 80 kg/ha.   Treatments included a nontreated control, Si applied alone and Si plus fungicides (edifenfos at 1L ha-1 and tricyclazole at 300 g ha-1) applied at the following growth stages: tillering (T), panicle initiation (PI), booting (B), 1% panicle emergence (1%), 50% panicle emergence (50%), PI, B, 1%, and 50%; B, 1% and 50%; 1% and 50%; B and 1%; PI and 1%; T (Figure 4).   Neck blast incidence was dramatically reduced using either Si alone or Si plus fungicides in comparison to the nontreated control (Figure 4). Silicon alone significantly reduced neck blast incidence by 40%. Si + one fungicide reduced neck blast between 75 to 90% while Si + two applications reduced neck blast between 76 to 94%. Silicon .+ three to five applications reduced neck blast between 94 to 98%. So, one application of the fungicide in combination with Si was as effective as two, although 3 to 5 provided better blast protection.

In another experiment, Si was incorporated prior to seeding at 0 and 5 Mg ha (Seebold et al, 1998). Two foliar applications of edifenfos were applied at 0, 10, 25 and 100% of recommended rates. Ratings of leaf blast for Si alone and Si plus edifenfos at various rates were 54-75% lower than in the nontreated control (Figure 5). The greatest leaf blast reductions were observed where Si plus the full rate of fungicide had been applied. Silicon + lower rates of fungicides (10% and 25%) were able to reduce leaf blast as effectively as a full rate of the fungicide. However, Si alone was just as effective as the fungicides alone or the fungicides + Si. The results from these experiments suggest that the number of fungicide applications and their rates may be reduced; consequently, saving growers either initial or additional application costs.

Figure 4. Effect of silicon and fungicide timings on neck blast incidence.
Fungicides timings are tillering (T), panicle initiation (PI), booting (B), 1% heading (1%), 50% heading (50%) and various combinations. Dark Bars represent FLSD value (P=0.05).

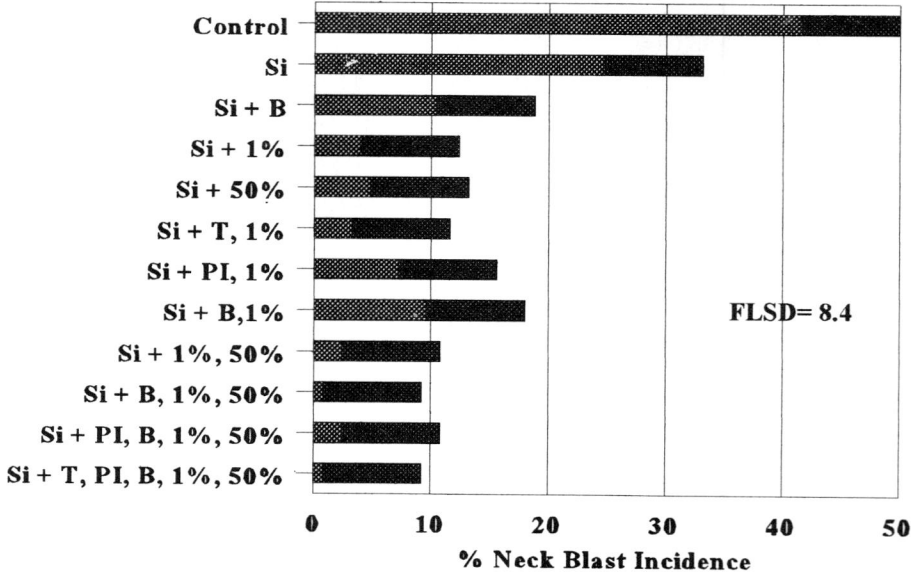

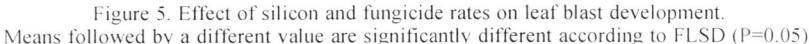

Figure 5. Effect of silicon and fungicide rates on leaf blast development.
Means followed by a different value are significantly different according to FLSD (P=0.05).

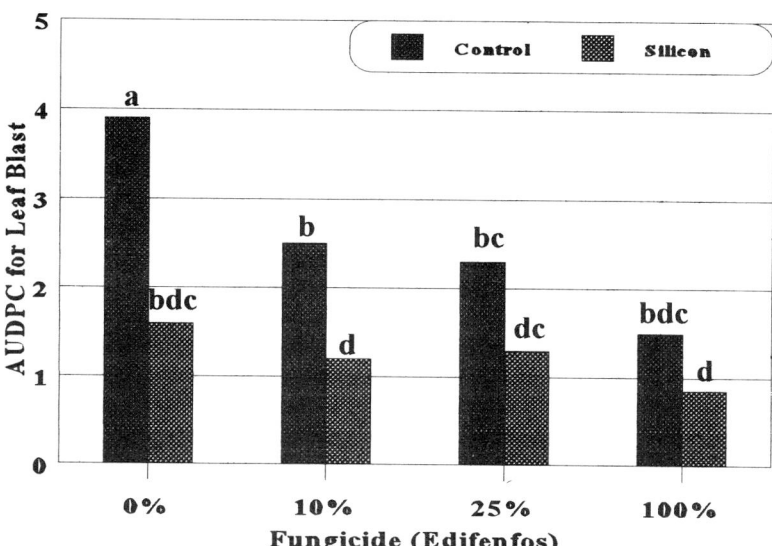

Silicon and Phosphorus Interactions

The efficacy of phosphorus has been reported to be enhanced when it is applied along with Si (Savant et al, 1997a). To study this potential interaction on rice blast, a factorial experiment was developed where Si was applied as wollastonite at 0, 1.3 and 2.6 Mg ha-1 and P was applied at 25 and 50 kg P ha-1 (Correa-Victoria et al, 1994). Due to the basic properties of wollastonite, chemically equivalent amounts of calcite lime were applied to the 0 and 1.3 Mg ha-1 Si treatments to equalize the lime value and Ca across treatments. There were no significant differences between the P rates for blast incidence (Figure 6). Blast incidence significantly decreased from a high of 30% in the control to as low as 15% with the addition of Si, however, there were no differences between Si rates Although P levels increased in rice tissue from 0.053% for 25 kg ha-1 to 0.062% for 50 kg P ha -1, there were no significant differences in total yields (data not shown). Yields however did increase over the control between 421 to 676 kg ha-1 with the addition of Si. These results suggest lower levels of P may be used in combination with Si for controlling rice blast, even though Si applications may enhance rice yield responses to higher applications of applied P.

Figure 6. Effect of phosphorus and silicon on neck blast development ('IAC 165').
Means followed by a different value are significantly different according to FLSD (P=0.05).

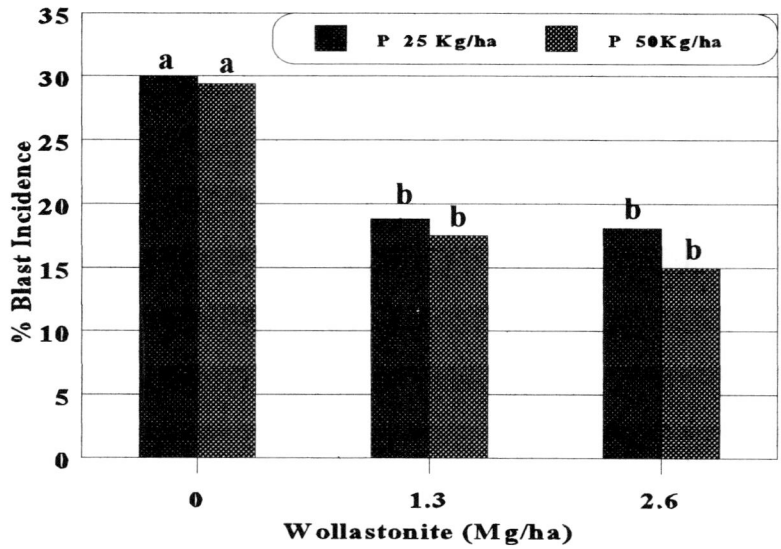

Outlook and Future Research Needs

Silicon fertilization of rice, especially where natural soil levels of Si are deemed less than optimum, offers promising results with respect to blast control and improved yields. Silicon sources have residual activity that persist over time, raising the possibility that applications need not be applied annually. Also, after the first initial Si amendment, subsequent application rate requirements might be considerably lower due to these residual effects. Silicate slags however are considered to be expensive sources so there is a need to find or develop cheaper and more efficient Si sources (Savant et al, 1997a). Recycling of rice hulls and/or straw may be one possible alternative. Silicon reduces susceptibility in rice to blast. Silicon can control blast to the same general degree as a fungicide and reduce the amount and frequency of fungicides needed. Silicon also appears to improve plant utilization of nutrients such as phosphorous. This implies that fertilizers (macro- and/or micro- nutrients) might be better managed if Si soil amendments are used thus helping to defray application costs. Therefore, Si sources and their management practices should be developed and practiced to improve rice production.

References

Correa-Victoria, F. J., Datnoff, L. E., Winslow, M. D., Okada, K., Friesen, J. I., Sanz, J. and Snyder, G. H. (1994) Silicon deficiency of upland rice on highly weathered Savanna soils of Colombia. II. Diseases and grain quality. IX Conferencia Internacional de Arroz para a America Latina e para o Caribe. V Reuniao Nacional de Pesquisa de Arroz, Goiania, Goias, Brazil, p. 65.

Datnoff, L. E. (1994) Influence of plant mineral nutrition on rice disease development. Pp. 89-100. IN: Teng, P. S., Heong, K. L. and Moody, K., eds., Advances in Rice Pest Management, IRRI, Los Banos, Philippines.

Datnoff, L. E., Deren, C. W., and Snyder, G. H. 1997. Silicon fertilization for disease management of rice in Florida. Crop Prot. 16:525-531.

Epstein, E. (1994) The anomaly of silicon in plant biology. Proc. Natl. Acad. Sci. USA 91:11-17.

Savant, N. K., Snyder, G. H., and Datnoff, L. E. (1997a) Silicon management and sustainable rice production. IN: Adv. Agron. Ed. by D. L. Sparks, 58:151-199, Academic Press, San Diego, CA, USA.

Savant, N. K., Datnoff, L. E., and Snyder, G. H. (1997b) Depletion of plant-available silicon in soils: A possible cause of declining rice yields. Comm. Soil Sci. & Plant Analysis 28:1245-1252.

Seebold, K., Datnoff, L., Correa-Victoria, F., and Kucharek, T. (1997) Effects of silicon and fungicide timing on foliar disease control and yield in upland rice. Phytopathology 87:S87 (Abstr).

Seebold, K. W., Datnoff, L. E., Correa-V., F. J., Kucharek, T., Snyder, G. H. and Tulande, E. (1998) Effect of calcium silicate plus fungicides at reduced rates on the control of leaf and neck blast in upland rice. 27th Rice Technical Working Group, Reno, NV.

Seebold, K., Datnoff, L., Correa-V., F., and Snyder, G. (1995) Effects of silicon and fungicides on leaf and neck blast development in rice. Phytopathology 85:1168 (Abstr).

THE EFFECT OF PLANT AGE ON THE SPATIAL AND TEMPORAL DYNAMICS OF RICE BLAST IN ARKANSAS

D.H. LONG, D.O. TE BEEST and J.C. CORRELL
University of Arkansas, Department of Plant Pathology
Plant Science 217, Fayetteville, Arkansas 72701

1. Abstract

Field experiments were conducted in 1997 to determine how plant maturity affected the rate and distance of rice blast development from foci. A sulfate non-utilizing (*sul*) mutant of *P.grisea* was used as a marked strain that could be distinguished from wild-type (background) inoculum. Plants, 28 or 42 days old, of the susceptible cultivar M204' were inoculated with *P. grisea* in the greenhouse. Inoculated plants were then transplanted as foci in the centers of 12.5-m^2 plots of M204 which also were either 28 or 45 days old (660 and 1090 degree day thermal units, respectively). Lesion formation was detected three days after transplanting on greenhouse inoculated plants and the focal strength was determined to be 45 lesions / focus in the younger plants and 90 lesions / focus in the older plants. The number of lesions per plant were recorded weekly at distances of 0.3, 1.6, 3.1, 4.7, and 6.3-m from the focal center (0.09-m^2). Disease developed rapidly at the focus within 10 days in both the younger and older plants (6 and 5 lesions / plant respectively). However, disease spread from foci differed considerably between the different aged plants whereby disease occurred earlier and was more severe in the younger plants. For example, a mean of 15 lesions / plant were observed on the younger plants 1.6-m from the foci after 18 days, while only 1 lesion / plant was observed at the same distance in the older plants. After 30 days, the mean number of lesions observed on the younger plants at all distances (45, 32, 24, 14 and 12 lesions / plant, respectively) was significantly higher than on the older plants (12, 12, 4, 2, and 0.5 lesions / plant, respectively). The *sul* mutant was recovered from over 90% of the lesions collected 10 and 40 days after transplanting, indicating inoculum introduced at foci at the beginning of the season was responsible for initiating and perpetuating the epidemic.

2. Introduction

Rice blast is one of the most destructive diseases of rice (*Oryza sativa* L) worldwide and can result in significant reductions in yield (Kingsolver et al.,1984; Torres and Teng, 1993). The pathogen, *Pyricularia grisea* Cav., infects rice from seedling stages through grain formation, causing symptoms on leaves, collars and necks. Rice blast epidemics often are more severe in temperate and sub-tropical ecosystems, but also can occur in other environments, if blast susceptible cultivars are widely grown or if effective management strategies are not implemented.

The epidemiology of rice blast disease has been thoroughly investigated in many

D. Tharreau et al. (eds.), Advances in Rice Blast Research, 188–195.

countries (Kim, 1987; Kingsolver et al., 1984; Suzuki, 1975), but development of rice blast in Arkansas has not been described and may differ because of different environmental conditions, host resistance, pathogen races, and different management and production strategies. Characterization of the spatial and temporal dynamics of rice blast is important in developing and recommending suitable and effective disease management strategies. Integrated pest management (IPM) strategies have been emphasized for many years, but many areas concerning rice blast development must be addressed before their implementation. For example, the risk and potential of disease spread of rice blast must be addressed before implementing an IPM management program that utilizes spot treatments with fungicides and the effects of various management strategies (i.e. water and nitrogen fertilization management) and plant maturity on disease spread of rice blast must be examined carefully before such management strategies are implemented. Therefore, the objectives of this research were to determine the how plant maturity affected the rate and distance of rice blast development from foci.

3. Materials and Methods

3.1 FIELD EXPERIMENTS

The experiment was conducted at the Pine Tree Branch Experiment Station in Colt, Arkansas in 1997. The experimental area was precision leveled for optimum water management and was bordered by trees on the northern side. The soil type was a Crowley silt loam. A susceptible cultivar M204' was drill-seeded at 18.2 Kg/ ha in 12.5-m^2 plots of 72 rows spaced 0.2-m apart. Treatments were applied in a randomized complete block design with six replications per treatment. Plants, 28 and 45 days old, of M204 were inoculated with *Pyricularia grisea* in the greenhouse. Inoculated plants were then transplanted as foci (0.09-m^2) (~144 plants) in the centers of the 12.5-m^2 plots that were also either 28 days (planted 23 May 1997) or 45 days (planted 03 May 1997) old. Accumulations of degree day thermal units (DD50's)(Helms, 1990) of the rice in the plots at the time of focal transplanting was 1090 DD50's for the older plants (~panicle initiation stage) and 660 DD50's for the younger plants (5th to 6th leaf stage). Accumulations of DD50's can be positively correlated with physiological maturity of the rice plant (Helms, 1990). Kaybonnet, a cultivar resistant to rice blast, was used as a barrier to minimize the interplot spread of *P. grisea*.

3.2 INOCULATION OF FOCI
Six to eight seeds of M204 were sown into each (6 x 6 -cm) cell of a 12 celled peat pot strip. Soil obtained from the Pine Tree Branch Experiment Station was combined with vermiculite at a 2:1 ratio. Seedlings were grown under greenhouse conditions at 24 to 30C with 16-hr photoperiods. Ammonium sulfate (5 g/ m^2) was applied to the seedlings 2-weeks after sowing and each week thereafter.

A sulfate non-utilizing (*sul*) mutant of *P. grisea* was used as a marked strain that could be distinguished from wild-type (background)inoculum (Harp and Correll, 1998). Cultures of the *sul* mutant, isolate 18/1, were maintained on rice bran agar (20-g rice bran

(Riceland); 15-g agar (Sigma) and 1000-ml of distilled water) for 7 to 10 days at 25C with a 12 hour photoperiod. Conidia from cultures of the *sul* mutant were harvested with distilled water and adjusted to 80,000 conidia / ml using a hemacytometer. Silwet L77 (0.01%) was added to the spore suspension prior to inoculation. One hundred milliliters of this conidial suspension was sprayed onto 160 (6 x 6-cm) cells using an air-assist spray apparatus. Plants were placed in a dew chamber following inoculations for 16 hours at 21C. Plants were removed from the dew chamber and placed on benches in the greenhouse and incubated for 3 days at 28C before being transplanted in the field.

3.3 DATA COLLECTION AND ANALYSIS

Foci were established by transplanting inoculated seedlings into plots containing the young plants (23 May planting date) and the old plants (03 May planting date). The number of lesions per plant was determined every 6 days at various distances from the initial focus (0.3, 1.6, 3.1, 4.7, and 6.3-m from the focus). The mean number of lesions for each treatment was used to construct disease progress curves. Apparent infection rates (Zadoks and Schein, 1979) and peak disease levels were analyzed using the GLM and the protected LSD procedures in SAS (SAS Institute, 1990). Fifty leaf blast lesions were collected 10 and 40 days after transplanting and were screened for the presence of the *sul* mutant. Initial isolations of *P. grisea* from these leaf blast lesions were made on RBA. After a seven day incubation period (25C), *P. grisea* isolates were transferred to RBA amended with 0.1% sodium selenate. *P. grisea* isolates that grew approximately 5 to 10 cm after 4 days were characterized as *sul* mutants. Wild-type isolates (background inoculum) are greatly restricted on this medium.

4. Results and Discussion

Disease symptoms developed on the focal transplants within 3 days after transplanting, resulting in 45 lesions / focus (~0.3 lesions / plant) in the younger plants (second planting) and 90 lesions / ocus (~0.6 lesions / plant) in the older plants (first planting). Disease increased rapidly on the focal transplants between 3 and 10 days in both treatments (6 and 5 lesions / plant, respectively).

There was much higher disease development in the younger plants than in the older plants in that disease levels (mean number of lesions / plant) were significantly higher in the younger plants at all five sampling points (0.3 to 6.3-m) from the disease focus (45, 32, 24, 14 and 12 lesions / plant, respectively) than on the older plants (12, 12, 4, 2, and 0.5 lesions / plant, respectively; Figure 1). The initial disease at distances of 4.7 and 6.3 m from the foci occurred earlier (days after transplanting) in the younger plants than in the older plants. Disease was observed after 18 to 24 days and after 30 to 36 days in the younger and older plants, respectively. Kim (1987) monitored rice blast epidemics from focal centers and calculated isopath (disease movement in a given direction) rates between 0.2 and 0.4-m /day. The rate of isopath movement from the focus to the farthest sampling point (6.3 m) in this study was 0.35-0.52 m / day in the younger plants and 0.21-0.26-m / day in the older.

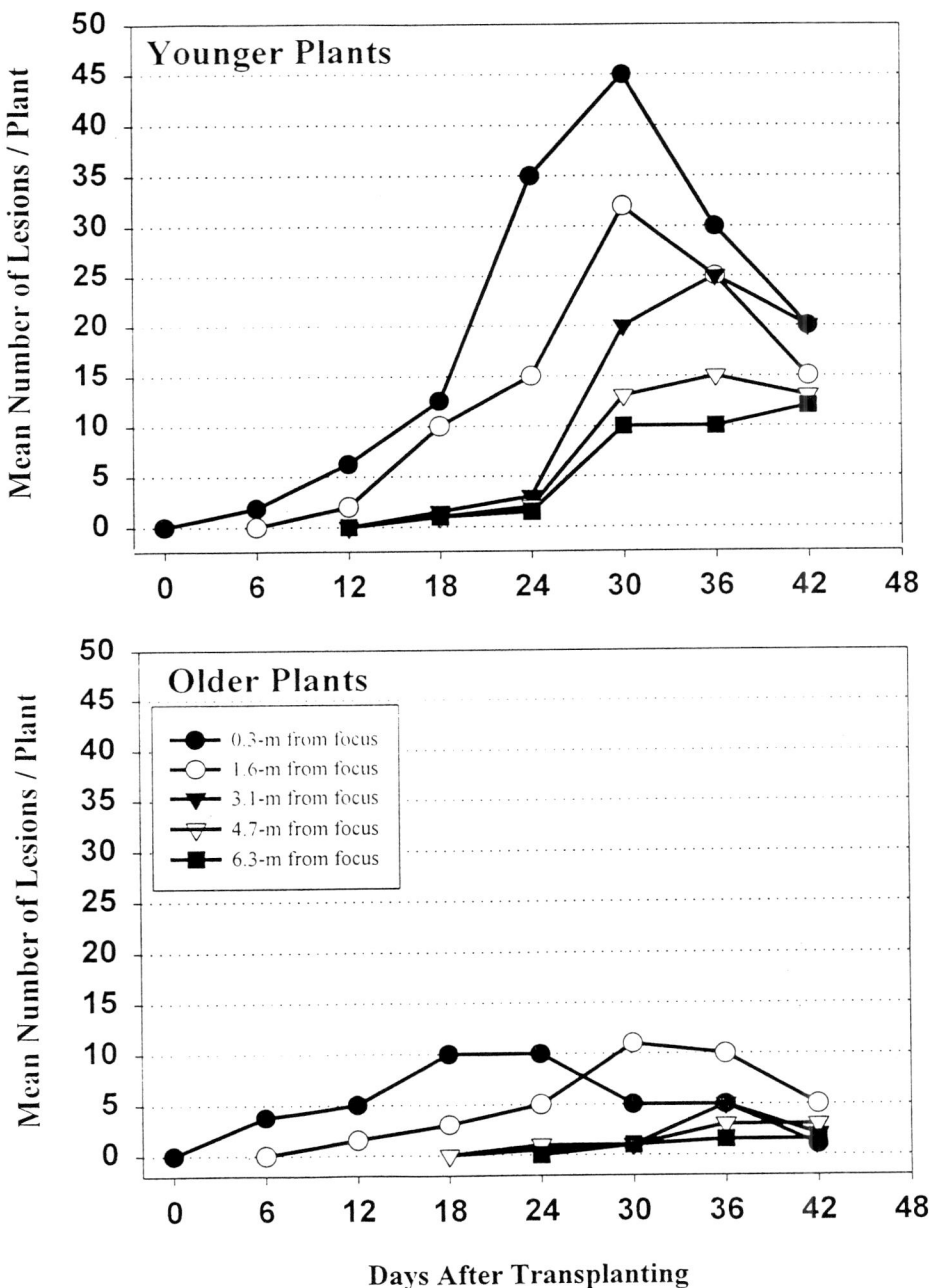

Figure 1. Disease progress curves for mean number of lesions per plant (leaf blast) on M204 planted on 23 May 1997 (younger plants) and 03 May 1997 (older plants) at distances 0.3, , 1.6, 3.1, 4.7 and 6.3-m from the foci. The x-axis is days after transplanting.

TABLE 1. Apparent infection rates for the first 18 days of disease development at the various distances from the foci in both the younger and older plants.

Distance[U]	Younger plants[V]	Older plants[W]
	Rate[X]	Rate
0.3-m	0.139	0.112
1.6-m	0.48	0.100
3.1-m	0.160	0.058
4.7-m	0.128	0.037
6.3-m	0.120	0.031
LSD[Y]	0.03	0.03
LSD[Z]	0.04	

[U] Distance from the focus (m).
[V] Rice plants were planted on 23 May 1997 (second planting).
[W] Rice plants were planted on 03 May 1997 (first planting).
[X] Apparent infection rate for the first 18 days of disease development.
[Y] Protected LSD's for each column.
[Z] Protected LSD's for both columns and rows.

The initial rate of disease increase was similar at 0.3, 1.6 and 3.1-m (comparing apparent infection rates) over the first 18 days of disease development for the each treatment (Table 1). Apparent infection rates for the same treatments also were similar in the younger plants, however, significant differences were observed at 0.3, 1.6 and 3.1-m from the foci when comparing peak disease ratings (Table 2). Similar infection rates were observed at 4.7 and 6.3-m in the younger plants and 0.3 and 1.6-m in the older plants. Initial disease development at these distances from the foci occurred while the plants were at similar physiological ages (1000 to 1250 DD50) (Figure 2), however, peak lesion numbers at 4.7 and 6.3-m in the younger plants were slightly higher than those observed on the older plants. Source strength behind the focal front was higher in the younger plants (~4 to 6 radial meters at 10 lesions / plant) than that observed in the older plants (~ 0 to 0.3 radial meters at 5 lesions / plant. Source strength at the focus early in the season is a key component of the disease cycle that determines the trend and shape of the disease progress curve throughout the season (Bastiaans, 1993; Pinnschmidt et al., 1993).

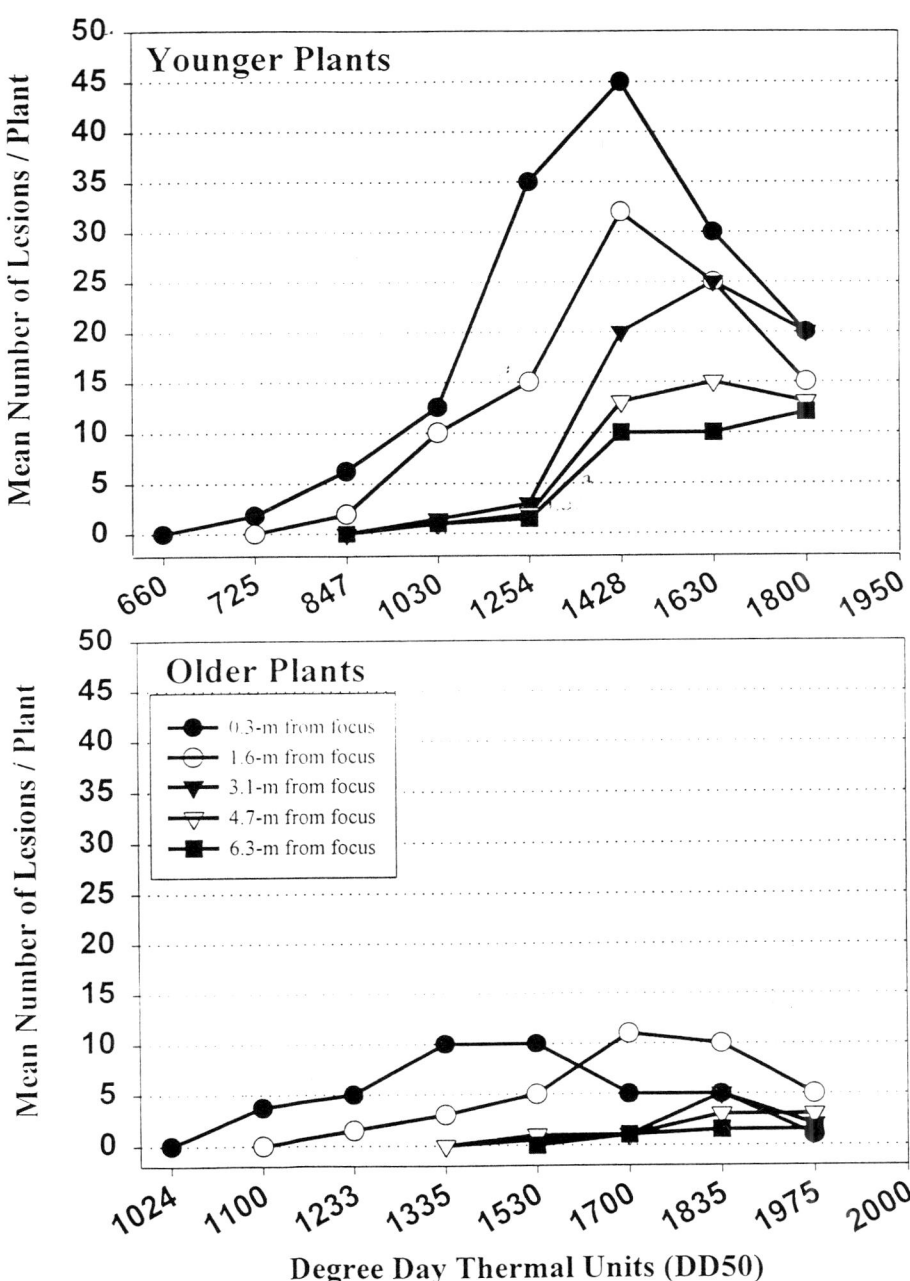

Figure 2. Disease progress curves for mean number of lesions per plant (leaf blast) on M204 planted on 23 May 1997 (younger plants) and 03 May 1997 (older plants) at distances 0.3, 1.6, 3.1, 4.7 and 6.3-m from the foci. The x-axis is the accumulation of degree day thermal units (DD50's).

TABLE 2. Maximum number of lesions/plant at the various distances from the foci in both the younger and older plants.

Distance	Lesion number[X]	
	Younger plants	Older plants
0.3-m	46	12
1.6-m	32	12
3.1-m	24	4
4.7-m	14	2
6.3-m	12	0.5
LSD[Y]	5	3
LSD[Z]	4	

[X] Maximum number of lesions per plant (mean lesions) for each treatment.
[Y] Protected LSD's for each column.
[Z] Protected LSD's for both columns and rows.

Disease levels at the end of the season also were higher on the younger plants (24, 22, 15, 12 and 11 lesions / plant, respectively) than in the older plants (0.5, 5, 0.5, 2 and 0.5 lesions / plant, respectively), however these differences were not as large as those observed when measuring peak lesion numbers for each treatment. A simple unimodal disease progress curve described leaf blast development, in that the mean number of lesions per plant increased to the highest levels at panicle primordial stage (1250 to 1300 DD50's), then declined gradually thereafter (reproductive growth stages). The decline in disease level has been attributed to adult resistance and the senescence of infected leaves (Bastiaans, 1993; Long et al., 1998). The decline in disease (number of lesions / plant) was significantly higher in those rice plants that had more disease at mid-season (1300 to 1440 DD50) than those that had maximum disease thereafter (1450 to 1600 DD50). Leaf senescence on plants that had maximum disease (mean number of lesions / plant) at mid-season (panicle primordial) had an average loss of 10 lesions / plant, while plants that had maximum disease after mid-season only lost 2 lesions / plant (data not shown).

The *sul* mutant was recovered from over 90% of the lesions collected 10 and 40 days after transplanting (100 and 91% respectively). These data confirmed that inoculum introduced at foci at the beginning of the season was responsible for initiating and perpetuating the epidemic

5. References

Bastiaans, L. 1993. Understanding yield reduction in rice due to leaf blast. Ph.D. Thesis, Wageningen Agricultural University, The Netherlands. Pages 1-127.

Harp, T.L. and Correll, J.C. 1998. Recovery and characterization of spontaneous, selenate-resistant mutants of *Magnaporthe grisea*, the rice blast pathogen. Mycologia **90**, (in press).

Helms, R.S. ed. 1990. Rice production handbook. University of Arkansas Cooperative Extension Service. Misc. Pub. 192. 90 pages.

Kim, C. H. 1987. Disease dispersal gradients of rice blast from a point source. Korean J. Plant Pathol. **3**, 131-136.

Kingsolver, C. H., T.H. Barksdale, and M.A. Marchetti. 1984. Rice blast epidemiology. Bulletin 853. The Pennsylvania State University, College of Agriculture, Agricultural Experiment Station, University Park, Pennsylvania. 33 pages.

Long, D.H., TeBeest, D.O. and Lee, F.N. 1998. Effect of Nitrogen Fertilization on Disease Progress of Rice Blast on Susceptible and Resistant Cultivars. Plant Disease **83**, (Submitted).

Pinnschmidt, H.O., Teng, P.S. Bonman, J.M., and Kranz, J. 1993!. A new assessment key for leaf blast (Bl). IRRN **18**, 45-46.

Suzuki, H. 1975. Meteorological factors in the epidemiology of rice blast. Ann. Rev. of Phytopathol. **13**, 239-256.

Torres, C.Q. and P.S. Teng. 1993. Path coefficient and regression analysis of the effects of leaf and panicle blast on tropical rice yield. Crop Prot. **12**, 296-302.

Zadoks, J.C. and Schein, R.D. 1997. Epidemiology and Plant Disease Management, Oxford University Press, New York. 427 pages.

EFFECT OF VARIETAL FIELD RESISTANCE FOR CONTROL OF RICE BLAST

M. YAMAGUCHII[1], H. SAITOH[2] AND T. HIGASHI[3]

[1]*Tohoku National Agricultural Experiment Station, Omagari, Akita 014-0102, Japan;* [2]*National Agricultural Research Center, Tsukuba, Ibaraki 305-8666,Japan;* [3]*Chugoku National Agricultural Experiment Station, Fukuyama, Hiroshima 721-8514, Japan*

1. Introduction

Rice blast caused by *Pyricularia oryzae* Cav. (telemorph: *Magnaporthe grisea* (Hebert)Barr) is the most widely distributed and most serious disease in Japan. Blast resistance in rice plants is the most important goal for rice breeders. The resistance is classified into two groups: true resistance and field resistance. True resistance is qualitative, race-specific and controlled by a small number of gene(s) with major effect; while field resistance is quantitative, non-specific and controlled by several multiple genes and/or polygenes, which have additive effect. Kiyosawa (1970) defined that true resistance completely inhibits sporulation and that field resistance reduces sporulation to some extent under field condition.

For control of rice blast, it is more important to use field resistance than true resistance , because breakdown in true resistance often occurs several years after

Table 1. Examples of blast resistant varieties broken down after release in Japan (Yaegashi, 1994)

Variety	Genotype for true resistance	Number of years from release to break down
Kusabue	*Pi-k*	1-6
Yuukara	*Pi-a Pi-k*	2
Teine	*Pi-a Pi-k*	2
Ugonishiki	*Pi-k*	2
Tachihonami	*Pi-k*	2
Minehikari	*Pi-a Pi-k^m*	3
Shimokita	*Pi-a Pi-ta*	2-7
Fukunishiki	*Pi-z*	5
Yamatenishiki	*Pi-i Pi-z*	1
Hamaasahi	*Pi-a Pi-i Pi-k Pi-b*	2
Akiyutaka	*Pi-k Pi-z*	3

D. Tharreau et al. (eds.), Advances in Rice Blast Research, 196–202.

Table 2. Leading varieties in Japan planted in 1996

Variety	Genotype for true resistance	Level of field resistance	Level of eating quality	Planted area (%)
1 Koshikikari	$Pi\text{-}k^{j}$	Susceptible	Excellent	30.6
2 Akitakomachi	$Pi\text{-}a\ Pi\text{-}i$	Susceptible	Excellent	7.4
3 Hitomebore	$Pi\text{-}i$	Susceptible	Excellent	7.1
4 Hinohikari	$Pi\text{-}a\ Pi\text{-}i$	Susceptible	Excellent	6.4
5 Kirara397	$Pi\text{-}i\ Pi\text{-}k$	Resistant	Medium	5.3
6 Sasanishiki	$Pi\text{-}a$	Susceptible	Good	3.6
Total				60.4

inducing a resistant variety. After the first breakdown of a line, Pi 5 in 1962, the breakdown of other varieties were reported from various part of Japan. Some of them are shown in Table 1 (Yaegashi, 1994). Since about 1967, it has become common to check the field resistance of newly developed varieties or strains in paddy fields and/or blast nursery and/or glasshouse (Ezuka, 1979). After checking, severe breakdown has not been observed on the field resistant varieties to the present. Therefore, the improvement of field resistance is fundamental in breeding program of blast resistance in Japan.

Table 2 shows recent leading varieties in Japan. Sixty percent of total planted area consists of only six cultivars. Although they have excellent eating quality, which is in commercial demand, their field resistance is susceptible, except for Kirara397. Therefore to control the blast, rice farmers use fungicides two to three times or more in a season even when the disease is not severe. As a result, Japan becomes the largest fungicide market. In recent years, however, the reduction of fungicide use is being desired to decrease chemical and labor costs, and to prevent environment.

How can we reduce chemicals use? Probabily by cultivation of resistant varieties instead of the susceptible varieties, we aim to clarify the effect of field resistance to control blast and to improve breeding for resistant varieties.

2. Blast Control with Varietal Field Resistance

To find the relationship between the degree of field resistance and fungicide treatments, four varieties with different field resistance - highly resistant Chubu 32, resistant Toyonishiki, medium Kiyonishiki and susceptible Sasanishiki - were planted in the paddy field at Tohoku National Agricultural Experiment Station in 1992 to 1993. Chubu 32 does not have true resistance genes, and the other three varieties have $Pi\text{-}a$ gene only, which is not available for native pathogenic races in Japan.

Four different blasticide treatments (0, 1, 2, 3 times) were carried out for each variety. Single treatment; application of probenazole (a week before the outbreak

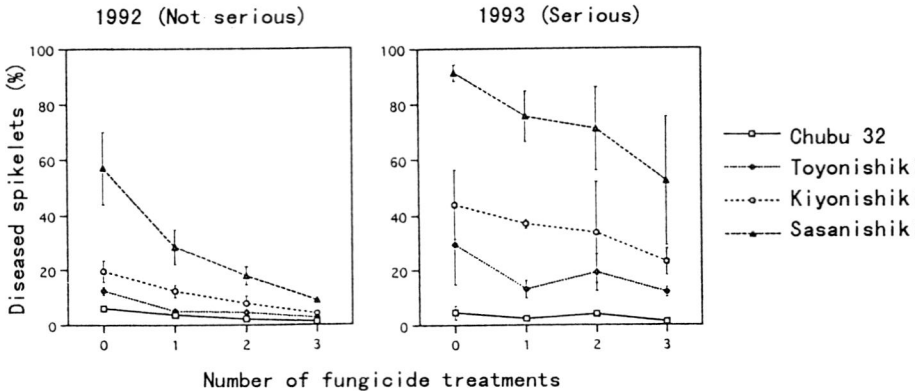

<figure>Figure 1.   Panicle blast severity in four varieties subjected to fungicide treatments.</figure>

of leaf blast in the field). Double treatment; application of probenazole, and isoprothiolane (three weeks before the heading time). Triple treatment; application of probenazole, isoprothiolane, and ferimzone + fthalide (at the full heading stage).

For the trials, 18 square meter-plots with three replications in split-plot experiment were conducted.

2.1. PANICLE BLAST SEVERITY IN DIFFERENT VARIETIES

Figure 1 shows the panicle blast severity in four varieties subjected to fungicide treatments. The ratio of damaged area by rice blast in Akita prefecture, where these trials were conducted, was 19.9 percent (not serious) in 1992; while in 1993 it was 63.3 percent (serious).

In 1992, the percentage of diseased spikelets of susceptible Sasanishiki with triple treatment was nearly the same as that of resistant Toyonishiki which was not treated. In 1993, Toyonishiki untreated, was more effective than Sasanishiki with triple treatment. Highly resistant Chubu 32 was more effective than Toyonishiki, and the resistance of Kiyonishiki was the average between Toyonishiki and Sasanishiki during both years. The varietal difference of panicle blast severity was bigger in 1993, than that in 1992.

2.2. YIELD IN DIFFERENT VARIETIES

In both years, yield loss caused by the blast disease, that is, the ratio of missing yield in fungicide-untreated varieties was highest in Sasanishiki, being 53.9 percent in 1992 and 71.7 percent in 1993. That of Kiyonishiki was 22.5 percent (1992) and 21.6 percent (1993), which was much the same as Toyonishiki, 23.0 percent (1992) and 16.8 percent (1993). Blast disease did not influence the yield of Chubu 32 (Figre 2).

In the case of yield , the result was similar to that of panicle blast severity ;

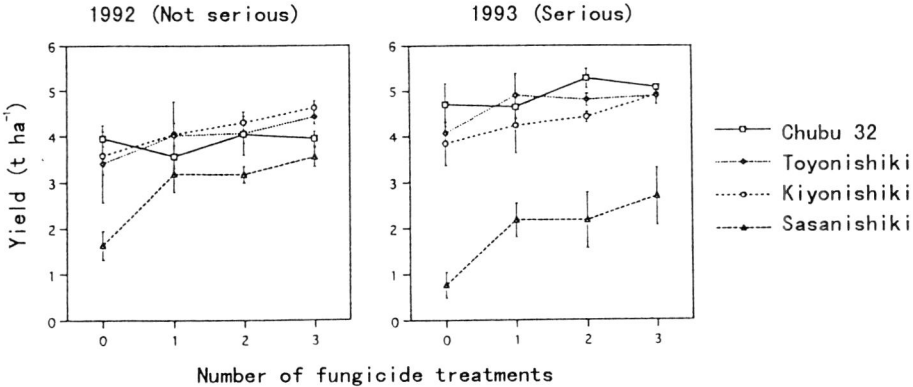

Figure 2. Yield in four varieties subjected to fungicide treatments.

Sasanishiki, treated thrice, yielded nearly the same as the untreated Toyonishiki in 1992. Untreated Toyonishiki, in 1993, yielded more than thrice-treated Sasanishiki.

All and all, the resistance effect of Toyonishiki was equal to the thrice-treated, susceptible Sasanishiki. The field resistance of Kiyonishiki somewhere between that of Toyonishiki and Sasanishiki. And that of higly resistant Chubu 32 was more effective than Sasanishiki with triple treatment. Based on the results, it can be concluded that breeding for field resistance to blast should focus on resistance at least equal to that of Toyonishiki.

2.3. COMPARISON OF BLAST CONTROL BETWEEN VARIETAL FIELD RESISTANCE AND FUNGICIDE TREATMENTS

To control blast, we determined if there are any advantages to use resistant varieties instead of chemicals. In 1992, although both panicle blast severity and yield were significantly influenced by varieties and number of fungicide treatments, varietal field resistance was more effective than fungicide treatments in both traits because contribution (the ratio of each variance to total variance) of varieties was about twice as much as that of number of fungicide treatments (Table 3).

In 1993, when blast was serious, panicle blast severity was influenced by varieties; however, it did not correlate with number of fngicide treatments, indicating blasticide treatments that year were not effective to panicle blast control, while yield was significantly influenced by varieties and blasticides (Table 4). The effect of varieties on panicle blast control and yield was more than 80 percent in 1993, and that was about twice as much as the effect in 1992.

All concsidered, for control of rice blast, utilization of resistant varieties was stable and more effective than chemical treatments in both years, especially in the year when blast was serious. In the serious blast season, the rainfall was generally above normal season, and the rain sometimes washes the fungicides off rice plants and interferes with the effect of chemicals for blast control. In such cases, farmers

M. YAMAGUCHI *et al.*

Table 3. Analysis of variance for panicle blast severity and yield in 1992

Source of variation	DF	Panicle blast severity		Yield	
		F	Contribution (%)	F	Contribution (%)
No. of fungicide treatments (N)	3	54.20**	26.6	16.40**	23.2
Block	2	1.78	0.3	1.72	0.7
Varieties (V)	3	93.46**	46.2	29.05**	42.2
N × V	9	13.78**	19.2	3.27**	10.2
Error (1+2)	30		7.7		23.7
Total	47		100.0		100.0

DF = degree of freedom.

**, * : Significant at 1% and 5% levels, respectively.

Table 4. Analysis of variance for panicle blast severity and yield in 1993

Source of variation	DF	Panicle blast severity		Yield	
		F	Contribution (%)	F	Contribution (%)
No. of fungicide treatments (N)	3	3.35	4.4	5.96*	7.3
Block	2	0.50	0.0	1.14	0.1
Error (1)	6	3.66*	2.7	4.81**	2.3
Varieties (V)	3	160.31**	81.7	265.56**	81.0
N × V	9	1.96	1.5	3.75**	2.5
Error (2)	24		9.7		6.8
Total	47		100.0		100.0

Symbols in this table are same as in Table 3.

have to use fungicides anew. We should recognize these advantages to use resistant varieties instead of chemical treatments in Japan.

3. Combination of Field Resistance and Eating Quality

Up to the recent time, it has been understood that breeding rice cultivars with high

Table 5.   Resistance effect of Okiniiri in farmers' fields

Field	Variety	Yield (t ha^{-1})	Panicle blast severity (%)	Eating quality
Farmer A	Okiniiri	7.2	4.0	Excellent
	Akitakomachi	6.3	13.5	Excellent
Farmer B	Okiniiri	7.7	4.5	Excellent
	Akitakomachi	6.4	7.0	Excellent

field resistance and good eating quality is difficult, and actually such cultivars have not been develped in Japan. Therefore, almost all the recent cultivars with excellent eating quality are susceptible to blast.

Recently, some examples to change such a view were reported. Yamaguchi *et al.* (1997) proved the possibility to breed lines with high field resistance and good eating quality by selecting a line having both characteristics. In 1996, a new cultivar Okiniiri was developed at Tohoku National Agricultural Experiment Station (Higashi *et al.*, 1997). The word 'okiniiri' means 'favorite', and this cultivar was selected from the progeny of a cross Chubu 47/Ouu 313. The field resistance of Okiniiri is resistant, which is equivalent to Toyonishiki, and this cultivar also has excellent eating quality, similar to that of Koshihikari and Akitakomachi, the latter two varieties having the best eating quality in Japan.

The resistance effect of Okiniiri was investigated in two farmers' fields of Akita prefecture in 1997 (Table 5). Both farmers did not use blasticides. Farmer A cultivated rice with organic fertilizer and without chemicals; Farmer B used herbicide and insecticide once only. The panicle blast severity of Okiniiri was 36 to 70 percent less than that of Akitakomachi, and the yield of Okiniiri was 13 to 19 percent more than that of Akitakomachi. Such blast resistant varieties as Okiniiri can be easily grown without or with minimum chemicals, or can be used for organic farming.

4. Conclusions

Finally, we can stress that in the near future, the goal of blast resistance breeding in Japan should be to improve the field resistance of recent susceptible cultivars with excellent eating quality at least to the level of resistant Toyonishiki. Of course, new varieties with high field resistance and excellent eating quality are desired. If it realized, we can reduce the utilization of blasicides greatly. Fortunately, aside from Okiniiri, some resistant cultivars with excellent or good eating quality have been developed at other experiment stations in Japan recently: like Koigokoro at National Agricultural Research Center in 1995; Manamusume at Miyagi Prefectural Furukawa Agricultural Experiment Station in 1997. Many more resistant varieties are desired, and to be developed later.

REFERENCES

Ezuka, A. (1979) Breeding for and genetics of blast resistance in Japan, in *Proceedings of The Rice Blast Workshop*, International Rice Research Institute, Los Banos, Philippines, pp. 27-48.

Higashi, T., Yamaguchi, M., Sunohara, Y., Oyamada, Z., Kowata, H., Tamura, Y., Yokogami, N., Saito., Ikeda, R., Inoue, M., and Matsumoto, S. (1997) Breeding of a new rice cultivar "Okiniiri", *Bull. Tohoku Natl. Agric. Exp. Stn.* 92, 15-33 (in Japanese, English summary).

Kiyosawa, S. (1970) A view on true resistance and field resistance, *J. Agric.Sci.* 25, 21-25 (inJapanese).

Yaegashi, H. (1994) Use of resistant varieties and disease control for paddy rice, *Agric. Hortic.* 69(1), 149-154 (in Japanese).

Yamaguchi, M., Kowata, H., Sunohara, Y., and Higashi, T. (1997) Selection of rice lines having good eating quality and high field resistance to blast, *Bull. Tohoku Natl. Agric. Exp. Stn.* 92, 35-42 (in Japanese, English summary).

EXPLORING OPTIMUM APPLICATION PROGRAMS OF FUNGICIDE USING A SIMULATOR FOR LEAF BLAST EPIDEMICS

K. ISHIGURO and T. NANSEKI
Tohoku National Agricultural Experiment Station
Shimo-Kuriyagawa, Morioka, Japan 020-0123

1. Introduction

In Japan, the conventional goal of rice production has been high and stable yield and quality. Since rice blast disease, caused by *Magnaporthe grisea*, is one of the most serious threats to rice production, effective control of the disease, especially, though the use of fungicide application, has been regarded as an important issue in achieving this goal.

Recent social consensus emphasizes economically reasonable and environmentally sound approaches to disease control. Since this is a new concept in Japanese agriculture, we have to set a new goal for rice production. Integrated Pest Management (IPM) is the optimization of disease control in an economically and ecologically sound manner [3]. This is compatible with the new framework. Therefore, we believe that IPM is an appropriate approach to the new goal.

IPM emphasizes the thoughtful combination of resistant varieties, cultural control practices and fungicide to control target diseases. For rice blast disease, substantial efforts to develop resistant rice varieties have been made. Knowledge of cultural control measures against the disease has accumulated. However, fungicide application is still the most important component of the disease management of rice blast.

So far, more than ten active compounds have been registered for blast control. Some of them, which have systemic activity, are used as granular formulation types. The rest of them, which have protectant and/or curative (post-infection and/or post-symptom) activity, are sprayed as topdressing. If schedules of the fungicide application could be flexibly determined on the basis of disease assessment and forecast, the fungicides could be more effectively and efficiently used. However, granular fungicides have to be applied prior to the start of epidemic. Consequently, these fungicides are only suitable for prefixed scheduling. Sprayed fungicides, on the other hand, are theoretically compatible with flexible scheduling. Unfortunately, there have been few studies on the decision-support system for scheduling. So, most growers adopt either a pre-fixed schedule or those starting just after the onset of a leaf blast epidemic.

D. Tharreau et al. (eds.), Advances in Rice Blast Research, 203–208.
© 2000 *Kluwer Academic Publishers. Printed in the Netherlands.*

The objective of this research was to develop a scientific basis for flexible fungicide application strategies for efficient and effective control of leaf blast epidemics. We demonstrate the usefulness of a decision-support system (DSS) by using a simulation analysis.

2. Decision support system

2.1. SIMULATION MODEL USED

BLASTL, a simulator of leaf blast epidemics, can predict the disease progression in an auto-infection pathosystem, when parameters of rice growth, cultivation conditions and weather data are provided [1]. Later, a sub-model for fungicide application was constructed and incorporated into the original model. The sub-model describes the behavior of a melanin biosynthesis inhibitor on the rice leaves after spraying and its efficacy [2]. We used this version of BLASTL as a component of DSS.

2.2. ASSUMPTIONS

There were a few problems to be solved before the development of DSS. While repetitive validation tests have shown the BLASTL model can predict the tendencies of epidemics well, the predictions do not always quantitatively correlate with epidemics in the field. The economic threshold level (ETL) of leaf blast has not yet been determined, mainly because it causes yield loss indirectly. That is, leaf blast plays a role as an inoculum for the subsequent panicle blast that mainly causes yield
losses.

To overcome these problems, four assumptions, which in some cases may be unrealistic, were made: (1) BLASTL can predict disease progression of leaf blast (as changes of lesion numbers per plot) accurately and precisely. (2) Yield losses due to the disease are correlated with the lesion numbers at the end of the epidemic. Consequently, ETL of leaf blast is determined based on lesion number. (3) The ETL can be arbitrarily determined in advance by the decision-maker, based on simulated disease progression by BLASTL using particular historical weather data and her/his experiences. (4) ETL is equal to the cost of a fungicide spray.

2.3. ALGORITHM

DSS makes a decision on a daily basis on whether or not fungicides need to be applied on that day. After the decision, the procedure is repeated the next day until the end of the epidemic. The procedure of each day has three steps: (1) prediction of disease progression with simulation (disease forecasting), (2) simulating fungicide application timing, and (3) decision-making based on this simulation.

In the first step, simulations are carried out for the current season using the weather data up to the day in question together with the data from a past season until the end of the epidemic. By changing the past weather data sets, various predictions ranging from the most pessimistic to optimistic one are made. DSS employs the Îmaximiniì criterion to make a decision. The criterion chooses a decision, which earns the best economic profit in the worst-case scenario. That is, if the disease progression of the worst-case scenario exceeds the ETL, then the procedure will go on to the second step to explore optimum fungicide timing. If not, the procedure skips from the first to the third step and makes a decision that any fungicide needs to be applied at that point. In the second step, disease progressions with various fungicide timings are simulated and evaluated. When immediate fungicide application at the point is evaluated as the only option to suppress the disease below the ETL, a decision to spray will be made. When even the immediate application cannot control the disease severity below the ETL but its economic benefit is more than one spraying cost, an immediate fungicide application will be suggested. In other cases, a decision to apply no fungicide at that time is made

In DSS, a few conditions are set: (1) the maximum number of application times is two, (2) the interval between the first application and second one is more than five days, (3) a decision and the consequent action are madewithin the same day, and (4) weather data up to the previous day of the season in question are available at each decision time.

3. Simulation analysis

3.1. EXPERIMENTAL DESIGN
To evaluate the effectiveness of DSS, we made a hypothesis that the performance of DSS would be better than those of conventional fungicide schedules. To test the hypothesis, we compared the performance of DSS with those of several other strategies, using simulation analysis, and 19 historical weather data sets (1974-1993, 1991 was not available) at Koriyama (Japan) were prepared.

The onset of each epidemic season was determined with Kobayashils criteria [4] based on the weather data mentioned previously. That is, a favorable condition for infection occurs when there is a wet period longer than 10 hours at night during which the minimum air temperature was higher than 16oC. The disease was assumed to appear after the first favorable condition for infection followed by the subsequent incubation period (about one week). The beginning of epidemics in the 19 seasons ranged from June 24 to July 18, and the average was July 6.

At first, disease progressions of the 19 seasons were simulated. Based on the results and past experience, we set 3.5 leaf lesions per plot as the ETL. One plot consisted of one hill, and the density of the hills was 22.2 per square meter (15 by 30cm planting interval). This ETL value meant that we considered fungicide application was necessary in only 6 out of the 19 seasons in retrospect. This means that Koriyama can be considered as a moderate blast disease area.

3.2. FUNGICIDE SPRAYING STRATEGIES COMPARED
In DSS, the weather data set of the season in question up to the previous day of the decision date followed by that of the 18 other seasons are used as the input weather data. Decision making about fungicide application starts after the disease onset has been confirmed. Thus, fungicide will never be applied before the disease onset. In DSS, decisions are made based on the results of disease forecasting as described previously. Since forecasting in itself contains some uncertainty, here we can consider DSS as îthe partial forecastî.

If complete weather data of the season in question are already known at mid-season, decisions can be also made through similar procedures. We can easily expect that this procedure will achieve the best result. Of course, such decision-making is impossible in the real world, since complete and accurate weather forecasting is currently impossible. We use this type of decision-making only for comparison, and call it as Îthe perfect forecastÎ. There are at least two conventional approaches to scheduling fungicide applications. One is the so-called Îcalendar sprayingÎ scheduling. The application dates are pre-fixed before the season starts. For calendar spraying, we designed five different programs. The dates of fungicide application were determined based on the average disease onset date, July 6 (Table 1). The scheduling of the other approach depends on the time of disease onset. Kobayashi [4] proposed, based on empirical experiences, that the first fungicide application should be done within one week after the onset of leaf blast to prevent a severe outbreak. We call this strategy 'onset dependent spraying' scheduling. For onset dependent spraying, we set up four programs (Table 1). For some programs, spraying starts two days after the disease onset, while others start seven days or more after the onset.

The performance of the 12 strategies, including a non-fungicide use program, were compared by the simulation experiment. The results are shown in Table 1.

Table 1. Evaluation of various strategies of fungicide application schedule by simulation. The values are average of 19 seasons.

Strategies	No. of spraying	Disease severity[1]		Total loss[2]
		Mean	S.D.	
Perfect forecast	0.32	2.77	1.65	3.88
Partial forecast	**0.89**	**2.55**	**1.62**	**5.87**
Calendar spraying				
on July 6[3]	1	7.83	10.24	11.33
on July 6, 13	2	3.19	3.45	10.19
on July 13	1	3.19	3.45	6.69
on July 13, 20	2	3.03	3.38	10.03
on July 20	1	6.91	8.30	10.41
Onset dependent spraying				
+ 2[4]	1	4.76	6.82	8.26
+2, +9	2	4.75	6.83	11.75
+9	1	5.96	9.05	9.46
+7, +14	2	4.57	5.92	11.57
Non-spraying	0	7.83	10.24	7.83

1) No. of leaf lesions per plot (hill) at the end of a leaf blast epidemic.
2) No. of leaf lesions per plot + spraying cost (3.5 x No. of spraying).
3) Dates of fungicide spraying.
4) Dates of fungicide spraying. Each value means days after disease onset.

3.3. PERFORMANCE OF THE DECISION SUPPORT SYSTEM

The average number of fungicide sprayings recommended by the partial forecast strategy (DSS) was less than once per season, and disease severity was less than any conventional strategy we tested. The standard deviation of the severity of the partial forecast was also much less than those of any conventional strategies. This suggests that DSS performed very well in this simulation experiment.

The perfect forecast indicates the upper limit of DSS performance. In this experiment, the average number of fungicide sprayings in the partial forecast was higher than that in the perfect one, while the average disease severity and its standard deviation were almost the same. Consequently, the average total loss of the perfect forecast was much lower than that of the partial one. These differences can be overcome by improving DSS in the future. For example, weather forecasts [5] could be included in the input weather data.

Detailed results (not shown) illustrated interesting differences between the decisions of the two forecast schedules. While the perfect forecast suggested no fungicide application in 13 of 19 seasons, the partial forecast recommended spraying in eight of the 13 seasons. Notably, the partial forecast suggested spraying twice in two of these eight seasons when the perfect one recommended no fungicide spray. In two other seasons, the partial forecast suggested spraying twice, while the perfect one recommended only one spraying. In another season, the timing of spraying in the partial forecast was later than that recommended by perfect forecast. Thus, the partial forecast correctly suggested the necessity of fungicide spraying or its spray timing in eight of 19 seasons.

Surprisingly, the performances of the onset dependent spraying schedules were not as high as we had expected. This may have been because spraying at the early phase of on epidemic may not sufficiently suppress the disease progression at a late phase, especially when the epidemic goes on for long period. Also, fungicide use in this strategy may not be efficient because fungicides have to be sprayed whenever disease occurrence is detected. In other words, this spraying strategy may be used in areas where severe blast epidemics occur more frequently but their average duration are shorter than we assumed in this experiment. These possibilities can also be examined with the simulation experiment by changing the environmental conditions. The calendar spraying strategy did not generally perform well in this experiment. However, a particular timing of spraying (on July 13) showed a considerably good performance. In this experiment, this time is the equivalent of one week after the average date of disease onset. At present, there is no clear explanation why this timing showed such a good result or whether this timing is generally effective. Further research is necessary to reach a conclusion.

3. Discussion

There have been few attempts to optimize spraying schedules of fungicides using disease simulators. Shtienberg et al. [7] and Raposo et al. [5] evaluated the performances of several spraying programs for potato late blight using simulation. Raynolds [6] demonstrated the mathematical optimization of apple scab management using a disease simulator. That is, they used simulation analyses when they evaluated their strategies or tried to explore the optimum solution. They did not use simulators for daily basis decision-making.

We used a simulator in our decision support system (DSS) on a daily basis, and evaluated its performance using simulation analysis. In DSS, disease simulations are reiterated many times to suggest a daily action. Recently, the tremendous advances in computer technology make these procedures feasible. The simulation experiment in the study suggests that the performance of DSS is satisfactory. However, so far, no mathematical demonstration has shown that the algorithm of DSS makes the optimum solution. Further research has to be carried out to do this.

The difficulty in estimating the ETL of rice leaf blast has been considered as an obstacle to developing DSS, and applying it to IPM practices. We overcame it by setting and assuming ETL arbitrarily based on past experience. This assumption is quite subjective. However, the decision-makers can accept the assumption, because they can determine ETL themselves.

Modifications of DSS are possible. For example, we can assume a time lag between a decision and the subsequent action. This condition is compatible for growers who cannot spray immediately after their decision due to logistical delays. While we used a protectant fungicide in this DSS, we can change it to curative ones. Information from weather forecasts can be introduced into the weather data for simulation. These changes should improve DSS's recommendations.

This study was carried out only with computer simulation experiments. Computer simulation is an essential approach in researching investigations into optimization. Field experiments are practically impossible to achieve the purpose. However, we believe that a careful combination of computer simulation with confirmation experiments in the field is important, in order to keep reality in the simulation. We are currently planning research to do this.

5. References

Hashimoto, A., Hirano, K. and Matsumoto, K. (1984) Studies on the forecasting of rice leaf blast development by application of computer simulation. Sp. Bull. Fukushima Pref. Agric. Exp. Stan. 2:1-104

Ishiguro, K., Takechi, S. and Hashimoto, A. (1992) Behavior of tricyclazole residue on rice leaves and its efficacy for rice leaf blast control in the field. Ann. Phytopathol. Soc. Jpn. 58:259-266.

Jacobsen, B. J. (1997) Role of plant pathology in integrated pest management. Annu. Rev. Phytopathol. 35:373-391.

Kobayashi, J. (1984) Studies on epidemic of rice leaf blast, *Pyricularia oryzae* Ca., in its early time. Bull. Akita Agric. Exp. Stan. 26:1-84.

Raposo, R., Wilks, D. S., and Fry, W. E. (1993) Evaluation of potato late blight forecasts modified to include weather forecasts: a simulation analysis. Phytopathology 83:103-108.

Raynolds, K. L. (1993) Simulating disease development and economic impact and the application of mathematical optimization methods for plant disease management (abstract). Abstracts of 6th ICPP, 6th ICPP, Montreal, pp.25.

Shtienberg, D. S., Doster, M. A., Pelletier, J. R., and Fry, W. E. (1989) Use of simulation models to develop a low-risk strategy to suppress early and late blight in potato foliage. Phytopathology 79:590-595.

COMPARATIVE CONTINENTAL VARIATION IN THE RICE BLAST FUNGUS USING SEQUENCE CHARACTERIZED AMPLIFIED REGION MARKERS

O. SOUBABERE[1], D. THARREAU[1], W. DIOH[2], M.H. LEBRUN[3],
J.L. NOTTEGHEM[4].
[1] CIRAD, [2] Université Paris-Sud, [3] CNRS-RPA, [4] ENSAM
[1] BP5035, 34032 Montpellier, France;[2] 91405 Orsay, France; [3] BP9163, 69623 Lyon, France; [4] 2 place Viala, 34060 Montpellier, France.

1. Genetic tools and *Magnaporthe grisea* population studies

Magnaporthe grisea (Herbert) Barr, a fungal pathogen causing Blast disease on rice and other Poaceae, is present in all rice growing areas. Only the asexual form, *Pyricularia grisea*, of this haploid Ascomycete has been identified in the field. *In vitro*, sexual reproduction can be realized between isolates of opposite mating types (Mat1.1/Mat1.2), provided that one isolate is female fertile. Only a few female fertile isolates have been identified among worldwide collections of thousands of isolates [4].

Up to date, population studies of *M. grisea* (Hebert) Barr were carried out with RFLP [2, 6], RAPD [Jorge, unpublished] or rep-PCR markers [1]. A repeated sequence (MGR586) has been commonly used to generate fingerprints [2, 3, 6, 7]. These techniques brought important knowledge on populations' diversity and structure at the field and country level. They confirmed predominance of asexual reproduction in the fields by showing the existence of clonal lineages. Populations' genetic structures where compared to pathotypic structures [2, 3, 6, 7]. Based on these datas, different control strategies based on host resistance [8] might be developped. The chosen strategy may not be the same in all the rice growing areas. Knowledge on the genetic and pathotypic diversity and structure of the fungal population in one particular region is essential to determine the suitable host resistance to be used. Although well adapted to local or regional studies the molecular markers commonly used for blast population studies are not well suited to world populations' comparison. For example, MGR fingerprints of isolates from different countries are usually completely different (no common band), which do not allow to determine genetic relationships between isolates from different countries. RAPD markers can be exploited to exibit similarities but are not easy to transfer in different laboratories (repetability, complexity of profiles) and, because two bands of the same size are not necessarily the same allele of one locus, they can not be used in some genetic studies. Thus, we developped new molecular tools, i.e. SCAR markers (Sequence Characterized Amplified Regions; [5]), to study populations of *M. grisea* pathogenic to rice at the worldwide scale.

D. Tharreau et al. (eds.), Advances in Rice Blast Research, 209–213.

2. SCAR markers reveal genetic structure at the worlwide scale

Fourteen pairs of primers were designed and used to characterize a sample of 64 isolates collected on rice worldwide (Africa, North and South America, Asia and Europe). When available, molecular characterization datas from previous studies were used to choose the isolates. Thus, in each geographic zone, isolates belonging to different clonal lineages were used as representatives to maximize diversity.

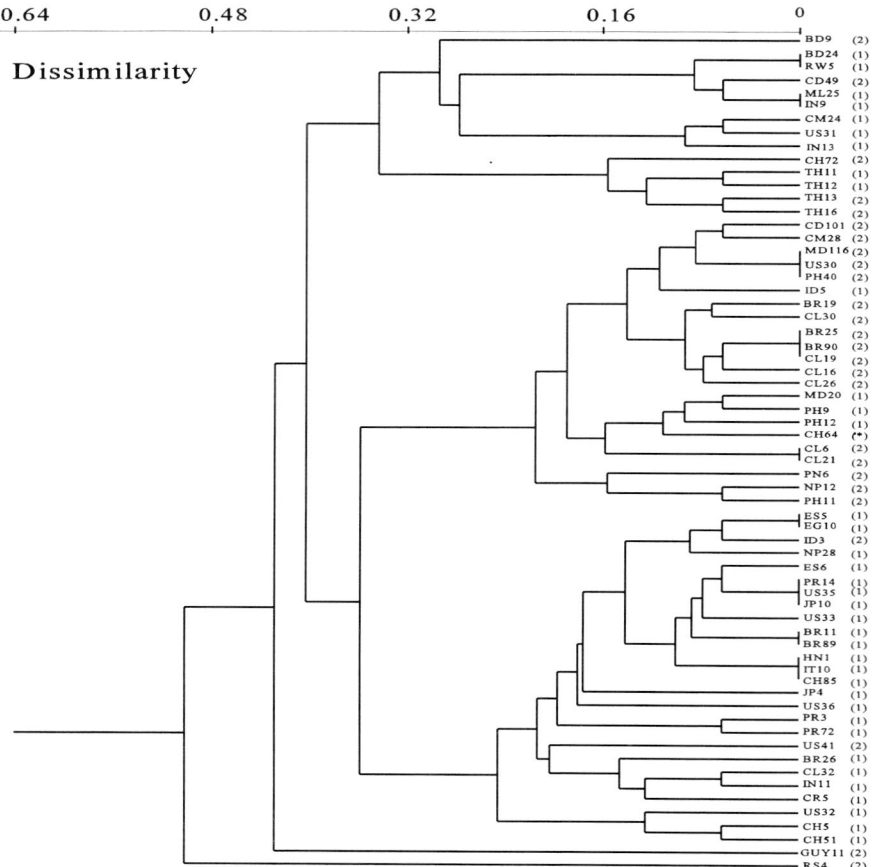

Figure 1 : UPGMA tree constructed with SCAR datas of 64 *Magnaporthe grisea* isolates pathogenic to rice from worlwide origin.
BD, Burundi; BR, Brazil; CD; Cote d'Ivoire; CH, China; CL, Colombia; CM, Cameroon; CR, Korea; EG, Egypt; ES, Spain; GUY, French Guyana; HN, Hungary, ID, Indonesia; IN, India; IT, Italy; JP, Japan; MD, Madagascar; ML, Mali; PH, The Philippines; PN, Panama; PR, Portugal; NP, Nepal; RS, Russia; RW, Rwanda; TH, Thailand; US, USA. (1)/(2): Mating type Mat1.1 and Mat1.2 respectively; * not determined.

After amplification, presence or absence of bands were scored, similarity between isolates was calculated and an UPGMA tree was constructed. This tree (figure 1) allowed to define five groups showing that structuration of the population exist at the worlwide level. Three isolates were unrelated to any other isolate (BD9, Guy11, RS4).

Most of the isolates (75%) belonged to groups 1 and 2. Group 1 was made of 26 isolates mainly of mating type Mat1.1 and originating from Asia, Americas and Europe (table 1). Group 2 gathered 22 isolates mainly of mating type Mat1.2 and originating from Asia, Americas and Africa. In group 3 only asian isolates were found; group 4 gathered isolates from all continents except Europe an Sout h America, group 5 contained african and asian isolates. Isolates from Asia were found in the five groups and almost all rare alleles identified in the populations from Africa, Europe or Americas were also found in the population from Asia.

TABLE 1. Geographic origin of isolates in each genetic group.

	Europe	S. America	N. America	Africa	Asia
Group 1	8	4	5	-	9
Group 2	-	10	1	4	7
Group 3	-	-	-	-	5
Group 4	-	-	1	1	1
Group 5	-	-	-	4	1
Not grouped	1	1	-	1	-

Haplotypic diversity was almost of one, indicating that each isolate was different from another and showing that sampling was as expected (maximum diversity). Allelic diversity was the highest in Asia (table 2) and the lowest in Europe.

TABLE 2. Genetic diversity.

	Number of isolates	Number of haplotypes	Nei's index	
			Haplotypic	Allelic
Africa	10	9	0.98	0.25
Europe	9	7	0.97	0.19
Americas	22	17	0.97	0.24
Asia	23	23	1	0.33

The two main groups of blast isolates (groups 1 and 2) could reflect specialization to two main genetic groups of rice (i.e. temperate japonicas and tropical indicas). The assumption of specialization of the pathogen is based on the geographic distribution of genetic groups of the pathogen that fit with the geographic distribution of the different rice types. For example, temperate japonicas are grown on all the sampled continents except Africa and, similarly, group 1 was found on all continents but Africa. In the same way, tropical indicas are not grown in North America and Europe, and group 2 was represented in these continents by only one isolate and none respectively. The correlation between genetic diversity of the host and genetic diversity of the pathogen in the regions could also be indicative of host specialization. In Europe, where only temperate japonicas are grown, diversity is the lowest. On the other hand, rice and blast diversity in Asia are high. Thus, rice could act as a major factor of population structuration of the pathogen.

TABLE 3. Genetic differenciation (Fst) among populations from different continents.

	Europe	Americas	Asia
Africa	0.538	0.280	0.140
	(0.397-0.679)	(0.174-0.386)	(0.077-0.203)
Europe	-	0.211	0.244
		(0.141-0.281)	(0.141-0.347)
Americas	-	-	0.064
			(0.016-0.112)

All the Fst values are significantly different from zero at the 1 % level (bootstrap, 1000 permutations).

Genetic differentiation between population of different continents was calculated using an estimator of the Fst parameter [9]. The population of each continent was significantly differentiated (table 3) suggesting rare exchanges between populations. The most differentiated populations were from Africa and Europe and the least from Asia and the Americas.

Results on levels of genetic diversity, distribution of isolates from Asia in all genetic groups and distribution of rare alleles suggest that the centre of diversification of the pathogen could be situated in Asia. Some isolates from Asia had the same haplotype as isolates from Americas, Europe and Africa. This could be indicative that migrations to other continents occurred. Transport of infected rice seeds or straw by men might have played a major role in the fungus spreading.

3. Conclusion

Preliminary studies on blast populations at the worldwide scale with SCAR markers have generated promising results. Several hypothesis have arisen that should be strengthened. As a prerequisite, these studies have to be extended to a greater number of isolates. The expected results are likely to bring new insights on the occurence of sexual reproduction in limited areas of Asia. For example, population diversity and structure has to be studied more intensively in some parts of China and foothills of the Himalaya. SCAR markers provide a tool to test migrations by studying genotype distribution but also allelic frequencies. The hypothesis that host selection is of importance for worldwide population structure has to be studied further, specially with pathogenicity tests.

4. References

1. George, M.L.C., Nelson, R.J., Zeigler, R.S., and Leung, H. (1998). Rapid genetic analysis of *Magnaporthe grisea* with PCR using endogenous repetitive DNA sequences. *Phytopathology* **88**, 223-229.
2. Levy, M.., Romao, J., Marchetti, M..A. and Hamer, J.E. (1991). DNA fingerprinting with a dispersed repeated sequence resolve pathotype diversity in the Rice Blast fungus. *The Plant Cell* **3**, 95-102.
3. Levy, M., Correa-Victoria, F. J., Zeigler, R. S., Xu, S., and Hamer, J. E. (1993). Genetic diversity of the Rice Blast fungus in disease nursery in Colombia. *Phytopathology* **83**, 1427-1433.
4. Nottteghem, J.L., and Silué, D. (1992). Distribution of mating types allele s in *Magnaporthe grisea* populations pathogenic to rice. *Phytopathology* **82**, 421-424.
5. Paran, I., and Michelmore, R.W. (1993). Development of a reliable PCR-based markers linked to downy mildew resistance genes in lettuce. *Theor. Appl. Genet.*, **85**, 935-993.
6. Roumen, E,. Levy, M., and Notteghem, J.L. (1997). Characterisation of the European pathogen population of *Magnaporthe grisea* by DNA fingerprinting and pathotype analysis. *European J. of Plant Pathology* **103**, 363-371..
7. Zeigler, R. S., Cuoc, L. X., Scott, R. P., Bernardo, M. A., Chen, D. H., Valent, B., and Nelson, R. J. (1995). The relationship between lineage and virulence in *Pyricularia grisea* in the Philippines. *Phytopathology* **85**, 443-451.
8. Zeigler, R.S., Thome, J., Nelson, R., Levy, M. and Correa-Victoria, F.J. (1994). Lineage exclusion : a proposal for linking blast population analysis to rice breeding, in R.S. Zeigler, S.A. Leong and P.S. Teng (eds.), The Rice Blast Disease, John Hopkins Press, Baltimore, Maryland, USA. pp. 267-291.
9. Weir, B.S. and Cockerham, C.C. (1984). Estimating F-statistics for the analysis of population structure. *Evolution* **38**, 1358-1370.

POPULATION DYNAMICS OF THE RICE BLAST PATHOGEN IN A SCREENING SITE IN COLOMBIA AND CHARACTERIZATION OF RESISTANCE

F.J. CORREA-VICTORIA, F. ESCOBAR, G. PRADO, and G. ARICAPA
Rice Project, Centro Internacional de Agricultura Tropical, CIAT
Apartado Aéreo 6713, Cali, Colombia

1. Introduction

Rice blast disease caused by *Pyricularia grisea* Sacc., the anamorph of *Magnaporthe grisea* (Hebert) Barr, is the main rice production constraint in Colombia. Development of resistant cultivars has been the preferred means of controlling this disease; however, development of resistant cultivars with durable blast resistance has been a very difficult task, especially in areas where rice is grown continuously such as in Colombia, and where the climatic conditions are highly conducive to blast. In general, blast resistance is defeated by the pathogen shortly after cultivar release (Correa-Victoria and Martinez, 1994). Major efforts are being made at CIAT to understand the high pathogen variation observed, often reported as the main cause of resistance breakdown (Correa-Victoria *et al.*, 1994). An extensive study of *P. grisea* diversity in Colombia was initiated in 1990 by analyzing the population structure in the country. We have analyzed extensively the virulence diversity and virulence spectrum of several hundred isolates collected from different rice cultivars. Isolates have been inoculated under greenhouse conditions onto rice cultivars with known resistance genes, rice commercial cultivars released in Colombia over a period of 30 years, sources of durable blast resistance, the international set of host differentials, and a set of near isogenic lines carrying known resistance genes.

MGR-DNA fingerprinting has been used to determine the genetic structure of the same blast populations. The pathogen complexity has been simplified into six genetic families, named SRL-1 to SRL-6 (Levy *et al.*, 1993). Studies on the relationship between virulence spectrum and genetic lineage suggest that perhaps the most important feature of lineage virulence spectrum is its limits, rather than the variation within. Developing stable blast resistance at CIAT is based on a strategy combining pathotyping and genetic characterization (MGR-DNA-fingerprinting) of local blast pathogen populations, and continuous evaluation and selection of rice breeding lines under high blast pressure and a diverse pathogen population. Segregating populations are evaluated for blast resistance under upland conditions to ensure heavy blast pressure (Correa-Victoria and Zeigler, 1993a, 1993b, 1995). Focusing on resistance genes to lineages of the pathogen, rather than on genes resistant to specific pathotypes, allow the identification of resistance gene combinations effective against entire targeted populations of the pathogen (Correa-Victoria and Martinez, 1994).

This paper summarizes data on the dynamics of virulence and genetic structure of a blast pathogen population in a breeding-screening site in Colombia in order to develop more effective strategies for the production of durable blast resistance.

D. Tharreau et al. (eds.), Advances in Rice Blast Research, 214–220.
© 2000 *Kluwer Academic Publishers. Printed in the Netherlands.*

2. Dynamics of Virulence and Genetic Structure of the Blast Pathogen in a Breeding Experiment Station in Colombia

The CIAT Rice Project has been breeding rice for blast resistance at a "hot spot" upland site, characterized by reliable and very high disease levels, as well as high pathogen diversity. This site is located in an important rice growing area, where rice blast is a major constraint. Uniform and high blast incidence is maintained in breeding plots during the entire season using spreader rows composed of a mixture of commercial cultivars and resistance sources with complementary susceptibility to different phenotypes and genetic families of the blast pathogen. Escapes are minimized by continuous exposure to the pathogen during the season, at all growth stages, and in several generations. Stability of resistance selected at this site has been associated with selection for high levels of multigenic resistance (Correa-Victoria and Zeigler, 1993b, 1995). However, stability and durability of resistance seem to be not only the function of host genotype, but also depend upon the dynamics of pathogenicity and virulence within the pathogen population. Then, establishment of a system for understanding pathogenic variability and its dynamics is essential for the development of feasible strategies to control blast with genetic resistance.

Continuous characterization of the blast pathogen in terms of race composition, compatibility with known resistance genes and frequency of virulent phenotypes has been established as routine work in the CIAT Rice Project in order to evaluate the effectiveness of this screening site for developing rice blast resistant lines. The population of *P. grisea* at this site has been found to be pathogenically very diverse. More than forty-five international races, representing all nine-race groups, were found. Virulence factors to all the international differentials and cultivars with at least 13 different genes for resistance have been identified at the site. Most of the isolates accumulate a large number of virulence factors to most known resistance genes, but none was virulent to all of them. Avirulence to all of the differentials is present in the pathogen population as well. Isolates within an international race may be further separated in different phenotypes according to their virulence on other rice cultivars known to have different sources of resistance. For example, six isolates each of races IA-103 and IA-128 could be further differentiated into 12 pathotypes on the basis of their virulence on commercial cultivars. This suggests that the international race designation do not fully described the entire pathogenic variability of the pathogen but merely a subset of this variability. The chronic epidemic environment and available hosts at our screening site apparently has promoted an accelerated and exaggerated version of the pathotype evolution that may occur in rice field monocultures. For a complete understanding of the pathogen diversity it is then very important to increase the number of differentials, including local commercial cultivars and other sources of resistance (Correa-Victoria and Zeigler, 1993a).

It is very important from an ecological, epidemiological, and breeding perspective to know how genetic diversity is maintained in the pathogen and how new, well adapted complex races arise. No evidence for sexual reproduction at this site has been observed (Correa-Victoria and Zeigler, 1993a). DNA-fingerprinting analysis of the blast pathogen population in Colombia's rice growing areas using the *P. grisea* repetitive DNA sequence MGR 586 suggests that the pathogen reproduces asexually being it grouped into six genetically different families. The genetic structure of the pathogen population in Colombia is relatively simple despite the great virulence diversity observed. The mean genetic similarity within linages was high, ranging from

92 to 98%. The mean similarity between lineages ranged from 37 to 85% (Levy *et al.*, 1993). The spectrum of virulence of isolates within each family is highly similar, differing mainly in single virulences. Although the six genetic families of the fungus share a high number of virulence factors, high specific interaction between some avirulence/virulence factors in the pathogen and resistance genes in the host has been observed (Table 1). This specific interaction is the basis for selecting the progenitors to be included in a breeding program aimed at obtaining more durable blast resistance. We have observed that the number of pathotypes within MGR-DNA fingerprint lineages is conditioned by a variety of factors and need to be assessed in a number of rice-growing areas under a variety of ecological conditions (Levy *et al.*, 1993). However, our results indicate that there is a close relationship between the lineage and virulence characteristics of the constituent individuals. This finding, together with the fact that the blast population in Colombia is composed of a limited number of lineages, indicate that population analysis at the virulence and lineage level could aid in directing resistance breeding projects that target the pathogen population in question.

TABLE 1. Virulence spectrum of six Colombian genetic lineages of *Pyricularia grisea*

Cultivar	Genetic Lineage						
	SRL-1	SRL-2	SRL-3	SRL-4	SRL-5	SRL-6A	SRL-6B
Near Isogenic Line							
C101 A 51 (Pi-2)	+	+		+		+	+
C101 LAC (Pi-1)					+		
C101 PKT (Pi-4a)		+		+	+	+	+
C104 PKT (Pi-3)		+		+	+		+
C105 TTP (Pi-4b)		+		+	+	+	+
Commercial Cultivars							
Oryzica Llanos 5							
Oryzica 1		+		+		+	+
Oryzica Yacu 9		+				+	+
Línea 2		+				+	+
Cica 8					+		+
Colombia 1				+			+
Oryzica Caribe 8				+			
Cica 9	+	+					+
Fanny	+	+	+	+	+	+	+

+ = Susceptible

Characterization of the genetic lineage structure, together with the virulence spectrum and the virulence frequencies within the whole blast pathogen population, should provide a more reliable estimate of the durability of cultivar resistance than only consideration of virulences or lineages alone. A summary of the combination of virulence assays and genetic lineages in the blast pathogen population at the screening site studied is shown in Table 1. This analysis is routinely conducted to identify individual or group of isolates within lineages representatives of the whole lineage spectrum of virulence in screening for blast resistance. One resistance gene may be defeated by isolates from different genetic lineages, suggesting that a common virulence factor can be shared between genetic lineages. This is the case of several of the isolines and for cultivar Fanny in Table 1. On the other hand, several virulence factors are shared among isolates within the same genetic lineage and may be accumulated in a single blast isolate. This is the case of isolates in lineages SRL-2, SRL-4, and SRL-6B (Table 1). Virulence frequency studies on 42 rice cultivars ranged between 0.0 and 0.86 (Correa-Victoria and Zeigler, 1993a). However, certain specific virulence-lineage-resistance gene interactions are observed where a resistance gene is effective against all the isolates of entire lineages yet is susceptible to most individuals of other lineages. Cultivars may have complementary resistance, which in combination could confer resistance to all the blast population described. The near isogenic lines C101 A51 and C101 LAC, carrying the resistance genes Pi-2 and Pi-1 respectively, have complementary resistance which, in combination, could confer resistance to all the blast pathogen population studied (Table 1).

TABLE 2. Compatibility frequency of six Colombian lineages of *Pyricularia grisea* on 201 Latin American rice cultivars

Genetic Lineage	Infected Cultivars (No.)	Compatibility Frequency (%)
SRL-3	12	6.0
SRL-4	34	16.9
SRL-2	52	25.9
SRL-1	61	30.3
SRL-5	138	68.7
SRL-6	162	80.6

Virulence and genetic lineage frequencies are driven by changes in the cultivated commercial cultivars that take place in farmer's fields. Genetic lineage frequencies change between years depending on the area planted with a rice cultivar and on the virulence spectrum of the lineage. Studies on the compatibility frequency reflecting the whole spectrum of virulence of the blast pathogen in Colombia was analyzed by inoculating representative isolates of each genetic lineage onto 201 Latin American rice cultivars (Table 2). Most cultivars were susceptible to genetic lineage SRL-6 (80.6%) followed by lineage SRL-5 (68.7). Lineage SRL-3 was compatible with very few cultivars. Pattern of compatibility observed with lineages SRL-6 and SRL-5 should be a result of the narrow and similar genetic base present in the Latin American rice germplasm. Lineage SRL-6 has been the most frequent segment of the population over the years in Colombia (Table 3). This lineage also expressed the greatest compatibility with the group of Latin American cultivars (Table 2). Lineages SRL-2 and SRL-4 have increased in frequency as planted area with susceptible commercial cultivars, Linea 2 and Oryzica Caribe 8 respectively, has also increased. The increase of these two lineages is not related to a capacity of infecting a large number of Latin American cultivars (Tables 2 and 3). Isolates of these two lineages exhibit however a wide spectrum of virulence on the Colombian commercial cultivars and near isogenic lines (Table 1). Frequency of lineages SRL-1 and SRL-5 has decreased (Table 3) in time as cultivars susceptible to these lineages have almost disappeared from farmer's fields. These two lineages exhibited a relatively large compatibility with the Latin American cultivars tested (Table 2). Frequency of lineages SRL-1 and SRL-3 has been near zero during the last years due to the absence of susceptible cultivars in commercial fields and the narrow virulence spectrum exhibited on the Colombian commercial cultivars (Table 1). A gain in virulence spectrum over time has been observed for most of the predominant lineages. This is illustrated by the lineages identified as SRL-6A and SRL-6B in Table 1. Monitoring gains in virulence is of practical use for identifying resistance donors that can be used in the event that resistance is defeated in the field by the new isolates. Although pathotype diversity per lineage is higher in a screening nursery site than commercial rice fields, amplification of the rate of evolution of new pathotypes leading to resistance breakdown may indeed originate under commercial fields. Blast isolates exhibiting initial steps of breakdown of the highly resistant cultivar Oryzica Llanos 5 were detected in commercial fields where lineage SRL-4 had increased in frequency. A false compatible interaction between lineage SRL-6 and Oryzica Llanos 5 had continuously been detected at our screening site. Nevertheless, virulence of compatible isolates on Oryzica Llanos 5 has not been stable and its resistance continues being durable under field conditions.

TABLE 3. Frequency of rice blast genetic lineages in a Colombian screening site

Genetic Lineage	Isolates (No.)	Frequency (%)
SRL-6	122	60
SRL-5	12	6
SRL-4	37	18
SRL-2	26	13
SRL-1	5	3
TOTAL	202	100

3. Identification of Blast Resistance Sources Effective Against all Members of Individual Genetic Lineages in Breeding for Durable Resistance

We have identified resistance sources expressing broad resistance against all lineages of the pathogen such as Oryzica Llanos 5 (Table 1) as well as complementary resistance to the different lineages such as C 101 A51, C101 LAC, Oryzica Yacu 9, Oryzica Caribe 8 (Table 1). Resistance can be expressed against all pathotypes of a lineage while being defeated by those of another lineage. Combinations of genes showing complementary resistance to different genetic families of the fungus should exclude any compatible interaction with a blast isolate.

TABLE 4. Rice lines (F_7) from the cross C101 LAC (Pi-1) x C101 A 51 (Pi-2) resistant to blast

Rice Lines	Field Reaction		Greenhouse	
	Leaf (1-9)	Neck (1-9)	SRL 6,4,3,2,1	SRL 5
CT 13432 (PL2)-1-1-M-M-M	2	1	R	R
CT 13432 (PL2)-4-2-M-M-M	2	1	R	R
CT 13432(PL2)-11-1M-M-M	2	1	R	R
CT 13432(PL4)-2-1-M-M-M	2	1	R	R
CT 13432(PL4)-2-2-M-M-M	2	1	R	R
CT 13432(PL4)-14-1-M-M-M	2	1	R	R
CT 13432(PL5)-1-1-M-M-M	2	1	R	R
CT 13432(PL5)-1-2-M-M-M	2	1	R	R
CT 13432(PL7)-5-2-M-M-M	2	1	R	R
CT 13432(PL7)-7-1-M-M-M	2	1	R	R
CT 13432(PL7)-10-1-M-M-M	2	1	R	R
CT-13432(PL8)-7-2-M-M-M	2	1	R	R
CT 13432(PL8)-9-1-M-M-M	2	1	R	R
CT 13432(PL8)-10-2-M-M-M	2	1	R	R
CT 13432(PL8)-15-3-M-M-M	2	1	R	R
C101 A51(Pi-2)	9	9	S(9)	R(1)
C101 LAC(Pi-1)	6	9	R(1)	S(7)

R = resistant (1-3); S = susceptible (>4). SRL = Blast Genetic Lineage 1 to 6

TABLE 5. Latin American rice cultivars exhibiting complementary resistance to six Colombian lineages of *Pyricularia grisea*

Cultivar	Origin	Genetic Lineage					
		SRL-1	SRL-2	SRL-3	SRL-4	SRL-5	SRL-6
MG-1	Brazil				+		
Rio Verde	Brazil				+		
O. Caribe 8	Colombia				+		
Sacia 2	Bolivia					+	
Amistad 82	Cuba					+	
IR 1529 ECIA	Cuba					+	
Perla	Cuba					+	
Juma 51	Dom. Republic					+	
Bamoa	Mexico					+	
Oryzica 3	Colombia						+
Capi 93	Honduras						+
PA-3	Peru						+
O. Yacu	Colombia		+				+
Line 2	Colombia		+				+
CICA 9	Colombia		+				+
Araure	Venezuela		+				+
Juma 61	Dom. Republic		+				+
O. Llanos	Colombia				+		+
IR 65	Philippines				+		+
Cuyamel 3820	Honduras				+		+
Icta Crispo	Guatemala					+	+
Juma 62	Dom. Republic		+			+	
Fanny	France	+	+	+	+	+	+

+= Compatible reaction

4. References

Correa-Victoria, F.J., and Martinez, C. (1994) Genetic structure and virulence diversity of *Pyricularia grisea* in breeding for rice blast resistance, in *Induced Mutations and Molecular Techniques for Crop Improvement*, IAEA, Vienna, pp. 133-145.

Correa-Victoria, F.J., and Zeigler, R.S. (1993a) Pathogenic variability in *Pyricularia grisea* at a rice blast "hot spot" breeding site in eastern Colombia, *Plant Disease* **77**, 1029-1035.

Correa-Victoria, F.J., and Zeigler, R.S. (1993b) Field breeding for durable rice blast resistance in the presence of diverse pathogen populations, in Th. Jacobs and J.E. Parlevliet (eds.), *Durability of Disease Resistance*, Kluwer Academic Publishers, Dordretch, pp. 215-218.

Correa-Victoria, F.J., and Zeigler, R.S. (1995) Stability of complete and partial resistance in rice to *Pyricularia grisea* under rainfed upland conditions in eastern Colombia, *Phytopathology* **85**, 977-982.

Correa-Victoria, F.J., Zeigler, R.S., and Levy, M. (1994) Virulence characteristics of genetic families of *Pyricularia grisea* in. Colombia, in R.S. Zeigler, S.A. Leong and P.S. Teng (eds.), *Rice Blast Disease*, CAB International, UK, pp. 211-229.

Levy, M., Correa-Victoria, F.J., Zeigler, R.S., Xu, S., and Hamer, J.E. (1993) Genetic diversity of the rice blast fungus in a disease nursery in Colombia, *Phytopathology* **83**, 1427-1433.

Zeigler, R.S., Tohme, J., Nelson, R.J., Levy, M., and Correa-Victoria, F.J. (1994) Lineage exclusion: a proposal for linking blast population analysis to resistance breeding, in R.S. Zeigler, S.A. Leong and P.S. Teng (eds.), *Rice Blast Disease*, CAB International, UK, pp. 267-292.

Zeigler, R.S., Cuoc, L.X., Scott, R.P., Bernanrdo, M.A., Chen, D.H., Valent, B., and Nelson, R.J. (1995) The relationship between lineage and virulence in *Pyricularia grisea* in the Philippines, *Phytopathology* **85**, 443-451.

CLONAL LINEAGES OF JAPANESE BLAST FUNGUS POPULATION
Variation of pathogenicity and karyotype

T. SONE and F. TOMITA
*Laboratory of Applied Microbiology, Faculty of Agriculture, Hokkaido
University*
Kita-9, Nishi-9, Kita-ku, Sapporo 060-8589, JAPAN

1. Introduction

In order to protect the rice crop from the blast which is one of the most seriously damaging disease of the rice crop, various approaches such as fungicides, resistant cultivar-breeding have been challenged. However, those approaches were not permanently effective because of the variability of the pathogen, *Magnaporthe grisea*. Many kinds of variations are found in the fungus. The most important variable feature is the pathogenicity (Ou, 1980). In Japan, the pathogenic race of the fungus is usually determined using nine differential varieties (Table 1, Yamada *et al.* 1976), and more than 40 pathogenic races has been identified. In addition, the pathogen is reported to be variable in fungicide resistance (Yoshino 1988). These variations should have been introduced genetically, by mutations or sexual / parasexual recombination.

The main objective of our project is to understand the molecular mechanisms that introduce such genetic variation into the fungal genome. For achieving this purpose, first of all, we should characterize genetically the fungal population where the fungus acquires the genetic variation. Molecular genetic tools, such as DNA fingerprinting and electrophoretic karyotyping have been used and found to be effective for such purpose, by uncovering clonal structure or CLPs (Chromosome Length Polymorphisms) among the fungal population.

In addition, molecular genetic analyses have made remarkable progress on more detailed genetic feature of the variability. Genetic maps with high resolution (Nitta *et al.* 1997) enabled us to resolve of genomic structure of the pathogen and chromosomal walking toward genes for host cultivar specificity (Valent *et al.* 1994, Mandel *et al*, 1997, Farman and Leong, 1998). A few avirulence genes have been cloned, and led us to find directly the mutations which is responsible for the host-specificity variation (Sweigard *et al.* 1995, Valent *et al.* 1994, Mandel *et al.* 1997, Farman and Leong, 1998). Those analyses have pointed out the importance of mobile genetic elements (Kempken and Kück, 1998) and genomic location (Valent *et al.* 1994) as factors inducing genetic instability. However, limited numbers of avirulence genes can be analyzed by such molecular analyses because the fertile cross is very rare, and the number of avirulence genes depends absolutely upon the strains used in the cross. Hence, it is necessary to analyze diverse strains of the fungus.

D. Tharreau et al. (eds.), Advances in Rice Blast Research, 221–233.
© 2000 *Kluwer Academic Publishers. Printed in the Netherlands.*

We chose Japanese strains of *M. grisea* as materials for studiying their variability for several reasons:

1. As genetic resource of avirulence genes. Each avirulence gene is pathogen's element that controls avirulence toward corresponding blast resistance gene in rice. Japanese blast pathogenic race differentiating system is based upon blast resistance genes in blast differential cultivars, and those resistance genes were analyzed using some differential strains of the pathogen, which is collected from Japanese field (Table 1, Kiyosawa and Ando 1997). Those differential strains are available for the analyses of avirulence genes.

2. As materials for the analysis of mutation. In Japan, some cases of the breakdown of resistance have been reported (Kushibuchi, 1997). Thus we can expect that the blast population should conserve the history of pathogenic variation in the genome. Furthermore, some field isolates were reported to be capable of inducing pathogenic race mutation in vitro (Namai *et al.* 1997), allowing us to extract that they have active mechanisms for the genomic variation.

3. Originality. Geographically, Japan is separated from other countries by sea. We can expect that Japanese blast-rice system has its original biological history of host-parasite interaction, which will not be found from the strains of other geographical area.

In this manuscript, we will present current knowledge about clonal lineages of Japanese blast fungus population which have been revealed to be rich in pathogenicity and karyotype diversity. Also we will show a model approach to uncover the mechanisms of chromosomal rearrangement.

2. DNA Fingerprinting of Japanese strains of *M. grisea*

DNA fingerprinting is a technique to study genetic relationship between individuals based upon information of DNA. The most popular type of the information is DNA banding pattern produced by Southern hybridization or PCR-based technology, such as RAPD (Random Amplified Polymorphic DNA) and AFLP (Amplified Fragment Length Polymorphism). In the analyses of the population genetics of *M. grisea*, Levy *et al.* (1991) first used pCB (MGR) 586 (Hamer *et al.* 1989) as the DNA fingerprinting probe which was hybridized to *Eco*RI-digested genomic DNAs of the fungus, and found some distinct patterns of fingerprints indicating the existence of clonal lineages of the fungus. The clear relationship between pathotypes and clonal lineages which was found in the study motivated researches on field isolates from other geographical area (Levy *et al.*, 1993, Xia *et al.* 1993, Zeigler *et al.* 1995, Chen *et al.*, 1995, Roumen *et al.* 1997). In those studies, a novel hypothesis "Lineage Exclusion" (Zeigler *et al.*, 1994 and 1995) was proposed and applied to breeding of rice cultivars. Moreover, Zeigler *et al.* (1997) uncovered the parasexual recombination of DNA of field isolates, and confirmed that the method can be applied to the analyses of the mutation event in the genome.

TABLE 1 Responces of differential varieties with true resistance genes to differential fungus strains (based on Kiyosawa and Ando, 1997)

Gene	Differential variety	Differential Strain (race)						
		Ina 72 (031)	Ina168 (101)	K. 53-33 (137)	K. 54-20 (003)	K. 54-04 (003)	Hoku 1 (007)	P-2b (303)
+	Shin 2	S	S	S	S	S	S	S
Pi-a	Aichi Asahi	R	R	S	S	S	S	S
Pi-i	Ishikari Shiroke	R	R	S	R	R	S	R
Pi-k	Kanto 51	S	R	S	R	R	R	R
Pi-k^m	Tsuyuake	S	R	S	R	R	R	R
Pi-z	Fukunishiki	R	R	R	R	R	R	R
Pi-ta	Yashiro-mochi	R	S	S	R	R	R	S
Pi-ta^2	Pi No.4	R	R	R	R	R	R	S
Pi-z^I	Toride 1	R	R	R	R	R	R	R

2.1. CLONAL LINEAGES OF JAPANESE STRAINS OF *M. GRISEA*

We conducted DNA fingerprinting to clarify the genetic relationship within Japanese strains. In the study, we isolated a repetitive DNA probe named pMG6015 from the genome of Ina72, one of Japanese differential strains. This probe contains a SINEs-like element named MGSR1 (Sone *et al.* 1993). The sequence of this probe was shown to be different from MGR586 by cross hybridization analysis.

Sone *et al.* (1997b) chose 7 differential strains and 17 field isolates for the first characterization of genetic relationships among Japanese strains of the fungus. In order to investigate the existence of a direct relationship between pathogenic race and genetic background, field isolates which have homologous pathogenic races as differential strains were chosen.

Figure 1 shows UPGMA-phylogenetic tree derived from the similarity of fingerprints with pMG6015 and MGR586. The bootstrapping analysis (Falsenstein, 1985) helped us to assign clonal lineages. First, at 0.68 similarity, we assigned the clonal lineages JBLA, JBLB and JBLC (for Japanese Blast Lineage A, B, and C) to the two robust clusters and P-2b. However, robust subclusters with the high 0.74 similarity level remained and these subclusters were not negligible because the clonal lineage assignment described above was made at lower similarity level than generally done. Moreover, if we assigned clonal lineages to these subclusters in addition to the three clusters at 0.68 similarity, we could detail a phylogenetic relationship of isolates. Thus we proposed five clonal lineages; JBLA-INA which contains Ina 72 & Ina 168, JBLB-K33 for Ken 53-33 and Ken 54-20, JBLC-P2B for P-2b, JBLB-HK1 for Hoku 1, and JBLA-K04 for Ken 54-04 and all field isolates.

According to this lineage assignment, we can conclude that there is no simple relationship between clonal lineages and pathogenic races among these strains because all the field isolates from various pathogenic races were revealed to be in a particular clonal lineage, JBLA-K04. However, the number of strains and pathogenic races considered in this part were not enough to make a conclusion about the relationships between genetic background and pathogenic race. Thus we used many more strains from various pathogenic races for the next step.

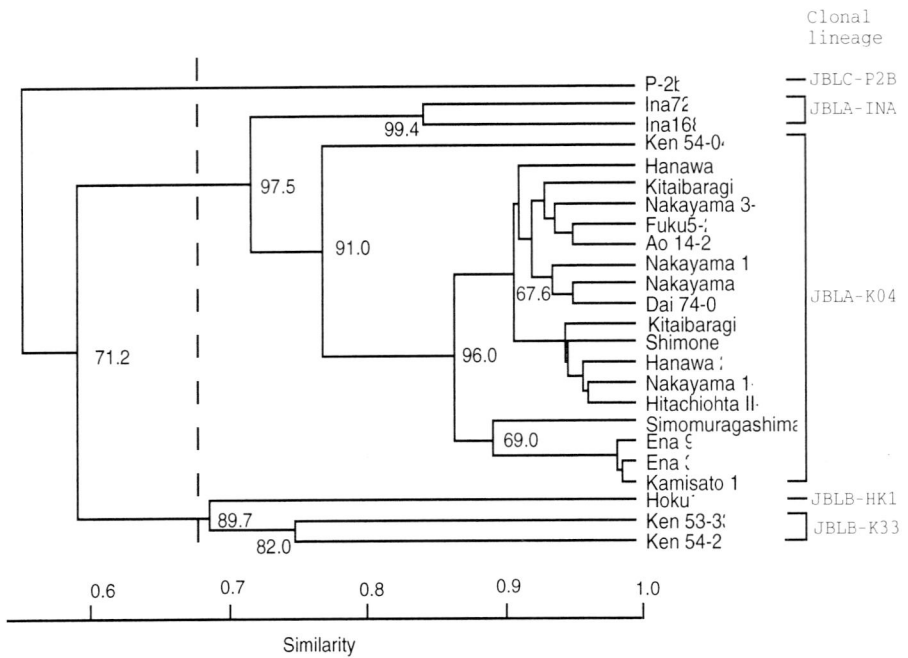

Figure 1. UPGMA-dendrogram of Japanese differential strains and field isolates based on DNA fingerprints derived with MGR586 and pMG6015. The numbers at forks indicate bootstrapping values, calculated with WinBoot (Yap and Nelson, 1996). Clonal lineages assignment was done at 0.68 similarity (dashed line). Assigned clonal lineages were indicated.

2.2. MAJORITY OF JBLA-K04 AND PATHOGENIC RACE VARIATION WITHIN THE LINEAGE

We collected strains under a clear perspective to make a population which can represent pathogenic races found in Japan as many as possible. The population used for the analysis finally contained 142 strains isolated from 1960s to 1994, and the locations of isolation covered almost all over Japan. Thirty-one kinds of pathogenic race from 001 to 477 were included in the population but larger numbers of strains of typical pathogenic races such as 001 or 003 were used due to the limitation of the material (Table 2).

The DNA fingerprints derived with both MGR586 and pMG6015 of the population contained only two pairs of identical pattern, thus the population were consisted of 140 haplotypes. A UPGMA phylogenetic tree was constructed based upon these data (Figure 2). Among numbers of clusters which appeared in the tree, two clusters at 0.8 similarity level were revealed to be robust by bootstrap analysis, a large cluster which contains 127 strains and a small cluster with 15 strains. By the analyses with differential strains, the large cluster were revealed to identical to the clonal lineage JBLA-K04 and the small one is the clonal lineage JBLB-K33. This suggests the population we used has a very simple clonal lineage construction, consisting of two clonal lineages.

Geographical and periodical distribution of each clonal lineage was distinct. The clonal lineage JBLA-K04 distributed all over the area and mostly from after 1980's. On the other hand, JBLB-K33 was specific to Tokai district, because 11 out of 15 JBLB-K33 strains were isolated from the area and mainly before 1980's.

The distribution of pathogenic races in clonal lineages was, on the other hand, very complex. The strains of JBLA-K04 were distributed to 28 kinds of pathogenic races, and JBLB-K33 were 10 kinds of races. Table 2 summarizes the distribution of clonal lineage toward each pathogenic race. The pathogenic races 001, 005, 017, 041, 043, 077, 301, and 303 were specific to the lineage JBLA-K04, while the race 303 was specific to JBLB-K33. Particularly, the races 001 and 005 were specific to the lineage JBLA-K04.

The pathogenic race of a strain is defined as the combination of the blast resistance genes in rice plant which the strain can invade. Thus we further analyzed the relationship between clonal lineages and resistance genes. Table 3 shows the ratio of the virulent isolates in each clonal lineage, toward each blast resistance genes of Japanese differential varieties. In the total of 127 strains of clonal lineage JBLA-K04, 3.1% to 99.2% invasive strains appeared toward each resistance genes, i.e. the clonal lineage contained both virulent and avirulent strains toward each differential cultivar. On the other hand, all the strains from the lineage JBLB-K33 were virulent toward $Pi\text{-}a$, and avirulent toward $Pi\text{-}ta^2$.

TABLE 2. Distribution of pathogenic races and clonal lineage among 142 field isolates.

Pathogenic races	Clonal lineage		Total
	JBLB-K33	JBLA-K04	
001	0	23	23
003	2	21	23
005	0	8	8
007	4	23	27
011	0	1	1
013	0	1	1
017	0	4	4
031	0	1	1
033	1	5	6
035	0	1	1
037	1	7	8
041	0	5	5
043	0	2	2
047	2	3	5
071	0	1	1
077	0	3	3
101	0	1	1
102	0	1	1
105	0	1	1
133	0	1	1
137	1	2	3
177	1	0	1
301	0	2	2
303	0	4	4
333	0	1	1
337	0	1	1
403	0	1	1
407	1	2	3
437	3	0	3
447	0	1	1
477	1	0	1

Figure 2. UPGMA-dendrogram od 142 field isolates. DNA fingerprints derived with MGR586 and pMG6015 were conbined and analyzed together. Clonal lineages were assiged using differential strains as internal controls. Bootstrapping percentages were indicated at forks.

TABLE 3. Frequency of invasive strains toward Japanese differencial cultivars in each clonal lineage

C.L.*	Total	Resistance Genes								
		$Pi\text{-}s^b$	$Pi\text{-}a$	$Pi\text{-}i$	$Pi\text{-}k$	$Pi\text{-}k^m$	$Pi\text{-}z$	$Pi\text{-}ta$	$Pi\text{-}ta^2$	$Pi\text{-}z^t$
K04	127	126	82	57	30	23	15	14	8	4
K33	15	15	15	13	7	7	4	1	0	5
Total	142	141	97	70	37	30	19	15	8	9

*clonal lineage.

Based upon these data, we concluded that a main factor for the pathogenic race diversity in Japan was the mutation of the avirulence genes, because virulent alleles of all putative avirulence genes toward blast resistance genes of Japanese blast differentiating system were distributed among the major lineage JBLA-K04. We must consider parasexual recombination as a factor to make diverse combination of the mutations of avirulence genes, and at least one mutation in an avirulence gene was necessary to make the donor of the new mutation allele. Sexual recombination would not be a process for the distribution of mutations, because most of rice-pathogenic field isolates were considered to be sterile in sexual cross. The absence of the intermediate lineage in the strains from the district where the both of clonal lineage exist might be an evidence for the absence of large scale recombinations between different clonal lineages.

Although the fungal population used in this study might not be suitable to draw conclusion on the field population of the fungus, the result indicated clearly that the clonal lineage JBLA-K04 is distributed all over Japan and constructs the main skeleton of it. From the study of Japanese differential strains which were isolated before 1960s, it was suggested that at least five clonal lineages of the pathogen existed in Japan, thus this lineage simplification would have occurred in the period from 1960 to 1980. During the period, many blast resistant cultivars were introduced in the field and some cases of breakdown were reported (Kushibuchi 1997). The difference in the pathogenicity profile between JBLA-K04 and JBLB-K33 might be the cause of the lineage simplification. On the other hand, however, population of rice cultivars in the field was also changed in the same period. The rate of high grain quality cultivars such as cultivar Koshihikari which have a few major blast resistance genes and low level of field resistance had increased remarkably, and this consequently required more fungicide treatments (Kushibuchi 1997). Thus consider some kinds of fungicides might be candidates for the cause of the lineage simplification.

3. Electrophoretic Karyotyping

Electrophoretic karyotyping is a useful method based on Pulsed-Field Gel Electrophoresis to know karyotypes of microbes whose chromosomes are not easy to be observed by microscope. Already, some features of electrophoretic karyotype of M. grisea have been identified (Valent et al. 1991, Talbot et al. 1993, Orbach et al. 1996). 1) usually n=6 or 7, 2) rice pathogens conserves mini-chromosomes smaller than 2Mb which does not segregate normally in the sexual cycle, and 3) highly polymorphic within clonal lineages. This CLPs (chromosome length polymorphisms) are also found in other pathogens of plants and animals (Zolan, 1995). In this chapter, we will present some data of electrophoretic karyotyping of Japanese field isolates and a model approach to the mechanism which introduce CLPs to the genome, using a karyotypic mutant.

3.1. KARYOTYPES OF FIELD ISOLATES

Figure 3 shows electrophoretic karyotyping of 17 strains belonging to JBLA-K04 (Sone *et al.* 1997b). The features of the karyotypes which described before were conserved in Japanese strains. The karyotypes were highly polymorphic but interestingly, some similarities existed among strains which belong to homologous or identical pathogenic races. For instance, two isolates in group 007 lacked a 4.7 Mb band. The two isolates of group 303 did not have any mini-chromosome band. Isolates in group 337 lacked a band of 3.5 Mb. This result might suggest the relationship between chromosomal rearrangement and pathogenic race. Further analysis with more strains is underway.

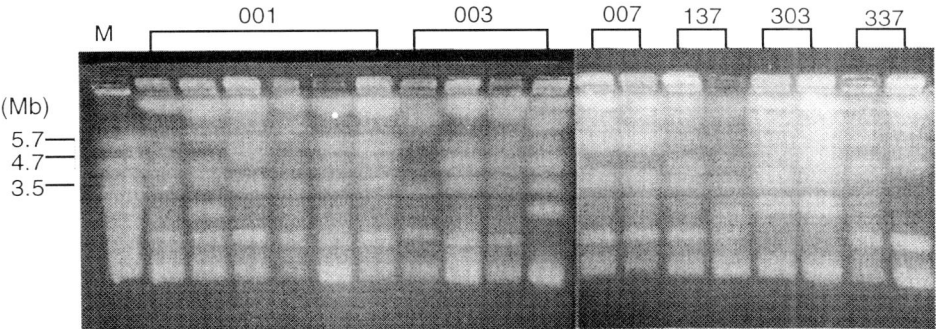

Figure 3. Chromosome length polymorphism of strians belong to JBLA-K04 (Sone *et al.* 1997b). Lane M contains *S. pombe* chromosomal sized DNA. Numbers avobe lanes indicates pathogenicity group, which contains strains of semi-homologous pathogenic races.

3.2. KARYOTYPIC MUTATION

During the analysis of electrophoretic karyotype, Sone *et al.* (1997a) found a reproducible karyotypic mutation from strain Ina168. The mutation is illustrated in Figure 4. In the mutational event, a chromosome (Band IIIb) in karyotype α suffered a deletion of 1Mb DNA to be chromosome-v in karyotype α'. This mutation occured at high frequency (>10%) in uni-direction; α to α'. Although the mutation was neutral with respect to the pathogenicity to Japanese differential cultivars, the process involved in the mutation might be common among other chromosomal rearrangements in the genome.

We found that rDNA cluster located on the same chromosome by PCR-amplified rDNA probe. However, the size of rDNA clusters were not different between the two karyotypes (Figure 5), indicating that the rearrangement in rDNA cluster was not the cause of the chromosomal deletion, unlike other rearrangements of rDNA-bearing chromosomes (Zolan, 1996). We should consider other elements for the DNA-deleting mechanism, such as transposons or other repetitive DNAs.

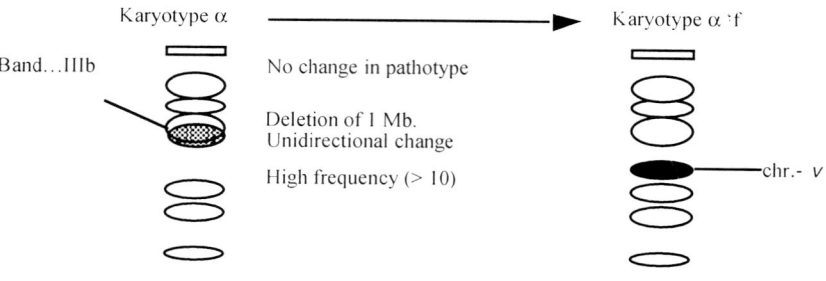

Figure 4. The α to α' karyotypic mutation in strain Ina168 (Sone *et al.* 1997a).

4. Conclusion

We have discussed about the pathogen's genetic diversity mainly by clonal lineage analysis of Japanese strains and its variation about pathogenicity and karyotype. Clonal lineage analysis, provides us many information not only on the population genetics of the pathogens, but also on the mechanisms introducing pathogenic variations. From DNA fingerprinting of Japanese strains, we concluded that the diversity of pathogenicity was caused primary by mutations in a particular clonal lineage JBLA-K04.

The detail mechanisms of such mutations are still unknown. From several reports of mutations in avirulence genes (Valent *et al.*, 1994, Sweigard *et al.* 1995, Mandel *et al.* 1997), mutation at polynucleotide level such as deletion of DNA, and insertion of transposons might be responsible for variability of genes in addition to simple base change. Some reports of unclonable DNA around host-specificity locus might indicate the existence of specific system which introduces such mutations (Mandel *et al.* 1997, Farman *et al.* 1998). It would be important to study not only on avirulence genes itself, but also on factors which introduce genetic rearrangement, such as transposons other repetitive DNAs.

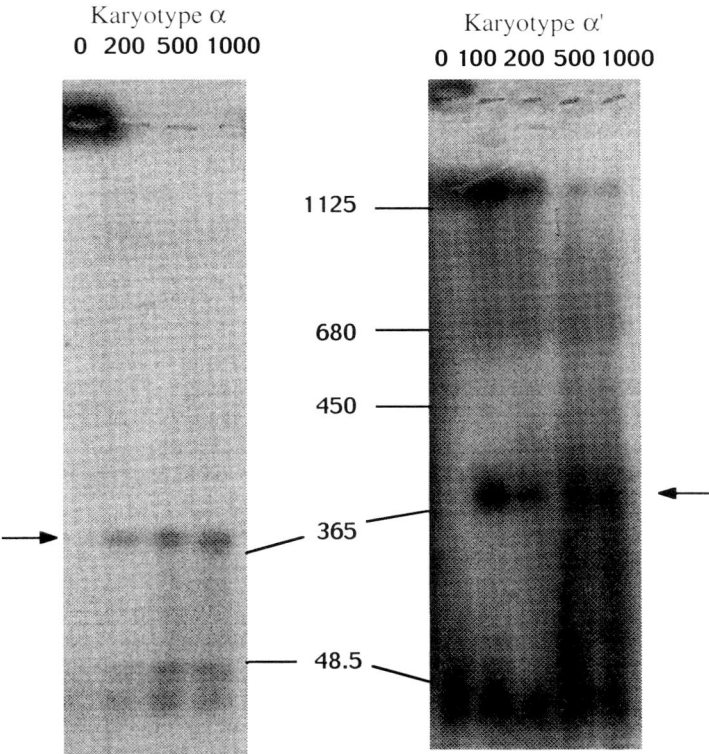

Figure 5. The size of rDNA cluster of strains of karyotypes α and α'. The numbers above lanes indicate units of *Bam* HI used for digestion of chromosomal-sized DNA samples. Samples were *Bam* HI (no cutting site in rDNA repeat unit) digested, CHEF-electrophoresed and hybridized with PCR-amplified rDNA probe. Arrows indicates the size of rDNA cluster (ca. 400kb). About 50 copies of rDNA repeat unit (ca.8 kb) included in the cluster. Sizes are given in kb.

   In this report, Japanese blast fungus population was revealed to be a rare example in which various pathogenic mutations had been accumulated, and thus they are suitable materials for the study on pathogenic mutations, its mechanisms, and dispersion of such mutations. Analysis of CLPs of field isolates and strain Ina168 will provide us an insight into the process of chromosomal rearrangements and its role in pathogenic mutations.

5. Acknowledgement

We thank Dr. S. Kiyosawa, Dr. M. Iwano, Dr. S. Koizumi, and Dr. Y Fujita for providing us fungal strains and many helpful suggestions. We are indebted to Dr. B. Valent for the kind gift of the probe MGR586. This work is partly supported by Grants-in-Aid from the Ministry of Education, Science and Culture of Japan (08456042 and 10-2369), and JSPS Research Fellowships for Young Scientists.

6. References

Chen, D. Zeigler, R.S., Leung, H. and Nelson, R.J.(1995) Population Structure of *Pyricularia grisea* at two screening sites in the Philippines. *Phytopathology* **85**, 1011-1020.

Felsenstein, J. (1985) Confidence limits on phylogenies: an approach using the bootstrap. *Evolution* **39**, 783-791.

Farman, M.L., and Leong, S.A. (1998) Chromosome walking to the *AVR1-CO39* avirulence gene of *Magnaporthe grisea*: discrepancy between the physical and genetic map. *Genetics* **150**, 1049-1058.

Hamer, J.E., Farrall, L., Orbach, M.J., Valent, B. and Chumley, F.G.(1989) Host species-specific conservation of a family of repeated DNA sequences in the genome of a fungal pathogen. *Proc. Natl. Acad. Sci. USA* **86**, 9981-9985.

Kempken, F. and Kück, U. (1998) Transposons in filamentous fungi-facts and perspectives. *BioEssays* **20**, 652-659.

Kiyosawa, S. and Ando, I. (1997) Inheritance of disease resistance, in M. Takane *et al* (eds.), *Science of the Rice Plant* vol. 3, Genetics, Food and Agriculture Policy Research Center, Tokyo, pp. 470-500.

Kushibuchi, K.(1997) Historical changes in rice cultivars, in M. Takane *et al* (eds.), *Science of the Rice Plant* vol. 3, Genetics, Food and Agriculture Policy Research Center, Tokyo, pp. 837-875.

Levy, M., Romao, J., Marchetti, M.A., and Hamer, J.E. (1991) DNA fingerprinting with a dispersed repeated sequence resolves pathotype diversity in the rice blast fungus. *The Plant Cell* **3**, 95-102.

Levy, M., Correa-Victoria, F.J., Zeigler, R.S., Xu, S. and Hamer, J.E. (1993) Genetic diversity of the rice blast fungus in a disease nursery in Colombia. *Phytopathology* **83**, 1427-1433.

Mandel, M.A., Crouch, V.W., Gunawardena, U.P., Harper, T.M. and Orbach, M.J. (1997) Physical mapping of the Magnaporthe grisea AVR1-MARA locus reveals the virulent allele contains two deletions. *Mol. Plant-Microbe Interact.* **10**, 1102-1105.

Namai, T., Fukushima, R., Ishibashi, Y., Ohba, J., and Togashi, J. (1997) Variation   in pathogenicity of incompatible races of rice blast fungus, Pyricularia oryzae Cavara on Panicles of Newly Developed Rice cvs. Haenuki and Domannaka. *Ann. Phytopathol. Soc. Jpn.* **63**, 37-43 (in Japanese).

Nitta, N.M., Farman, M.L. and Leong, S.A. (1997) Genome organization of *Magnaporthe grisea*: integration of genetics maps, clustering of transposable elements and identification of genome duplications and rearrangements, *Theor. Appl. Genet.* **95**, 20-32.

Orbach, M.J., Chumley, F.G. and Valent, B. (1996) Electrophoretic karyotypes of Magnaporthe grisea pathogens of diverse grasses. *Mol. Plant-Microbe Interact.* **9**, 261-271.

Ou, S.H. (1980) Pathogen variability and host resistance in rice blast disease, *Annu. Rev. Phytopathol.* **18**, 167-187.

Roumen, E., Levy, M. and Notthghem, J.L. (1997) Characterization of the European pathogen population of *Magnaporthe grisea* by DNA fingerprinting and pathotype analysis. *Euro. J. Plant Pathol.* **103**, 363-371.

Sone, T., Suto, M. and Tomita, F. (1993) Host species-specific repetitive DNA sequence in the genome of *Magnaporthe grisea*, the rice blast fungus. *Biosci. Biotech. Biochem.* **57**, 1128-1230.

Sone, T., Abe, T., Suto, M. and Tomita, F. (1997a) Identification and characterization of a karyotypic mutation in *Magnaporthe grisea. Biosci. Biotech. Biochem.* **61**, 81-86.

Sone, T., Yoshida, N., Suto, M. and Tomita, F. (1997b) DNA fingerprinting and electrophoretic karyotyping of Japanese isolates of rice blast fungus. *Ann.   Phytopathol. Soc. Jpn.* **63**, 155-163.

Sweigard, J.A., Carroll, A.M., Kang, S., Farrall, L., Chumley, F.G. and Valent, B. (1995) Identification, cloning, and characterization of *Pwl2*, a gene for host species specificity in the rice blast fungus. *The Plant Cell* **7**, 1221-1233.

Talbot, N.J., Salch, Y.P., Ma, M. and Hamer, J.E. (1993) Karyotypic variation within clonal lineages of the rice blast fungus, *Magnaporthe grisea. Appl. Environ. Microbiol.* **59**, 585-593.

Valent, B. and Chumley, F. G. (1991) Molecular genetic analysis of the rice blast fungus *Magnaporthe grisea*. *Annu. Rev. Phytopathol.* **29**, 443-467.

Valent, B. and Chumley, F. G. (1994) Avirulence genes and mechanisms of genetic instability in the rice blast fungus. in Zeigler, R.S. *et al.* (eds.), *Rice Blast Disease*, CAB International, Wallingford, pp. 111-134.

Xia, J.Q., Correll, J.C., Lee, F.N., Marchetti, M.A. and Rhoads, D.D. (1993) DNA fingerprinting to examine microgeographic variation in the *Magnaporthe grisea* (*Pyricularia grisea*) population in two rice fields in Arkansas. *Phytopathology* **83**, 1029-1035.

Yap, I. and Nelson, R. J. (1996) Winboot: a program for performing bootstrap analysis of binary data to determine the confidence limits of UPGMA-based dendrograms. IRRI Discussion Paper Series No. 14. International Rice Research Institute, P. O. Box 933, Manila, Philippines.

Yamada, M., Kiyosawa, S., Yamaguchi, T., Hirano, T., Kobayashi, T., Kushibuchi, K. and Watanabe, S. (1976) Proposal of a new method for differentiating races of *Pyricularia oryzae* Cavara in Japan. *Ann. Phytopathol. Soc. Jpn.* **42**, 216-219.

Yoshino, R. (1988) Present status of occurrence and control of blast disease in Japan. *Japan Pesticide Information* **52**, 3-8.

Zeigler, R.S., Tohme, J., Nelson, R., Levy, M. and Correa-Victoria, F.J. (1994) Lineage exclusion: a proposal for linking blast population analysis to resistance breeding. in Zeigler, R.S. *et al.* (eds.), *Rice Blast Disease*, CAB International, Wallingford, pp. 267-292.

Zeigler R.S., Cuoc, L.X., Scott, R.P., Bernardo, M.A., Chen, D.H., Valent, B. and Nelson, R.J. (1995) The relationship between lineage and virulence in *Pyricularia grisea* in the Philippines. *Phytopathology* **85**, 443-451.

Zeigler, R.S., Scott, P.P., Leung, H., Bordeos, A.A., and Nelson, R.J. (1997) Evidence of parasexual exchange of DNA in the rice blast fungus challenges its exclusive clonality. *Phytopathology* **87**, 284-294.

Zolan, M.E. (1995) Chromosome-length polymorphism in fungi. *Microbiol. Rev.* **59**, 686-698.

DIFFERENTIAL CHANGES IN HOST SPECIFICITY AMONG MGR586 DNA FINGERPRINT GROUPS OF *PYRICULARIA GRISEA*

J. C. CORRELL, T. L. HARP, J. C. GUERBER, and F. N. LEE
Department of Plant Pathology, University of Arkansas
Fayetteville, AR 72701 USA

1. Abstract

The development of new races of the rice blast pathogen continues to circumvent the best efforts of plant breeders to develop blast resistant cultivars. Sampling field populations of the rice blast fungus from 1990-1997 has indicated that four MGR586 DNA fingerprint groups (A, B, C, and D) occur in the contemporary rice blast pathogen population in Arkansas. All field isolates examined in fingerprint groups A and C had an IB-49 virulence phenotype whereas most isolates in fingerprint groups B and D had an IC-17 virulence phenotype. However, several field isolates in fingerprint group B had an IG-1 or an IC-1k virulence phenotype; race IC-1k was virulent on the more recently released resistant cultivar Katy. Because race diversity was greater within the B fingerprint group based on evidence from the field isolates, it was hypothesized that spontaneous mutations affecting host specificity could occur more readily among isolates in fingerprint group B. Five representative isolates, one each from DNA fingerprint groups A, C, and D, and two isolates (an IG1 and an IC-17 isolate) in fingerprint group B were inoculated on the cultivars M201, Newbonnet, and Katy. A marked strain (a sulfate nonutilizing mutant) of each isolate was used, along with DNA fingerprints, to rule out any possibility of cross-contamination during the inoculation tests. A single-spored isolate was recovered from 15-20 rare lesions observed on the resistant cultivar Katy after inoculation with all five isolates. None of the single-spore isolates recovered from the A or C isolate inoculations were virulent when re-inoculated onto Katy whereas 2 of 20 single-spored isolates from the D isolate inoculation were virulent when re-inoculated onto Katy. However, 13 of 16 single-spored isolates recovered from the B isolate (race IG1) inoculation and 17 of 19 single spored isolates recovered from the B isolate (race IC17) inoculation were virulent when re-inoculated onto Katy. The data indicate that spontaneous mutation to virulence on Katy occurred at a much higher frequency in the two isolates in the B fingerprint group than the isolates in fingerprint groups A, C, or D. Inoculations with additional isolates in all four fingerprint groups indicated that the frequency of the spontaneous host-specificity mutation was consistently higher among some, but not all isolates, in the B fingerprint group. Although circumstantial, the data indicate that virulent phenotypes capable of infecting Newbonnet and Katy originated from the resident population.

2. Introduction

The origin of new races of the rice blast pathogen continues to circumvent the best efforts of plant breeders to develop durable blast resistant cultivars. Races which can

D. Tharreau et al. (eds.), Advances in Rice Blast Research, 234–242.

overcome resistance can result from a.) new races which are introduced into a given geographical area; b.) indigenous races which occur at a low frequency in the population but increase upon cultivation of a new cultivar; or c.) changes in virulence among indigenous pathogen genotypes. The pathogenic variability of *P. grisea* has been examined in considerable detail (1,5,12,13). We now know that specific virulent mutants can be selected from rare lesions on resistant cultivars and arise due to specific changes in avirulence genes (8,14,15). The objectives of this study were to determine if changes in virulence could occur among isolates representing four contemporary MGR586 DNA fingerprint groups of *P. grisea* commonly found in Arkansas and determine if the frequency of these changes differed among the different groups.

Although eight MGR586 DNA fingerprint groups (A through H) have been recovered from rice in Arkansas, blast surveys conducted between 1990 and 1998 indicate that four groups (A, B, C, and D) occur in the contemporary population (18) (Fig. 1). Furthermore, there is a close correspondence between race and MGR586 group among isolates from Arkansas (3,16).

Changes in race frequency have occurred due to changes in host genotypes being widely cultivated (11). In Arkansas, race IG-1 was common on cultivars grown in the 1970's and early 1980's (9). By 1986, the high yielding cultivar Newbonnet became popular and was planted on approximately 70% of the acreage in the state. Newbonnet was highly susceptible to races IC-17 and IB-49 which became prevalent by the late 1980's. The cultivar, Katy which is resistant to races IC-17 and IB-49, was grown commercially in 1989 and by 1993 was being grown on 20 percent of the acreage in Arkansas. By 1994, a number of isolates were recovered which could cause severe disease on Katy (IC-1k) (2).

3. Materials and Methods

3.1 ISOLATES, MGR586 DNA FINGERPRINTS GROUPS, AND RACE

Isolates recovered from Arkansas, which belonged to one of four MGR586 DNA fingerprint groups and represented races IG-1, IC-17, IC-1k, and IB-49 were examined (Table 1, Fig. 1). Field isolates collected between 1975 and 1994 which belonged to DNA fingerprint group B indicated that race diversity was present within this group but not detected within the other three (Table 2). Thus, greenhouse tests were conducted to determine if differential changes in host specificity could occur among isolates from each of the four MGR586 DNA fingerprint groups (A, B, C, and D).

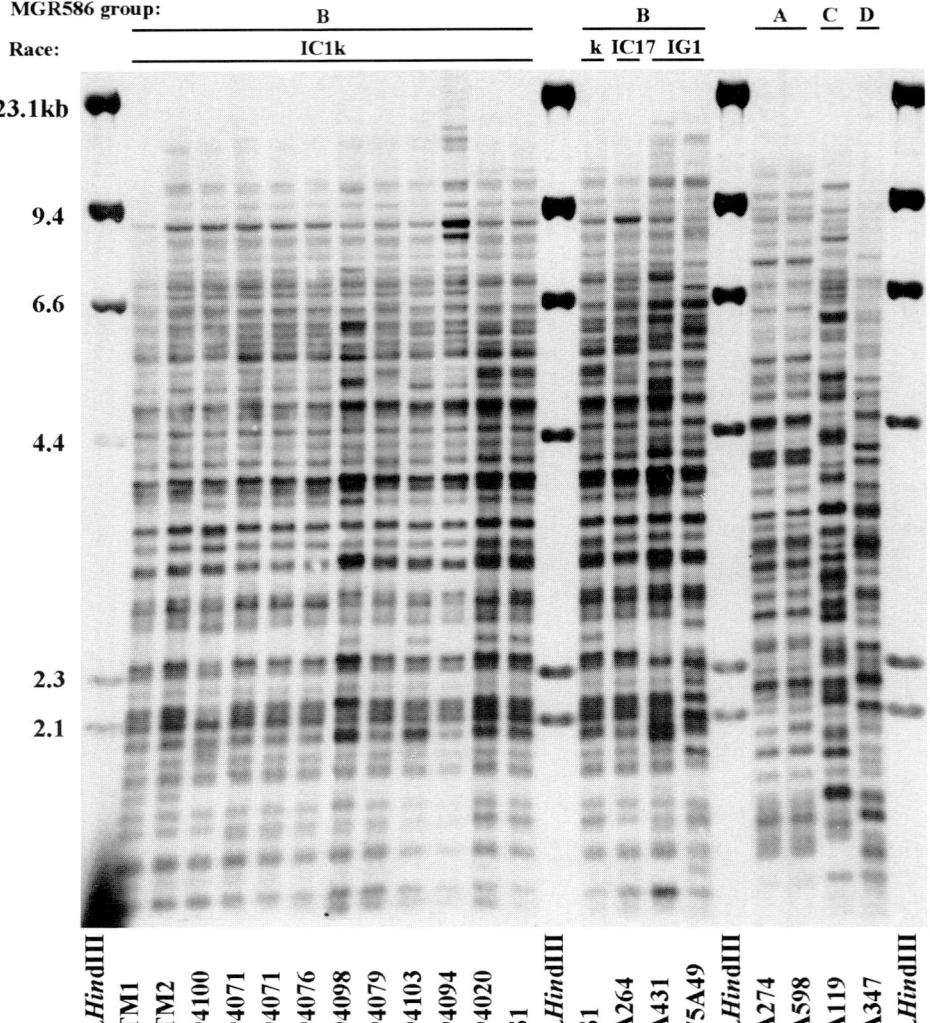

Table 1. Mean disease reactions of five isolates representing four races of *Pyricularia grisea* from Arkansas.

	Isolate (Race)									
	75A49(1G-		24(1G-1)		8(1B-49)		17(1C-17		S-1 (1C-1†	
Cultivar	Mean[a]	(Range[b])	Mean	(Range)	Mean	Range	Mean	Range	Mean	Range
M204	6.7	(5.3-7.8)	6.8	(5.8-7.5)	7.8	(5.8-8.8)	6.7	(5.3-7.5)	6.4	(5.3-7.8)
Starbonnet	7.0	(5.5-8.0)	7.0	(6.0-8.5)	7.0	(5.0-8.3)	7.3	(5.5-8.0)	6.8	(5.8-7.5)
Tebonnet	4.2	(3.5-4.8)	4.1	(2.8-5.0)	4.8	(3.7-5.5)	4.3	(2.3-5.3)	4.4	(3.7-5.3)
Newbonnet	1.6	(0.5-2.3)	1.0	(0.5-2.0)	6.9	(5.3-8.8)	7.0	(6.3-8.0)	6.7	(5.5-7.8)
Lemont	1.1	(0.5-2.5)	0.5	(0.0-1.3)	6.7	(5.5-8.5)	6.7	(5.8-7.3)	6.4	(5.5-7.0)
Zenith	1.6	(1.0-2.5)	1.2	(0.3-2.5)	6.1	(4.3-7.8)	1.4	(0.3-2.5)	1.3	(0.3-2.3)
Mars	2.3	(1.7-3.0)	2.7	(2.3-3.0)	7.8	(6.0-8.8)	2.7	(2.0-3.3)	2.3	(1.3-3.8)
Katy	1.7	(0.5-3.0)	1.9	(0.5-3.5)	2.2	(0.3-3.0)	1.9	(0.8-4.5)	6.2	(5.3-7.0)
Katy	1.3	(0.3-2.3)	1.3	(0.3-2.5)	2.0	(0.5-3.5)	1.7	(0.3-5.5)	6.2	(5.3-7.3)
Zenith	1.3	(0.0-2.0)	1.5	(0.7-2.0)	4.2	(3.3-5.8)	1.1	(0.0-2.8)	1.5	(0.3-4.0)
NP125	1.5	(0.3-2.5)	1.7	(1.0-2.8)	6.2	(5.3-7.5)	5.1	(3.0-7.3)	4.8	(3.8-6.3)
Usen	1.2	(0.3-2.8)	1.3	(0.7-3.0)	4.4	(3.5-5.3)	2.0	(1.0-4.8)	4.5	(3.5-5.3)
Dular	1.0	(0.3-2.0)	0.9	(0.0-1.5)	6.4	(5.5-7.3)	6.3	(5.3-7.3)	6.2	(5.5-6.8)
Kanto 51	4.2	(3.8-4.5)	3.9	(3.0-4.5)	6.1	(5.5-7.3)	5.9	(4.5-6.8)	5.8	(5.0-6.5)
Sha Tiao Tsao	4.0	(2.0-5.3)	4.2	(2.7-5.0)	4.7	(3.0-6.8)	4.2	(3.5-5.3)	4.4	(2.8-5.5)
Caloro	2.7	(1.3-4.0)	3.2	(1.5-4.3)	5.2	(4.3-6.5)	3.4	(2.0-4.3)	4.6	(3.5-5.3)

[a] Each number is the mean disease reaction of six inoculation tests on a scale of 0-9 where (0-3.0) = resistant; (3.1- 4.0) = intermediate; and (>4.0)= susceptible

[b] Range is the minimum and maximum mean disease rating for the six inoculation tests.

3.2 GREENHOUSE INOCULATION TESTS

Greenhouse inoculation tests were conducted on the cultivars M204, Newbonnet, and Katy. Each of the races could be easily distinguished based on disease reactions under standardized conditions as previously described (Table 2).

Table 2. MGR586 Disease reactions of the various isolates

MGR586 DNA grouplsolate	Race		Year Collected	Cultivar [a]		
				M204	Newbonnet	Katy
A	3/3	1B-49	1993			-
B	20/1	1G-1	1995		-	-
B	75A9/3	1C-17	1993			-
B	S-1	1C-1k	1994			
C	32/1	1B-49	1993			-
D	58/1	1C-17	1993		-	-

[a] Disease reaction was based on a scale of 0-9. "+" = disease rating 4.0; "-" disease rating ‹ 4.0.

To determine if changes in virulence could occur among various fungal genotypes, rare susceptible lesions (Fig. 2) were recovered from otherwise resistant cultivars as described by Valent (15). The lesions were incubated and a single monoconidial isolate was recovered per rare lesion and examined for virulence on the three cultivars in comparison with the parental isolate.

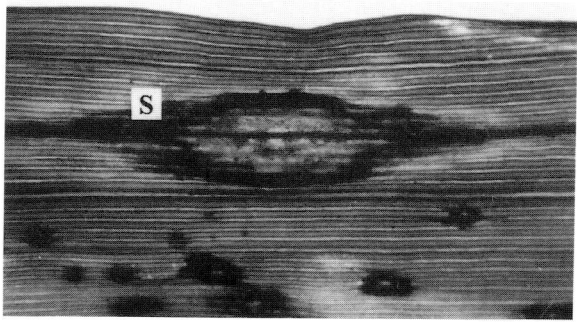

Figure 2. Katy showing a rare susceptible "S" lesion near several resistant reactions.

To confirm that rare lesions were caused by the original parental genotype, sulfate non-utilizing (*sul*) mutants of the wildtype strains were used. Sul mutations have previously been shown to be useful markers with *P. grisea* and do not affect pathogenicity (7,10). In addition, the MGR586 fingerprint (6) of isolates was also confirmed.

4. Results and Discussion

Monoconidial isolates were recovered from rare lesions on either Newbonnet or Katy when inoculated with either of two isolates of race IG-1 (75A49 and 24), both in MGR586 group B) collected in 1975 and 1993. The majority of the monoconidial isolates recovered were found to be virulent on the cultivar from which they were recovered (Table 3 and 4, Newbonnet data not shown). These data indicate that distinct host specificity changes can occur within a given fungal genotype. These "race-shifts" were observed in all seven isolates in MGR586 DNA group B, three of the six group D isolates, but none of the group A or C isolates (Table 3 and 4). In addition, the frequency of "race-shift" isolates was considerably higher among the isolates in MGR586 group B.

Table 3. Frequency by which monoconidial isolates recovered from rare lesions on Katy exhibited changes in host specificity.

MGR586 DNA group	Parental isolate[a]	Number of lesions recovered[b]	Number of isolates virulent[c]	%
A	3/3	20	0	0%
B	20/1	19	17	89%
B	75A49/3	16	13	81%
C	32/1	20	0	0%
D	58/1	20	2	10%

[a] *sul* mutant used.
[b] Number of "susceptible" lesions from which a single monoconidial isolate was recovered.
[c] Number of monoconidial isolates which showed a distinct change in host specificity (were virulent on Katy).

Three lines of evidence, historical, population, and experimental were examined to determine the origin of races of *P. grisea* which could cause disease on the cultivar Katy in Arkansas. Historical evidence indicated that there could have been a progression of races in Arkansas from IG-1 to IC-17, which could cause disease on Newbonnet (released in 1983), to IC-1k which could cause disease on Katy (released in 1989) based on the cultivars grown, the acreage planted, and when the cultivars were released (9).

Table 4. Frequency by which monoconidial isolates recovered from rare lesions on Katy exhibited changes in host specificity in five experiments.

MGR586 DNA group	Parental isolate[a]	Number of lesions recovered[b]	Number of isolates virulent[c]	%
A	6/1	0,3,0,5,2	0,0,0,0,0	0%
A	8/1	1,0,0,2,0	0,0,0,0,0	0%
A	9/101	0,0,0,3,2	0,0,0,0,0	0%
A	12/1	0,0,0,-,0	0,0,0,-,0	0%
A	15/1	0,1,0,-,0	0,0,0,-,0	0%
B	18/1	1,3,1,6,0	0,3,1,6,0	90%
B	23/1	0,0,0,1,0	0,0,0,1,0	100%
B	24/1	3,1,1,-,0	3,1,1,-,0	100%
B	25/104	12,3,6,3,4	9,2,5,3,4	82%
B	26/1	5,0,1,-,1	5,0,1,-,1	100%
C	34/1	0,0,0,0,0	0,0,0,0,0	0%
C	35/1	0,0,0-,0	0,0,0,-,0	0%
C	38/1	0,0,0,0,0	0,0,0,0,0	0%
C	44/1	0,0,0,-,0	0,0,0,-,0	0%
C	45/101	0,0,0,0,0	0,0,0,0,0	0%
D	47/1	0,0,0,-,0	0,0,0,-,0	0%
D	54/103	0,0,1,-,1	0,0,1,-,1	100%
D	56/2	0,0,0,1,0	0,0,0,0,0	0%
D	57/14	0,1,0,-,0	0,0,0,-,0	0%
D	60/1	0,3,0,2,6	0,3,0,2,6	100%

[a] *sul* mutant used.
[b] Number of "susceptible" lesions from which a single monoconidial isolate recovered from each of five experiments.
[c] Number of monoconidial isolates which showed a distinct change in host specificity (were virulent on Katy).

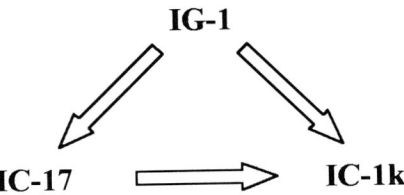

Figure 3. Changes in host specificity which were demonstrated with specific fungal genotypes on the cultivars M204, Newbonnet, Katy.

The population data indicated that all three races (IG-1, IC-17, and IC-1k all occur within fingerprint group B (Fig. 3). Furthermore, all IC1k isolates recovered from Arkansas in 1994 from different geographical areas belonged to fingerprint group B. Experimentally, changes in host specificity could be demonstrated among isolates in the B and D fingerprint groups, but the frequency of changes was considerably higher among isolates in the B group (Table 3 and 4). Thus, the three lines of evidence indicate that isolates capable of causing disease on Katy under field conditions likely originated from indigenous genotypes in Arkansas. However, in addition to changes in host specificity, additional selection for fitness would likely occur. Therefore, proof that changes in host specificity do not interfere with fitness characteristics needs to be demonstrated. Even if selection for fitness were severe, it appears likely that isolates which are pathogenic on Katy would originate from group B isolates provided there is not a direct linkage between changes in virulence and fitness.

The frequency by which changes in host specificity could occur was distinctly higher among isolates in the B group. These data are consistent with isolates in this group being hypervariable for a number of other genetic characteristics. Perhaps this lineage is more variable due to the number or frequency of movement of certain mobile genetic elements which also are capable of disrupting avirulence genes. However, the direct affect on avirulence genes and the movement of such genes within and between lineages need further investigation (4,17,18).

5. References

Bonman, J. M, Vergel de Dios, T. I., Bandong, J. M., and Lee, E. J. 1986. Pathogenic variability of monoconidial isolates of Pyricularia oryzae in Korea and in the Philippines. Plant Dis. **71**, 127-130.

Correll, J.C., Guerber, J.C., and Lee, F.N. 1995. Virulence instability of an MGR-DNA fingerprint group of the rice blast pathogen. Mycol. Soc. Newsletter **46**, 9.

Correll, J.C., Guerber, J.C., and Lee, F.N. 1995. Examination of virulence among isolates in MGR586 DNA fingerprint groups of the rice blast pathogen, *Pyricularia grisea*, in Arkansas. Phytopathology **85**, 1199.

Correll, J.C., Harp, T. L., Lee, F. N., and Zeigler, R. S. 1998. The use of vegetative compatibility to examine population structure in *Pyricularia grisea*. Phytopathology **88**, S19.

Giatgong, P, and Frederiksen, R. A. 1969. Pathogenic variability and cytology of monoconidial subcultures of *Pyricularia oryzae*. Phytopathology **59**, 1152-1157.

Hamer, J.E., Farrall, L., Orbach, M.J., Valent, B., and Chumley, F.G. 1989. Host species-specific conservation of a family of repeated DNA sequences in the genome of a fungal plant pathogen. Proc. Natl. Acad. Sci. USA **86**, 9981-9985.

Harp, T.L., and Correll, J.C. 1998. Recovery and characterization of spontaneous, selenate-resistant mutants of *Magnaporthe grisea*, the rice blast pathogen. Mycologia **90**, (in press)

Kang, S. Sweigard, J. A., and Valent, B. 1995. The PWL host specificity gene family in the blast fungus, *Magnaporthe grisea*. Mol. Plant-Microbe Interact. **8**, 939-948.

Lee, F.N. 1994. Rice breeding programs, blast epidemics and blast management in the United States. Rice Blast Disease 489-500 pp. Zeigler, R.S., Leong, S., and Teng, P.S. eds. Commonwealth Agricultural Bureaux International, Willingford, UK.

Long, D. H., TeBeest, D. O., and Correll, J. C. 1998. The effect of plant age on the spatial and temporal dynamics of rice blast in Arkansas. (this volume).

Marchetti, M.A. 1994. Race-specific and rate-reducing resistance to rice blast in U. S. cultivars. In Rice Blast Disease 231-244 pp. Zeigler, R.S., Leong, S., and Teng, P.S. eds. Commonwealth Agricultural Bureaux International, Willingford, UK.

Ou, S.H. 1980. Pathogen variability and host resistance in rice blast disease. Annu. Rev. Phytopathol. **18**, 167-187.

Ou, S. H., and Ayad, M. R. 1968. Pathogenic races of Pyricularia oryzae originating from single lesions and monoconidial cultures. Phytopathology **58**, 179-182.

Shull, V., and Hamer, J.E. 1994. Genomic structure and variability in *Pyricularia grisea*. In Rice Blast Disease 65-86 pp. Zeigler, R.S., Leong, S.A., and Teng, P.S. eds. Commonwealth Agricultural Bureaux International, Willingford, U.K.

Valent, B. and Chumley, F.G. 1994. Avirulence genes and mechanism of genetic instability in the rice blast fungus. Rice Blast Disease 111-134 pp. Zeigler, R. S., Leong, S., and Teng, P.S. eds. Commonwealth Agricultural Bureaux International, Willingford, UK.

Xia, J. Q., Correll, J. C., Lee, F. N., Rhoads, D. D., and Marchetti, M. A. 1993. DNA fingerprinting to examine variation in the *Magnaporthe grisea (Pyricularia grisea)* population in two rice fields in Arkansas. Phytopathology **83**, 1029-1035.

Zeigler, R. S. 1998. Recombination in Magnaporthe grisea. Ann. Rev. Phytopath. **36**, 249-275.

Zeigler, R.S., Scott, R. P., Leung, H., Bordeos, A. A., Kumar, J., and Nelson, R.J. 1997. Evidence of parasexual exchange of DNA in the rice blast fungus challenges its exclusive clonality. Phytopathology **87**, 284-294.

EVIDENCE FOR RECOMBINATION IN *MAGNAPORTHE GRISEA* RICE PATHOGENS: A CASE STUDY FROM THE INDIAN HIMALAYAS.

R. S. Zeigler[1], J. Kumar[2], R. J. Nelson[1,3] and H. Leung[1].
[1]*International Rice Research Institute, P. O. Box 933, Manila, Philippines.* [2]*Hill Campus, G. B. Pant University, Ranichauri 249199, Uttar Pradesh, India.* [3]*Present affiliation: International Potato Center, Lima, Peru*

The balance between asexual, clonal reproduction and recombination in microbial populations and their relative contributions to population structure and dynamics is a topic of hot debate (see Tibayrenc et al., 1991 and Maynard Smith et al., 1993). Many fungi are capable of both asexual and sexual reproduction, but it is not clear under which circumstances their sexual recombination potential is expressed (Lenski, 1993; Kohn 1995). Fungi that previously were thought to reproduce exclusively asexually in nature have been shown to have population structures more consistent with populations to which recombination contributes significantly (see Burt et al. 1996; Geiser, et al. 1994).

The often enormous asexual reproductive capacity of many fungi, in general, and *Magnaporthe grisea* in particular, can result in a few clones rapidly coming to dominate a population. Correcting for clonality is necessary to remove the bias imposed by differential success of recombinant progeny. However, this can reduce sample sizes and adversely affect the power of statistical tests for recombination. Ambiguity as to what, in fact, constitutes a population and the possibility that population admixtures can occur seriously hinder analyses as well. Because of the difficulties in applying any single test to determine if recombination contributes significantly to population structure, Lenski (1993) and Milgroom (1996) argue that a range of statistical and biological considerations be brought to bear on the question of the contribution of recombination to population structure.

It is important to consider that recombination is a characteristic of populations, rather than an attribute of a species. Global analyses that examine very few isolates from widely dispersed populations do not provide sufficient resolution to determine if recombination is important in some of them. Furthermore, in a species such as *M. grisea* that has undergone recent pan-global migration and has probably been subjected to extreme population bottlenecks, it is likely that some populations are indeed completely asexual, while others could retain the capacity for recombination. Indeed, Leslie and Klein (1995) argue that in fungal species such as *M. grisea*, where there is

D. Tharreau et al. (eds.), Advances in Rice Blast Research, 243–247.

both sexual and asexual capacity, female fertility will be lost in uniform environments, such as those presented by modern agriculture for crop pathogens. They further suggest that in heterogeneous environments, where the wide range of fitness variation maintained by recombination should favor its retention, female fertility should persist.

In this brief summary, we present results from an intensive multi-year study on *M. grisea* populations from rice in the Indian Himalayan region. This area was selected as the mountainous environment is very heterogeneous (across small distances) and variable (over time). Cropping systems are characterized by small fields and diverse traditional rice cultivars that in other Himalayan areas have been shown to be genetically heterogeneous for resistance to *M. grisea* (Thinlay, 1998). We argue that if sexual recombination is contributing to population structure a suite of characteristics should be observed. 1) Populations should be diverse; 2) Many unique genotypes should be found; 3) In a population where both clonal reproduction and recombination occur, they should show signs of "epidemic structure" (*sensu* Maynard Smith *et al.*, 1993); 4) Unlinked loci should be randomly associated (in gametic phase equilibrium); 5) Both mating types should be present in appreciable frequencies; and 6) Female fertility should be present. These topics will be addressed in order in the following sections, and will focus on two different populations in the region: from Matli and Ranichauri, ~ 1 hectare, each. Because of space limitations, supporting data must be drastically summarized; however, more detail can be found in Kumar et al., 1999)

Basis for Analyses. Several hundred isolates from both populations were analyzed using MGR586 (Hamer et al., 1989) and single locus sequences found on different chromosomes (Skinner et al., 1993). Isolates were grouped into putative lineages based on MGR586 similarity (Chen et al., 1995) of 70% or greater similarity. Analysis with single locus probes showed that isolates within putative lineages had identical, or near-identical, single locus profiles that differed among lineages. Cluster analyses of pair-wise comparisons among isolates from the two probe systems yielded phenograms that were essentially identical. Isolates within lineage shared apparently recent and common ancestry that differed among lineages.

1) *Population Diversity.* Shannon's Index (normalized) was used to estimate diversity at the haplotype and lineage level. As in other studies haplotypic diversity was near its maximum of 1.0 at both sites; however, the lineage diversity was much higher than in other studies. Comparing similar sampling time frames (two years in Matli) with published data from two sites in the Philippines (Chen *et al.*, 1995) and one in Colombia (Levy *et al.*, 1993) yielded the following Shannon's Index values: Matli = 0.68; Cavinti, Phil. = 0.24; Los Banos, Phil. = 0.10; Santa Rosa, Colombia = 0.29. Rarefaction analysis showed that lineage diversity at Matli was significantly greater than that observed either in the Philippines or in Colombia. Diversity similar to that of Matli was observed in Ranichauri. For these comparisons it is instructive to note that in Matli almost three times the number of lineages was detected with fewer than one third the number of isolates sampled in Colombia. Furthermore, the Colombian and Philippine sites are breeding nurseries managed to generate and maintain pathotypic diversity, while Matli samples are from farmers' fields.

2) *Unique Genotypes*. In our overall study of the Himalayas, of 45 lineages identified in our 1992-93 collection (222 isolates) 31 (69%) were found at only one site and were represented by only one or two isolates. In our continuing samples from Matli from 1992-1996, 36 lineages (unique haplotypes) have been identified, two thirds of which were detected as only one or two isolates. In 1997 an additional 12 lineages were identified, of which 11 were represented by only one isolate. In Ranichauri, 13 lineages were identified over a two year period, 8 of which were represented by a single isolate. Haplotypic diversity among isolates of well-represented lineages was very low in Ranichauri, suggesting very recent common origin. These data are in stark contrast to the few lineages reported from Colombia (Levy *et al.*, 1993), the US (Xia *et al.*, 1993) and Europe (Roumen, *et al.*, 1997), and suggest that further sampling would yield many more lineages. In a recombining population alleles of multiple locus sequences such as MGR586 should randomly mix, if they are well-distributed across the genome and meiosis functions normally. The MGR586 fingerprints of the unique products of recombination would tend towards increasing similarity. In a comparison of a similarity matrix of matching coefficients of MGR586 of all lineages from the first two years sampling of the Matli population show this to be the case. Of the 756 comparisons, over 90% had similarities $\geq$ 50%. This is similar to the observed $\sim$ 95% with similarity $\geq$ 50% in another putatively recombining population (Geiser, *et al.*, 1994) and very different from the putatively non-recombining US *M. grisea* population, where only 30% of the comparisons were $\geq$ 50% similar. Despite the presence of many unique genotypes, unique, or "private" alleles indicative of genetic isolation (Slatkin, 1985) were uncommon among the lineages.

3) *Population Dynamics Consistent with Epidemic Structure*. In Matli, samples from 1992, 1993, and 1994 each yielded new lineages, while only a few persisted from year to year. One lineage showed a steady trend of increasing in frequency in our samples. By 1995 only two lineages were represented, with the frequency of one >95%. In 1996 intensive sampling of one field (< 400 m^2) all isolates from one area of the field yielded only the predominant isolate detected the previous year. However, at the other side of the field 12 different and previously undetected lineages were observed, of which 11 were represented by only one isolate each. In Ranichauri, initial samples yielded only two lineages in the population, but in 1995 an additional 10 lineages were detected, eight of which were represented by a single isolate.

4) *Gametic Phase Equilibrium*. To eliminate the effect of clonality, we only analyzed one isolate per lineage (Chen and MacDonald, 1996; Milgroom, 1995; Maynard Smith *et al.*, 1993) for the equilibrium analyses. We applied two tests to determine if single locus sequences were randomly associated and in equilibrium. Fisher's Exact Test (Weir, 1990) allowed us to ask how many locus pairs were in equilibrium and a multilocus variance test (Brown et al., 1980) allowed us to ask if the overall association of loci differed from that expected if they were unlinked. For Fisher's exact test analyzing all alleles, as well as the more rigorous approach

of just analyzing the most common alleles (Burt, 1996), showed that the distribution of alleles for >90% of the locus pairs was consistent with random association. The multilocus variance test yielded an index of association not significantly different from zero for the lineage corrected set of isolates, which means the hypothesis of random association of loci cannot be rejected. For both tests analysis of the full data set uncorrected for clonality yielded significant association among alleles, as expected.

5) *Presence of Both Mating Types.* Mating type was determined by crossing with locally developed tester strains of opposite mating type. These testers isolates where collected from *Eleusine coracana* in Ranichauri. Mating type of the testers and of the isolates were calibrated with international nomenclature by PCR analysis (Xu and Hamer, 1995) using primers generously provided by J-R Xu. Mating type was assigned as the opposite of the tester isolate with which perithecia were developed. For approximately 100 isolates the MAT1-1:MAT1-2 ratio was 68:32 for Matli isolates and 70:30 for Ranichauri isolates. Differences among populations were striking, as in two other populations the ratios were ~ 95:5. In populations such as these where asexual reproductive capacity is high and where there can be strong selection pressure, there is no reason to expect that mating types should be in a 1:1 ratio in samples not corrected for clonality.

6) *Presence of Fertility.* Mating type assays allowed a simultaneous assessment of fertility in the two populations. In Matli and Ranichauri 57% and 73%, respectively, of the approximately 100 isolates tested from 1996 collections were sexually fertile, producing abundant fertile perithecia with tester isolates. In Matli 42% were male fertile, only, with these testers, 5% were female fertile, and the remaining 10% were hermaphrodites. In Ranichauri 46% were male fertile, 14% were female fertile, and 13% were hermaphrodites. The frequency of male only fertility is not surprising; however, female fertility combined with male sterility is interesting. Fertility appears to be a within-population characteristic. Fertile isolates wold frequently behave as sterile when paired with hermaphrodites of the opposite mating type from different populations. Thus, fertility estimates using tester isolates derived from *Eleusine* may be low. These results suggest several open fields of inquiry into the genetics and physiology of sexual fertility in this fungus.

Our results strongly suggest that recombination has played an important role in the structure of populations in the Indian Himalayas. The dynamics of these populations and the presence of female fertility suggest that the source of this recombination is sexual, and that it is recent or, perhaps, contemporary. Considering the recent reports of sexual fertility in Southern China and in highly diverse populations from Thailand (Mekwatanakarn, et al. (submitted)) sexual fertility and recombination may be rather common across the area of South and Southeast Asia where rice was domesticated. Our data suggest that in these diverse environments populations may be mosaics of clonal and recombining subpopulations, and the frequency of recombination in a population may fluctuate dramatically over only seasons.

References Cited

Brown, A. H. D., M. W. Feldman and E. Nevo, 1980 Multilocus structure of natural population of *Hordeum spontaneum*. Genetics 96: 523-536.

Burt, A., A. C. Carter, G. L. Koenig, T. J. White, and J. W. Taylor, 1996 Molecular markers reveal cryptic sex in the human pathogen *Coccidioides immitis*. Proc. Natl. Acad. Sci. USA. 93: 770-773.

Chen, D., R. S. Zeigler, H. Leung, and R. J. Nelson, 1995 Population structure of *Pyricularia grisea* at two screening sites in the Philippines. Phytopathology 85: 1011-1020.

Chen, R. S., and B. A. McDonald, 1996 Sexual reproduction plays a major role in the genetic structure of populations of the fungus *Mycosphaerella graminicola*. Genetics 142: 1119-1127.

Geiser, D., M. Arnold, and W. Timberlake. 1994 Sexual origins of British *Aspergillus nidulans* isolates. Proc. Natl. Acad. Sci. USA. 91: 2349-2352.

Hamer, J. E., L. Farrall, M. J. Orbach, B. Valent and F. G. Chumley, 1989 Host species specific conservation of a family of repeated DNA sequences in the genome of a fungal plant pathogen. Proc. Natl. Acad. Sci. USA. 86: 9981-9985.

Kohn, L. M., 1995 The clonal dynamic in wild and agricultural plant populations. Can. J. Bot. 73(Suppl. 1): S1231-S1240.

Kumar, J., Nelson, R. J., and Zeigler, R. S. 1999. Population structure and dynamics of *Magnaporthe grisea* in the Indian Himalayas. GENETICS 152:971-984.

Lenski, R. E., 1993 Assessing the genetic structure of microbial populations. Proc. Natl. Acad. Sci. USA. 90: 4344-4336.

Leslie, J. F., and Klein, K. K., 1996 Female fertility and mating type effects on effective population size and evolution in filamentous fungi. Genetics 144:557-567.

Levy, M., F. J. Correa-Victoria, R. S. Zeigler, S. Xu and J. E. Hamer, 1993 Genetic diversity of rice blast fungus in a disease screening nursery in Colombia. Phytopathology 83: 1427-1433.

Mekwatanakarn, P., K   , W., and Zeigler, R.S. 1999. Sexually fertile field isolates of *Magnaporthe grisea* from North and Northeast Thailand. Plant Disease (In Press).

Maynard Smith, J., N. H. Smith, M. O'Rourke and B. G. Pratt, 1993 How clonal are bacteria? Proc. Natl. Acad. Sci. USA. 90: 4384-4388

Roumen, E., M. Levy and J. L. Notteghem, 1996. Characterization of the European pathogen population of *Magnaporthe grisea* by DNA fingerprinting and pathotype analysis. European J. Pl. Path. 00: 000 - 000.

Skinner, D. Z., A. D. Budde, M. L. Farman, J. R. Smith, H. Leung and S. A. Leung, 1993 Genome organization of *Magnaporthe grisea*: genetic map, electrophoretic karyotype, and occurrence of repeated DNAs. Theor. Appl. Genet. 87: 545 - 557.

Slatkinf, M. 1985. Rare alleles as indicators of gene flow. Evolution 39:53-65.

Thinlay. 1998. Rice blast, caused by *Magnaporthe grisea*, in Bhutan and development of strategies for resistance breeding and management. PhD Thesis, ETH (#   ), Zurich.

Tibayrenc, M., F. Kjellberg, J. Arnaud, B. Oury, S. Frederique Breniere, M. Darde, and F. J. Ayala, 1991 Are eukaryotic microorganism clonal or sexual? A population genetics vantage. Proc. Natl. Acad. Sci.USA. 88: 5129-5133.

Weir, B. S., 1990 *Genetic Data Analysis*. Sinauer Associates, Sunderland, MA.

Xia, J. Q., J. C. Correll, F. N. Lee, M. A. Marchetti and D. D. Rhoads, 1993 DNA fingerprinting to examine microgeographic variation in the *Magnaporthe grisea* (*Pyricularia grisea*) population in two rice fields in Arkansas. Phytopathology 83: 1029-1035.

Xu, J.-R. and Hamer, J. E. 1995. Assessment of *Magnaporthe grisea* mating type by spore PCR. Fung. Genet. Newslet. 42:80

BIOCHEMICAL AND BIOMECHANICAL ASPECTS OF APPRESSORIAL DEVELOPMENT IN *MAGNAPORTHE GRISEA*

DIANA J. DAVIS
Department of Chemistry
College of Mount St. Joseph
Cincinnati, Ohio 45233, USA

CHRIS BURLAK
AND NICHOLAS P. MONEY
Department of Botany
Miami University
Oxford, Ohio 45056, USA

ABSTRACT. This article is concerned with the developmental processes that prime the appressorium of *Magnaporthe grisea* for mechanical penetration of the host surface. Experiments have shown that melanized appressoria generate enormous turgor pressure following the accumulation of high concentrations of cytoplasmic osmolytes, and thereby deliver substantial invasive force at the tip of a slender penetration hypha. Recent molecular genetic studies provide authoritative evidence for the central role played by melanin biosynthesis in the operation of the appressorium, and still other studies have revealed the identity of some of the osmolytes responsible for turgor generation. However, our picture of appressorial function remains limited by a number of problems: the maximum level of appressorial pressure is still questionable, the metabolic source of the cytoplasmic osmolytes is a mystery, and the processes that control melanization of the appressorial wall are poorly understood. Progress toward the resolution of these issues is discussed.

1. Introduction

The appressorium of *Magnaporthe grisea* is a cellular platform that achieves mechanical penetration of the host cuticle and epidermal cell wall [1-4]. The necessary invasive force is derived from turgor pressure within the appressorium, and is applied at the tip of a penetration hypha that develops from the base of the cell. During the infection process, the necessary tight contact between the fungus and substrate is maintained by an adhesive secreted by the appressorium [5]. Penetration proceeds only when the force exerted at the tip of the penetration hypha exceeds the mechanical resistance offered by the underlying cuticle and cell wall; the role of exoenzymes in reducing the mechanical resistance offered by host tissues is unknown [6]. In this article we explore both the

D. Tharreau et al. (eds.), Advances in Rice Blast Research, 248–256.

magnitude and metabolic source of the turgor pressure generated by appressoria, and also consider the process of appressorial melanization.

2. Appressorial turgor pressure

To date, there have been no direct measurements of appressorial turgor pressure, and the available indirect data rest on estimates of the osmolality, or osmotic pressure, of the appressorial cytoplasm. Three different approaches have been used to estimate the osmotic pressure of the appressorium; these are referred to as the cytorrhysis, stabilization, and melting point methods. In the study published by Howard and colleagues [7], appressoria were incubated in a series of aqueous solutions of polyethylene glycol (PEG)-300 providing a defined range of osmotic pressures. Exposure to PEG induces water loss from the appressorium, and at a critical concentration of PEG the cell collapses; this phenomenon is described as cytorrhysis and is explained in detail elsewhere [8]. Cytorrhysis occurs only when the osmotic pressure of the solution bathing the cell exceeds the osmotic pressure of its cytoplasm. By scoring the number of appressoria that collapse over a wide range of PEG concentrations, it is possible to estimate the medium osmotic pressure that matches the osmotic pressure of the cytoplasm (Fig. 1).

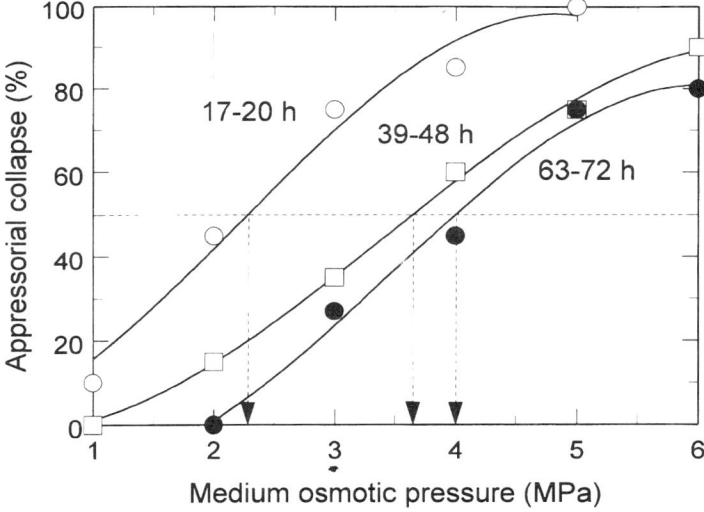

Fig. 1. Relationship between external osmotic pressure (controlled with PEG-300) and appressorial collapse in *M. grisea*. Each data point represents the percentage of 20 randomly-selected appressoria that collapsed within a 15 min exposure to PEG. Appressoria formed on plastic coverslips, and separate coverslip cultures were incubated in each PEG solution. Plot shows results from appressoria of three different ages reflecting the time following conidial harvest. Estimates of appressorial turgor in Howard et al. [7] were based on this type of experiment.

This method provides estimates of extraordinary levels of turgor in melanized appressoria; since 50% of mature appressoria collapse in a PEG solution with an osmotic pressure of 4.0 MPa, it is reasonable to suggest that the cytoplasm of these cells has a comparable osmotic pressure. To arrive at a turgor estimate we assume that the osmotic differential between the appressorium and the water on a dew-covered leaf surface (the *in vivo* environment) results in the development of a turgor pressure equal in magnitude to the osmotic pressure (since pure water has zero osmotic pressure). Figure 1 shows that a smaller proportion of the mature appressoria do not collapse even when the medium osmotic pressure is raised to 6.0 MPa. This is consistent with the proposition that appressoria may generate enormous turgor pressures (perhaps even in excess of 8.0 MPa [7]).

The turgor estimates derived from the cytorrhysis experiments are supported by the results from a second type of experiment in which the cells were again incubated in a series of PEG solutions and deflated to atmospheric pressure. However, instead of scoring the population for collapse, cells in each of these PEG solutions were punctured with the tip of a micropipet and observed for forcible expulsion of cytoplasm. The principle is very simple: if the contents of the cell are at or below atmospheric pressure they will not explode when the cell wall is damaged with the pipet. Therefore, the estimate of turgor from the second method is based on the proportion of appressoria that are stabilized against bursting (Fig. 2).

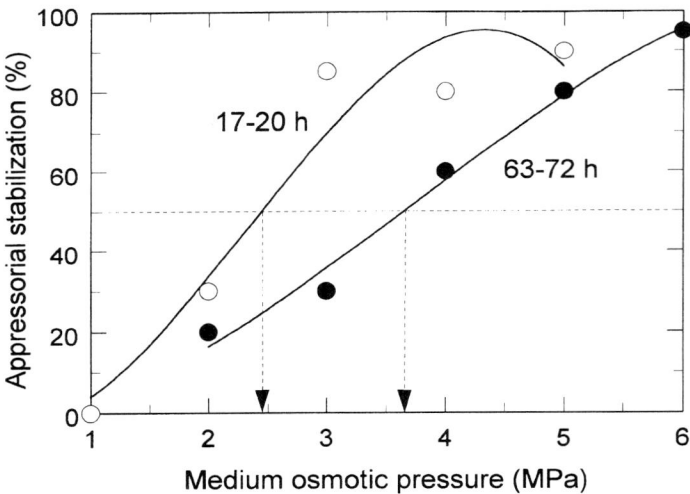

Fig. 2. Effect of external osmotic pressure (controlled with PEG-300) on appressorial stabilization following puncture with a micropipet. Each data point represents the percentage of 20 randomly-selected appressoria that did not burst in response to penetration with a micropipet, after a 15 minute exposure to PEG. Appressoria formed on plastic coverslips, and separate coverslip cultures were incubated in each PEG solution. Plot shows results from appressoria of two different ages reflecting the time following conidial harvest. The results of this experiment affirmed the estimates of high turgor pressure derived from Fig. 1.

It is useful to bear in mind that the experimental approaches illustrated in Figs. 1 and 2 have some important limitations. When a cell is incubated in a solution with an osmotic pressure in excess of its cytoplasm (i.e. the solution is hyperosmotic or hypertonic with respect to the cell), the cell will usually shrink before it collapses. Therefore, we may actually overestimate cellular osmotic pressure if we base our measurement on the osmotic pressure of the solution that induces 50% of the appressoria to collapse, because shrinkage concentrates the solutes dissolved in the cytoplasm. Precisely the same argument applies to the stabilization experiments. But in both cases, the data can be corrected to provide an accurate measure of cytoplasmic osmotic pressure by accounting for the reduction in cell volume at the critical concentration of PEG that causes collapse or stabilizes the cells against bursting. In the cytorrhysis experiments, this is determined by measuring the reduction in volume of cells underlined preceding collapse. In the study by Howard et al. [7] no changes in appressorial volume were observed, but it should be noted that even quite small changes in diameter have a profound effect upon cell volume and osmotic pressure. For example, a 1.0 μm reduction in the diameter of a 10.0 μm sphere will cause a 37% increase in osmolyte concentration. There are other potential sources of error in the cytorrhysis and stabilization approaches for determining cytoplasmic osmotic pressure. If the cell wall does not completely exclude the test solute molecule, influx of the solute can result in a tremendous overestimate of the concentration of the matching solution. The efficacy of the cytorrhysis method also hinges on the assumption that the cell wall offers negligible resistance to collapse under compressive stress; once turgor has been reduced to zero (equal to atmospheric pressure), collapse should follow any further increase in external osmotic pressure. The fact that the cytorrhysis and stabilization experiments on *M. grisea* provide very similar estimates of appressorial osmotic pressure is encouraging, but other fungal cells do show great resistance to collapse under osmotic stress [8]. Appreciation of the limitations of the cytorrhysis method is important not only in relation to our understanding of the biomechanics of host penetration, but also because the technique has been adopted as a swift screen for mutants that may be defective in host penetration [9].

Relative to the various sources of error inherent in the cytorrhysis and stabilization experiments, somewhat lower estimates of appressorial osmotic pressure and turgor have been made by comparing the melting points of intracellular and extracellular ice on a microscope cooling stage [10]. So, while we are certain that the pressures generated by appressoria play an essential role in pathogenesis, authoritative data on the magnitude of the turgor pressure will require direct mechanical measurements from individual cells. In relation to the penetration process itself, the invasive force at the tip of the penetration hypha is the critical variable rather than the pressure within the appressorium. The reader is referred to a separate paper [6] for further details of this issue.

3. Osmolyte accumulation

Glycerol has been identified as a major constituent of the appressorial cytoplasm with estimates of concentrations in excess of 3.0 M [11]. The claim that glycerol is a dominant osmolyte within the appressorium is particularly interesting in relation to fine structural studies on developing appressoria In 1990, Bourett and Howard [12] showed that characteristic glycogen inclusions called rosettes are present in high concentrations within young appressoria, but disappear in mature cells. They suggested that the glycogen rosettes may be a major source of the osmolytes responsible for the increase in appressorial turgor. With recognition of the importance of glycerol in appressorial physiology, we can provide a partial test of this hypothesis by posing the following specific question: if all of the glycogen rosettes in an appressorium were metabolized to form glycerol through the standard metabolic pathway, how much glycerol would be generated? The answer is derived through a series of simple calculations.

3.1. HOW MANY ROSETTES ARE PRESENT IN AN APPRESSORIUM?

Tim Bourett and Rick Howard [12] counted the number of glycogen rosettes within their fabulously preserved, freeze-substituted sections of appressoria and found a maximum of 4 rosettes per square μm (4.02 ± 1.23, mean $\pm$ standard error) 8 hours following conidial harvest (by which time the appressorium is isolated from the conidium). These figures for number per square μm can be converted to number per unit volume, with reference to the section thickness of 100 nm. Therefore, an average of 4 rosettes were found in a slab with a volume of 1.0×10^{-19} m^3.

The remainder of the calculations, in Sections 3.2-3.7, serve to predict the maximum concentration of glycerol that would be produced if the glycogen rosettes within a volume of 1.0×10^{-19} m^3 were metabolized to form glycerol. By modeling concentration changes in a small volume the need to estimate the volume of the appressorium is obviated, but nevertheless the calculations yield a glycerol concentration that applies to the entire appressorial cytoplasm.

3.2. WHAT IS THE VOLUME OF A SINGLE GLYCOGEN ROSETTE?

The average diameter of a rosette in *M. grisea* is 100 nm (the same as the thickness of the EM section [12]), which is similar to the size of rosettes in mammalian liver.

Volume $= 4/3\pi r^3 = 4/3\pi \times (50 \times 10^{-9}$ m$)^3 \times 1000$ L/m$^3 = 5.2 \times 10^{-19}$ L

3.3. WHAT IS THE DENSITY OF GLYCOGEN?

To calculate the mass of glycogen in a rosette we must know the density of glycogen. Published work [13] provides a figure of 1.3 g per mL, and this was verified by making our own measurements using glycogen purified from invertebrate and mammalian sources (Sigma, St. Louis, MO).

3.4. WHAT IS THE MASS OF GLYCOGEN IN 4 ROSETTES?

The mass of glycogen is given by the number of rosettes x volume of a single rosette x density of glycogen:

4 rosettes x 5.2×10^{-19} L x 1000 mL/L x 1.3 g/mL $= 2.7 \times 10^{-15}$ g glycogen.

3.5. HOW MANY MOLES OF GLUCOSE CAN BE GENERATED FROM THIS MASS OF GLYCOGEN?

Glycogen consists of α-(1$\rightarrow$4) linked glucose units. Therefore,

2.7×10^{-15} g glycogen/162.10 g per mol glucose units $= 1.7 \times 10^{-17}$ mol glucose units.

3.6. HOW MANY MOLES OF GLYCEROL CAN BE GENERATED FROM THIS MASS OF GLUCOSE?

Each mole of glucose can be metabolized to form 2 moles of glycerol. Therefore,

1.7×10^{-17} mol glucose units x 2 mol glycerol per mol glucose $= 3.4 \times 10^{-17}$ mol glycerol.

This is the amount of glycerol that can be produced from 4 glycogen rosettes within the volume of 1.0×10^{-19} m^3.

3.7. WHAT IS THE PREDICTED GLYCEROL CONCENTRATION WITHIN THE APPRESSORIUM BASED ON THE ROSETTE SOURCE?

The molarity of glycerol in the appressorium is then equal to the number of moles divided by the volume of the slab of the EM section.

3.4×10^{-17} mol glycerol/1.0×10^{-19} m^3 x 1000 L per m^3 = **0.34 M glycerol.**

This is 10-fold lower than the estimate based on the enzymatic assay of glycerol from appressoria grown on synthetic film [11]. If a glycerol concentration in excess of 3.0 molar is confirmed, the present calculations show that the glycogen rosettes are not likely to be a major source of these osmolytes, and that other sources of glycerol (e.g. lipid metabolism) must be explored. Alternatively, it is possible that glycerol levels may be much lower than the available measurements suggest.

4. Relationship between glycerol accumulation and turgor pressure

Measurements of osmotic pressure using vapor pressure deficit osmometry show that a 3.3 M solution of glycerol (the maximum level of appressorial glycerol measured in the study by De Jong et al. [11]) generates an osmotic pressure of 11.0 MPa. Therefore, a cell that contained this much glycerol would have the potential to generate a turgor pressure of 11.0 MPa in pure water, which is higher even than the maximum levels estimated by Howard et al. [7]. By contrast, a more modest estimate of turgor is derived from the calculations based on glycogen availability; 0.34 M glycerol generates an osmotic pressure of 0.8 MPa. Irrespective of the actual glycerol concentration, the presence of other osmolytes such as sugars and inorganic ions will further increase cytoplasmic osmotic pressure and turgor.

5. Melanin synthesis as a determinant of osmolyte accumulation and pressure generation

Some years ago, C. P. Woloshuk and colleagues [14] established that melanin biosynthesis is an essential determinant of pathogenicity in rice blast disease, and demonstrated that non-melanized appressoria are unable to penetrate the host surface. More recently, authoritative evidence for a link between melanin biosynthesis and pathogenicity has come from the work of Kawamura and colleagues [15] who were able to restore pathogenicity to melanin-deficient mutants of *M. grisea* by transformation with a cosmid clone containing genes that are essential for melanin biosynthesis. This was a particularly elegant experiment, because the complementary genes were cloned not from *M. grisea*, but from *Alternaria alternata*. In a related study [16], transformation of an albino strain (Pks1⁻) of *Colletotrichum lagenarium* with a single gene from *A. alternata* (*ALM*, that probably encodes a polyketide synthase) also restored melanization and penetrative ability to its appressoria.

Physiological analysis of pigmentation mutants of *M. grisea* has shown that appressorial melanin is essential for osmolyte accumulation, and for the development of high turgor pressures [10,11]. A model linking melanin synthesis to osmolyte accumulation has been proposed [17] which suggests that the appressorial plasma membrane and melanized cell wall act as a dual permeability barrier to limit osmolyte leakage. This awaits confirmation, emendation, or rejection through further experiment. But even though light microscopic observations show that the "blackening" of the appressorium precedes and predicts the development of high turgor pressure [10], we cannot rule out the possibility that melanin exerts its effect upon pathogenicity via a less direct mechanism. For example, the expression of genes controlling the melanization process could simply stimulate the metabolic reactions that lead to osmolyte synthesis. According to this frightening scenario, the melanin within the appressorial wall could prove to be a red herring for those of us intent on understanding appressorial function!

Further data on the metabolic activity within melanized and non-melanized appressoria could resolve this issue. However, the conventional methods of "bucket biochemistry" are almost useless in studying the development of an asynchronous population of cells that develop in relatively low densities on hydrophobic surfaces. With this in mind, we have made some progress in understanding the pigmentation process in appressoria by collecting spectral information from individual cells using a technique called Raman microspectroscopy [18,19]. Experiments performed on appressoria developing on coverglasses and silicon wafers show that the highly-characteristic Raman spectrum obtained from samples of purified DHN-melanin (Fig. 3 inset) can also be obtained from single appressoria (Fig. 3C). Other experiments are consistent with the idea that intermediates in the polyketide pathway (such as 1,3,8-trihydroxynaphthalene) may be present in the conidium and are transported into the developing appressorium (Fig. 3). Following this transfer, melanization presumably proceeds with the polymerization of these compounds within the appressorial wall. Clearly then, microspectroscopy offers the level of spatial resolution that will be essential for examining biochemical processes within individual living cells.

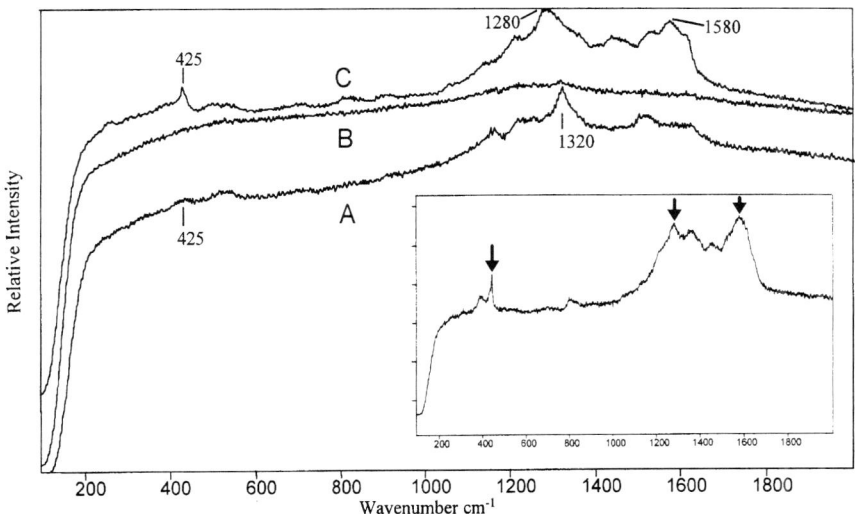

Fig. 3. Raman spectrum obtained from (A) a single conidium, (B) the young appressorium attached to the same conidium, and (C) the same appressorium after maturation. Conidia of *M. grisea* strain 0-42 were collected on a coverglass and soaked with distilled water. Spectra were obtained with a Renishaw spectrometer (Wooton-under-Edge, United Kingdom) coupled to a standard Olympus microscope platform. The light source was a 2.7 mW helium-neon laser with a 760 nm excitation source; data were collected using a holographic notch filter with a spectral slit width of 40 μm and a wavenumber resolution of 4 cm^{-1}. Each spectrum represents the average of 10 scans of a 3 μm diameter sampling area (middle of the conidium or appressorium). Note that the spectra from both the conidium (A) and mature appressorium (C) contain some peaks at identical wavenumbers, while the spectrum of the young appressorium is relatively featureless. Inset shows Raman spectrum of laccase-derived DHN-melanin; note peaks (arrows) at 425, 1280, and 1580 cm^{-1}.

6. Future prospects

The discussion of appressorial chemistry provided in this paper is most useful if it has conveyed a picture of profound uncertainty. Despite a number of important advances in the last decade, we still know very little about the chemical composition of the appressorial cytoplasm. This is a critical deficit in our knowledge, and in the absence of even a partial inventory of appressorial osmolytes we cannot expect to understand the metabolic events that precede mechanical penetration of the host. In terms of biomechanical studies on host infection, prospects for future research are very encouraging because a number of new techniques have been developed recently. For example, silicon beam force transducers have already provided measurements of the microNewton forces exerted by individual hyphae of other fungi [Money, unpublished data], and the atomic force microscope may prove of real utility in studying changes in cellular pressure. Both approaches also have the potential to provide a definitive measure of that most intractable and important variable, appressorial turgor pressure.

Acknowledgments

The authors thank Dr. Andy Sommer, Director of the Molecular Microspectroscopy Laboratory at Miami University, for assistance with the spectroscopic analyses.

References

1. Talbot, N. J. (1995) Having a blast: Exploring the pathogenicity of *Magnaporthe grisea*, *Trends Microbiol.* **3**, 9-16.
2. Howard, R. J., and Valent, B. (1996) Breaking and entering: host penetration by the fungal rice blast pathogen, *Magnaporthe grisea*, *Annu. Rev. Microbiol.* **50**, 491-512.
3. Dean, R. J. (1997) Signal pathways and appressorium morphogenesis, *Annu. Rev. Phytopathol.* **35**, 211-234.
4. Howard, R. J. (1997) Breaching the outer barriers - cuticle and cell wall penetration, in G. Carroll and P. Tudzynski (eds.), *The Mycota*, Vol 5, Part A, *Plant Relationships*, Springer-Verlag, Berlin Heidelberg New York, pp. 43-60.
5. Ebata, Y., Yamamoto, H., and Uchiyama, T. (1998) Chemical composition of the glue from appressoria of *Magnaporthe grisea*, *Biosci. Biotechnol. Biochem.* **62**, 672-674.
6. Money, N. P. (1998) Mechanics of invasive fungal growth and the significance of turgor in plant infection, in K. Kohmoto and O. C. Yoder (eds.), *Molecular Genetics of Host-Specific Toxins in Plant Disease*, Kluwer Academic Publishers, Dordrecht, The Netherlands, pp. 261-271.
7. Howard, R. J., Ferrari, M. A., Roach, D. H., and Money, N. P. (1991) Penetration of hard substrates by a fungus employing enormous turgor pressures, *Proc. Natl. Acad. Sci. U.S.A.* **88**, 11281-11284.
8. Money, N. P., Caesar-Tonthat, T., Frederick, B., and Henson, J. M. (1998) Melanin synthesis is associated with changes in hyphopodial permeability, wall rigidity, and turgor pressure in *Gaeumannomyces graminis* var. *graminis*, *Fungal Genet. Biol.* **24**, 240-251.
9. Balhadere, P. V., Foster, A. J., and Talbot, N. J. (1998) Identification of pathogenicity mutants of the rice blast fungus *Magnaporthe grisea* by insertional mutagenesis, *Molec. Plant Microbe Interact.*, in press.
10. Money, N. P., and Howard, R. J. (1996) Confirmation of a link between fungal pigmentation, turgor pressure, and pathogenicity using a new method of turgor measurement, *Fungal Genet. Biol.* **20**, 217-227.
11. De Jong, J. C., McCormack, B. J., Smirnoff, N., and Talbot, N. J. (1997) Glycerol generates turgor in rice blast, *Nature* **389**, 244-245.
12. Bourett, T. M., and Howard, R. J. (1990) *In vitro* development of penetration structures in the rice blast fungus *Magnaporthe grisea*, *Can. J. Bot.* **68**, 329-342.
13. Meyer, K. H. (1943) The chemistry of glycogen, *Adv. Enzymol.* **3**, 109-135.
14. Woloshuk, C. P., Sisler, H. D., and Vigil, E. L. (1983) Action of the antipenetrant, tricyclazole, on appressoria of *Pyricularia oryzae*, *Physiol. Plant Pathol.* **22**, 245-259.
15. Kawamura, C., Moriwaki, J., Kimura, N., Fujita, Y., Fugi, S-i., Hirano, T., Koizumi, S., and Tsuge, T. (1997) The melanin biosynthesis genes of *Alternaria alternata* can restore pathogenicity of the melanin-deficient mutants of *Magnaporthe grisea*, *Mol. Plant-Microbe Interact.* **10**, 446-453.
16. Takano, Y., Kubo, Y., Kawamura, C., Tsuge, T., and Furusawa, I. (1997) The *Alternaria alternata* melanin biosynthesis gene restores appressorial melanization and penetration of cellulose membranes in the melanin-deficient albino mutant of *Colletotrichum lagenarium*, *Fungal Genet. Biol.* **21**, 131-140.
17. Money, N. P. (1997) Mechanism linking cellular pigmentation and pathogenicity in rice blast disease: A commentary, *Fungal Genet. Biol.* **22**, 151-152.
18. Hester, R. E., and Girling, R. B. (1991) *Spectroscopy of Biological Molecules*, The Royal Society of Chemistry, Cambridge, U.K.
19. Greve, J., and Puppels, G. J. (1993) Raman microspectroscopy of single whole cells, in R. J. H. Clark and R. E. Hester (eds.), *Biomolecular Spectroscopy*, Part A, Vol 20, Wiley and Sons, New York, pp. 231-265.

RECENT STUDIES ON PATHOGENICITY GENES IN RICE BLAST FUNGUS, *MAGNAPORTHE GRISEA*

K. ADACHI, M. URBAN*, T. BHARGAVA**, F. TENJO, J.E. HAMER
Department of Biological Sciences, Purdue University
West Lafayette, Indiana 47907-1392, U.S.A.

Current address:
* *Monsanto Europe S.A., Parc Scientifique Fleming,*
Rue Laid Burniat 5, B-1348 Louvain-La-Neuve, Belgium
** *Boyce Thompson Institute, Cornell University,*
Ithaca, New York 14853-1801, U.S.A.

1. Introduction

Rice blast is the most damaging disease in rice growing areas throughout the world. The causal agent of rice blast disease is an ascomycete fungus *Magnaporthe grisea*. *M. grisea* is highly suited for research on fungal pathogenicity, since many experimental tools including classical and molecular genetic technique are available. Several genes affecting pathogenicity have been isolated from *M. grisea* which can be called 'pathogenicity genes' in the broad sense (see Schafer, 1994). Those genes are divided into three groups: genes for plant-specific nutrition, genes for infection-related morphogenesis and genes for symptom development and disease establishment. The first group of genes provide important information for understanding the physiology of invasive fungal growth. Insertional mutagenesis have identified that *PTH3* encoding a histidine biosynthesis enzyme is necessary for full pathogenicity (Sweigard *et al.*, 1998) and that a mutant exhibiting methionine auxotrophy shows reduced pathogenicity (Balhadère *et al.*, 1999). The second group is currently represented by genes encoding signal transduction components involved in infection-related morphogenesis. Recently, we identified the first example of a gene which may be classified into the third group (Urban *et al.*, 1999). Here, we focus on the second and third groups of pathogenicity genes.

D. Tharreau et al. (eds.), Advances in Rice Blast Research, 257–266.

2. Genes Involved in Signaling Pathways during Appressorium Formation

At the early stage of the infection cycle, *M. grisea* differentiates a dome-shaped melanized structure, called an appressorium, in which a high concentration of glycerol accumulates so as to generate turgor pressure for penetration into rice leaves. *M. grisea* forms appressoria not only on plant surfaces, but also on artificial hydrophobic surfaces. Self-assembly of the hydrophobin protein Mpg1 is likely to play a roll in surface sensing mechanisms to trigger appressorium formation (Beckerman and Ebbole, 1996). With the knowledge that eukaryotic signal transduction systems are highly conserved, several *M. grisea* genes encoding signal transduction pathway components have been isolated using the polymerase chain reaction (PCR). Gene knockout experiments have confirmed that some of these candidate genes are essential for either appressorium formation or function.

2.1. cAMP SIGNALING

cAMP is a ubiquitous secondary messenger in prokaryotic and eukaryotic cells. In fungi, cAMP plays a central role as a signaling molecule in growth, morphogenesis and pathogenesis (Kronstad, 1997). Pharmacological studies show that cAMP signaling is involved in appressorium formation of *M. grisea* (Lee and Dean, 1993). A key enzyme in the cAMP signaling pathway is adenylate cyclase which converts ATP to cAMP. A gene encoding a putative adenylate cyclase (*MAC1*) was cloned and disrupted (Choi and Dean, 1997; Adachi and Hamer, 1998). *MAC1* can complement a classically defined *Neurospora crassa* adenylate cyclase mutant (*cr-1*), indicating that *Magnaporthe* adenylate cyclase is functionally related to that of *Neurospora*. Δ*mac1* mutants show pleiotropic defects in growth, conidiation, mating ability and appressorium formation. The deficiency in appressorium formation is restored by the addition of exogenous cAMP. We identified bypass suppressor mutations of the Mac1⁻ phenotype, designated *sum* (for suppressors of Mac1⁻). *sum* mutations completely restore the defects of Δ*mac1* mutants, although suppressed mutants do not recover full pathogenicity. Moreover, Δ*mac1 sum* mutants make appressorium not only on hydrophobic surfaces, but also on hydrophilic surfaces which are noninductive substrates for appressorium formation of wild type strains. Molecular cloning shows that one suppressor mutation (*sum1-99*) changes a conserved leucine residue at position 211 to arginine in cAMP binding domain A of the PKA regulatory subunit gene (*SUM1*). PKA assays demonstrate that Δ*mac1 sum1-99* mutants have constitutively activated PKA activity, suggesting the point mutation in *sum1-99* changes the structure of the PKA complex and alters regulation of PKA activity. Interestingly, mutations in a PKA catalytic subunit gene (*CPKA*) have very specific effects on appressorium penetration and, unlike mutations in *MAC1*, do not affect vegetative growth, sexual or asexual morphogenesis (Mitchell and Dean, 1995; Xu *et al.*, 1997). These data suggest that cAMP signaling pathway in *M. grisea* diverges at the level of PKA regulation. That is, *M. grisea* must have other PKA catalytic subunit (s) than CpkA which must be associated with Sum1 and play specific roles in growth and morphogenesis (Figure 1).

Yeast extract and *Saccharomyces cerevisiae* α factor inhibit appressorium formation of *M. grisea* in a mating-type-dependent manner (Beckerman *et al.*, 1997). It is proposed that yeast extract and α factor interact with a membrane-associated Mat1-2 receptor and lead to inhibition of cAMP signaling. A strain with the *sum1-99* mutation is free from this inhibition, suggesting that the inhibition acts upstream of cAMP generation (Figure 1).

Heterotrimeric G protein coupled receptors are known to transduce cell surface signals. G protein α subunits often regulate adenylate cyclase in eukaryotic cells. Three G protein α subunit genes were cloned and characterized from *M. grisea*, designated *magA*, *magB* and *magC* (Liu and Dean, 1997). Deletion of *magB* significantly reduces appressorium formation and pathogenicity, whereas deletion of *magA* or *magC* does not. Δ*magB* mutants also have pleiotropic defects in vegetative growth, conidiation and sexual development, highly similar to Δ*mac1* mutants. Moreover, exogenously added cAMP restores appressorium formation in Δ*magB* mutants. These data suggest that MagB is a positive regulator of Mac1 either directly or indirectly (Figure 1).

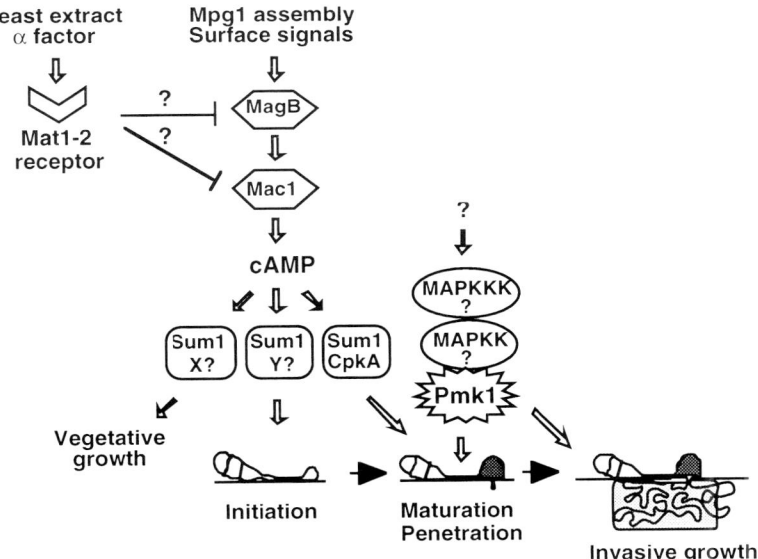

Figure 1. A model for cAMP and MAP kinase signalings during appressorium formation of *M. grisea*. X and Y represent PKA catalytic subunits other than CpkA.

2.2. MAP KINASE SIGNALING

Mitogen-activated protein (MAP) kinase signaling is also a well-conserved pathway among eukaryotes. In *S. cerevisiae*, several MAP kinase pathways have been identified which regulate responses to high external osmolarity, mating, cell wall integrity, spore formation and pseudohyphal growth (reviewed in Banuett, 1998). Three MAP kinases were cloned from *M. grisea* and examined for involvement in pathogenicity.

Pmk1, a homolog of yeast Fus3 and Kss1, has a specific role in appressorium maturation and acts downstream or independently from the cAMP signaling pathway for appressorium initiation (Figure 1; Xu and Hamer, 1996). Δ*pmk1* mutants never make appressoria and therefore are nonpathogenic. Pmk1 is also necessary for invasive growth in plants.

Mps1 is a homolog of yeast Slt2 and regulates cell wall integrity (Xu and Hamer, 1998). Δ*mps1* mutants form normal appressoria, but can not penetrate into plant cells. In addition, Δ*mps1* mutants show reduced conidiation and female fertility, but also display acute sensitivity to cell wall-digesting enzymes and undergo wall lysis in older regions of the fungal mycelia. These defects suggest an inability to remodel cell wall growth during specific development processes such as appressorium formation, conidiation and prolonged hyphal growth.

Osm1/Pmk2 is a homolog of yeast MAP kinase Hog1. Although Hog1 controls glycerol metabolism in response to high osmotic stress, Osm1 is not involved in glycerol accumulation during appressorium formation and *osm1* mutants do not show major reductions in pathogenicity (see Talbot's chapter).

3. Genes Required Specifically for Invasive Growth in Plants

Plants have various mechanisms to protect themselves against microorganisms. Among numerous microorganisms only plant pathogens can successfully penetrate and colonize into plant cells and finally cause diseases. Thus, plant pathogens have particular tactics to overcome plant defense mechanisms and survive inside plant tissues. One of the chemical defense mechanisms deployed by plants is a production of antifungal proteins (Dixon *et al.*, 1994) and more specialized antibiotics, called phytoalexins (reviewed in Osbourn, 1996). It has been demonstrated that some fungi produce phytoalexin-detoxifying enzymes (Van Etten *et al.*, 1994) or nonphytotoxic compounds, called suppressors, which disturb plant defense responses (reviewed in Shiraishi *et al.*, 1994). However, other fungal mechanisms to ensure survival in plant cells are still largely unknown.

3.1 THE ABC TRANSPORTER, ABC1, IS ESSENTIAL FOR INVASIVE GROWTH

In an effort to identify genes required for invasive growth and survival in plant cells, we performed an insertional mutagenesis screen using DNA mediated transformation of *M. grisea* strain Guy11 and screened transformants for defects in pathogenicity by leaf sheath injection (Xu and Hamer, 1996; Lau and Hamer, 1998). Our mutant screen identified a mutant strain TF7-3131 with dramatically reduced pathogenicity on rice cultivar CO39. The plasmid insertion in strain TF7-3131 identified a novel gene, named *ABC1*. Molecular analysis reveals that the plasmid insertion of strain TF7-3131 is located in the putative promoter region upstream of the start codon. *ABC1* shows high similarity to members of the ABC protein superfamily. The secondary structure and domain organization are typical for ABC-type-transporter proteins. The Abc1 amino acid sequence is composed of two homologous halves, each comprising one N-terminal hydrophilic domain followed by a C-terminal hydrophobic domain. Each hydrophilic domain contains an ATP binding sequence cassette (ABC) and each hydrophobic domain encodes six transmembrane spanning a helices. ABC transporters are widely conserved proteins found from bacteria to man and function as ATP-driven efflux pumps of small hydrophobic molecules. To analyze the function of Abc1 we constructed an *ABC1* deletion strain by one-step-gene replacement. Spray inoculation assays on rice and barley show that both *abc1* insertional and deletion mutants are dramatically reduced in pathogenicity to the same extent, and produce rare lesions (one lesion for every five leaves). Quantitative microscopic analysis (Figure 2) on detached barley leaves shows that the ability to form appressorium and penetrate plant cells is not affected. However, only 4% of *abc1* mutants formed invasive hyphae compared with Guy11 (65%) in a detached leaf assay on barley. Invasive hyphae formed by 4% of the *abc1* mutants were short and did not proliferate extensively within plant cells. Furthermore, viable *abc1* mutants could not be recovered from inoculated plant. This suggests that *abc1* mutants fail early in pathogenicity, most likely just after penetration.

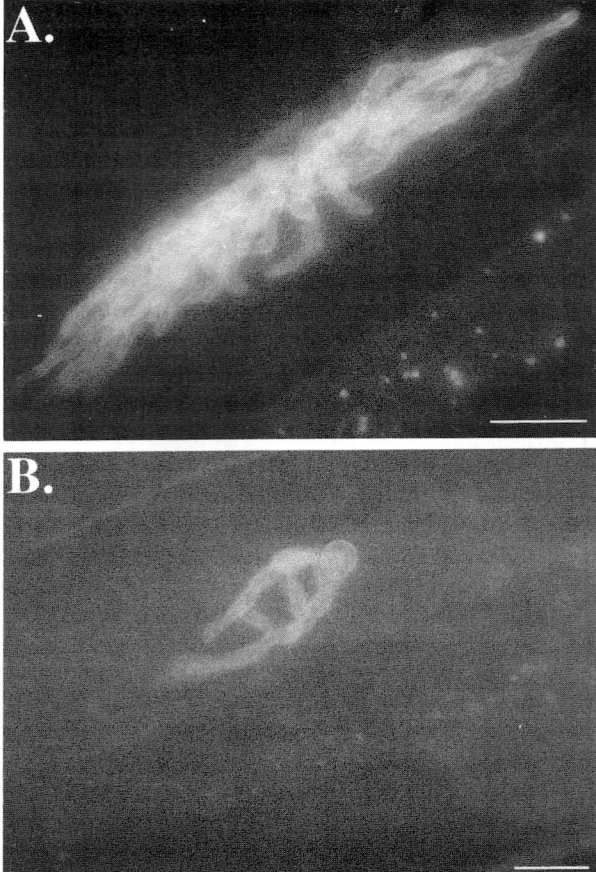

Figure 2.   Penetration assays on detached barley leaves after 60 h incubation.   (A) Anillin blue stained wild type strain Guy11 penetrates barley epidermis cells and differentiates dense spreading infection hyphae made visible with epifluorescense microscopy.   (B) Only in rare instances the *abc1* deletion mutant forms small infection hyphae.   Bar = 10μm.

The most related proteins to Abc1 include two yeast ABC transporters Pdr5 and Cdr1.   Both yeast transporters confer multidrug resistance (MDR) against xenobiotics (Balzi *et al.*, 1994; Hirata *et al.*, 1994; Sanglard *et al.*, 1996).   To determine if Abc1 plays a role in multidrug resistance, we tested the effects of a variety of candidate drugs on *abc1* mutants.   The tested drugs included protein synthesis inhibitors, sterol biosynthesis inhibitors, mutagenic agents and the rice phytoalexin, sakuranetin. However, our assays did not detect any difference in drug sensitivities between *abc1* mutants and wild type strain Guy11. This finding suggests that Abc1 may not play a general role in multidrug resistance, but may have a more specialized role in pathogenicity.   MDR-type ABC transporters, such as Pdr5 and the homologous transporter Snq2 in *S. cerevisiae* are transcriptionally up-regulated by exposure to metabolic poisons and by heat shock (Miyahara *et al.*, 1996).   Interestingly, *ABC1* mRNA levels can be induced by cycloheximide, rice phytoalexin sakuranetin and various azole antifungals.   In these experiments, it is also demonstrated that the plasmid insertion in TF7-3131 abolishes *ABC1* inducibility by various cytotoxic drugs. This finding suggests that the up-regulation of Abc1 is essential for invasive growth in plants.

Our findings provide the first evidence that ABC transporters may play a specific role in fungal pathogenicity. *M. grisea* Abc1 seems to be required specifically for the symptom development process. At least two hypothesis can be proposed to explain the function of Abc1. First, Abc1 may be required to export a fungal toxin during an early stage of pathogenesis. An alternative hypothesis is that Abc1 provides a defense function during early stages of pathogenesis, by acting as an efflux pump to provide resistance to antimicrobial compounds produced in rice cells.   Either way, the identification of the substrate of Abc1 will give a major insight in how the rice blast fungus establishes disease.

4.   Conclusions

Recent progress in fungal molecular biology is accelerating to accumulate an understanding of fungal pathogenicity mechanisms.   Additional insertional mutant hunts are being conducted by several laboratories. Differential or subtractive cDNA screening is being used to identify fungal genes expressed at specific stages in pathogenesis. Genomic and EST sequencing projects are making it easier to obtain particular genes through homology searches with DNA databases available on the World Wide Web. For large-scale functional analysis of fungal genes, more efficient gene knockout strategies will be necessary. The identifications of an increasing number of pathogenicity genes should provide new targets for designing antifungal drugs and may create new strategies for engineering disease resistant crops.   Both results will be of enormous importance for control of rice blast disease.

Acknowledgments

We thank all members in Hamer laboratory for useful and helpful discussions. K.A. was supported by a Japan Society for Promotion of Science Postdoctoral Fellowship for Research Abroad.   M.U. was a recipient of a postdoctoral fellowship from the Deutsche Forschungsgemeinschaft. The rice blast research in Hamer laboratory has been supported by a National Science Foundation grant.

References

Adachi, K. and Hamer, J.H. (1998) Divergent cAMP signaling pathways regulate growth and pathogenesis in the rice blast fungus *Magnaporthe grisea*, *Plant Cell* **10**, 1361-1373.

Balhadère, P.V., Foster, A.J. and Talbot, N.J. (1999) Identification of pathogenicity mutants of the rice blast fungus *Magnaporthe grisea* by insertional mutagenesis, *Mol. Plant-Microbe Interact.* **12**, 129-142.

Balzi, E., Wang, M., Leterme, S., Van Dyck, L., and Goffeau, A. (1994) PDR5, a novel yeast multidrug resisance conferring transporter controlled by the transcription regulator PDR1, *J. Biol. Chem.* **269**, 2206-2214.

Banuett, F. (1998) Signaling in the yeasts: An informational cascade with links to the filamentous fungi, *Microbiol. Mol. Biol. Rev.* **62**, 249-274.

Beckerman, J.L. and Ebbole, D.J. (1996) *MPG1*, a gene encoding a fungal hydrophobin of *Magnaporthe grisea*, is involved in surface recognition, *Mol. Plant-Microbe Interact.* **9**, 450-456.

Beckerman, J.L., Naider, F., and Ebbole, D.J. (1997) Inhibition of pathogenicity of the rice blast fungus by *Saccharomyces cerevisiae* α-factor, *Science* **276**, 1116-1119.

Choi, W. and Dean, R.A. (1997) The adenylate cyclase gene *MAC1* of *Magnaporthe grisea* controls appressorium formation and other aspects of growth and development, *Plant Cell* **9**, 1973-1983.

Dixon, R.A., Harrison, M.J., and Lamb, C.J. (1994) Early events in the activation of plant defense responses, *Annu. Rev. Phytopathol.* **32**, 479-501.

Hirata, D., Yano, K., Miyahara, K., and Miyakawa, T. (1994) *Saccharomyces cerevisiae* YDR1, which encodes a member of the ATP-binding cassette (ABC) superfamily, is required for multidrug resistence, *Curr. Genet.* **26**, 285-294.

Kronstad, J.W. (1997) Virulence and cAMP in smuts, blast and blights, *Trends Plant Sci.* **2**, 193-199.

Lau, G. and Hamer, J.E. (1998) *Acropetal*: a gene required for pathogenicity and termination of spore differentiation in the rice blast fungus, *Magnaporthe grisea*, *Fungal Genet. Biol.* **24**, 228-239.

Lee, Y.-H. and Dean, R.A. (1993) cAMP regulates infection structure formation in the plant pathogenic fungus *Magnaporthe grisea*, *Plant Cell* **5,** 693-700.

Liu, S. and Dean, R.A. (1997) G protein α subunit genes control growth, development, and pathogenicity of *Magnaporthe grisea*, *Mol. Plant-Microbe Interact.* **10,** 1075-1086.

Mitchell, T.K. and Dean, R.A. (1995) The cAMP-dependent protein kinase catalytic subunit is required for appressorium formation and pathogenesis by the rice blast pathogen *Magnaporthe grisea*, *Plant Cell* **7,** 1869-1878.

Miyahahara, K., Hirata, D., and Miyakawa, T. (1996) yAP-1- and yAP-2-mediated, heat shock-induced transcriptional activation of the multidrug resistance ABC transporter genes in *Saccharomyces cerevisiae*, *Curr. Genet.* **29,** 103-105.

Osbourn, A.E. (1996) Preformed antimicrobial compounds and plant defense against fungal attack, *Plant Cell* **8,** 1821-1831.

Sanglard, D., Ischer, F., Monod, M., and Bille, J. (1996) Susceptibilities of *Candida albicans* multidrug transporter mutants to various antifungal agents and other metabolic inhibitors, *Antimicr. Agents and Chemother.* **40,** 2300-2305.

Schafer, W. (1994) Molecular mechanisms of fungal pathogenicity to plants, *Annu. Rev. Phytopathol.* **32,** 461-477.

Shiraishi, T., Yamada, T., Saitoh, K., Kato, T., Toyoda, K., Yoshioka, H., Kim, H. -M., Ichinose, Y., Tahara, M., and Oku, H. (1994) Suppressors: Determinants of specificity produced by plant pathogens, *Plant Cell Physiol.* **35,** 1107-1119.

Sweigard, J.A., Carroll, A.M., Farrall, L., Chumley, F.G., and Valent, B. (1998) *Magnaporthe grisea* pathogenicity genes obtained through insertional mutagenesis, *Mol. Plant-Microbe Interact.* **11,** 404-412.

Urban, M., Bhargava, T., and Hamer, J.E. (1999) An ATP-driven efflux pump is a novel pathogenicity factor in rice blast disease, *EMBO J.* **18,** 512-521.

VanEtten, H.D., Soby, S., Wasmann, C., and McCluskey, K. (1994) Pathogenicity genes in fungi, in   (eds.), *Advances in Molecular Genetics in Plant-Microbe Interactions*, Kluwer Academic Publishers, Dordrecht, pp.163-170.

Xu, J.-R. and Hamer, J.E. (1996) MAP-kinase and cAMP signaling regulate infection structure formation and pathogenic growth in the rice blast fungus *Magnaporthe grisea*, *Genes Dev.* **10,** 2696-2706.

Xu, J.-R., Staiger, C.J., and Hamer, J.E. (1998) Inactivation of the mitogen-activated protein kinase mps1 from the rice blast fungus prevents penetration of host cells but allows activation of plant defense responses, *Proc Natl Acad Sci USA* **95,** 12713-12718.

Xu, J.-R., Urban, M., Sweigard, J.A., and Hamer, J.E. (1997) The *CPKA* gene of *Magnaporthe grisea* is essential for appressorial penetration, *Mol. Plant-Microbe Interact.* **10,** 187-194.

INHIBITION OF SIGNAL TRANSDUCTION LEADING TO APPRESSORIUM FORMATION IN *MAGNAPORTHE GRISEA* BY GLISOPRENINS

E. THINES[1], F. EILBERT[2], H. ANKE[2] & O. STERNER[3]
[1] *School of Biological Sciences, University of Exeter, Washington Singer Laboratories, Perry Road, Exeter EX4 4QL, United Kingdom*
[2] *Dept. of Biotechnology, University of Kaiserslautern, Paul-Ehrlich-Str. 23, D-67663 Kaiserslautern, Germany*
[3] *Organic Chemistry 2, University of Lund, S-22100 Lund, Sweden*

Abstract

From submerged cultures of the deuteromycete *Gliocladium roseum*, strain HA190-95, glisoprenins A, C, D and E were isolated as inhibitors of appressorium formation in *Magnaporthe grisea*. The compounds inhibited formation of infection structures on hydrophobic, stimulating surfaces. Glisoprenin E was ten times less active compared to the other glisoprenins. Formation of appressoria on non-stimulating surfaces, induced by the cAMP analogue 8-(4-chloro-phenylthio)-adenosine-3',5'-monophosphate or by 1,16-hexadecanediol, a plant wax component, was not affected by glisoprenins, indicating that at least two signal transduction pathways are involved in appressorium formation in *M. grisea*. Inhibition of appressorium development by glisoprenins on hydrophobic surface could be reversed in a competitive manner by 1,2-dioctanoylglycerol, a known activator of protein kinase C (PKC) but not by 1-oleoyl-2-acetylglycerol, the most effective inducer of PKC in mammalian cells.

Introduction

Hydrophobicity of the plant surface has been shown to induce appressorium formation in the rice blast fungus *Magnaporthe grisea*, in addition to individual wax components and cAMP[1-4]. Test systems were therefore set up for screening to isolate inhibitors of appressorium formation on different surfaces in the presence and absence of inducers. Extracts used in this screening were obtained from submerged cultures of higher fungi and showed no antifungal activity. Three hundred extracts were tested and four exhibited inhibitory activity. Isolation of the active compounds from submerged cultures of the deuteromycete *Gliocladium roseum*, HA190-95, led

D. Tharreau et al. (eds.), Advances in Rice Blast Research, 267–270.
© 2000 *Kluwer Academic Publishers. Printed in the Netherlands.*

to four active glisoprenins. All four compounds inhibited appressorium formation exclusively on hydrophobic surfaces.

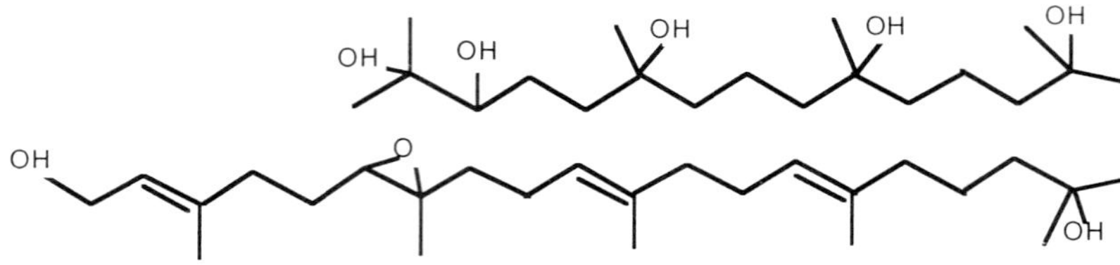

Figure 1. Structure of glisoprenin C

Results

        The fungal strain HA190-95 was isolated from the decaying fruiting body of a basidiomycete identified as *Gliocladium roseum* Bainier. Mycelia cultures of *Gliocladium roseum*, HA 190-95, were cultivated in 20 1 scale and bioactivity-guided fractionation yielded glisoprenin A, C, D, and E as inhibitors of appressorium formation. Whereas glisoprenin A had been already known from a *Gliocladium* species,[7] compounds C, D, and E were new fungal secondary metabolites[5,8]. Compared to glisoprenins A,C and D, glisoprenin E was ten times less active[5]. Under treatment with 4 µg/ml glisoprenin A, C or D, 80 % of the germinating conidia failed to form appressoria, whereas 40 µg/ml glisoprenin E were needed to obtain similar levels of inhibition.

All four compounds inhibited appressorium formation in *Magnaporthe grisea* on hydrophobic surfaces, whereas appressorium formation on hydrophilic surfaces induced by 1,16-hexadecanediol or chlorophenylthio-cAMP was not affected. Inducers of appressorium formation on hydrophilic surfaces, such as 1,16-hexadecanediol and chlorophenylthio-cAMP, failed to undergo the inhibition on hydrophobic surfaces after treatment with glisoprenins. Even at the tenfold concentration, needed to induce appressorium formation on the hydrophilic surface, 1,16-hexadecanediol and chlorophenylthio-cAMP failed to restore appressorium formation under treatment with the inhibitor. However 1,2-dioctanoylglycerol, a known inducer of PKC in mammalian cell lines[10], not only induced appressorium formation on hydrophilic surfaces, but was able to restore appressorium formation after treatment with glisoprenin. Inhibition by glisoprenins could be reversed in a competitive manner by 1,2-dioctanoylglycerol but not by 1-oleoyl-2-acetyl-glycerol[5]. We therefore suggest, that PKC is involved in the signal transduction leading to appressorium formation in *Magnaporthe grisea* on hydrophobic surface. The notion that PKC plays an important role in appressorium formation of filamentous fungi has also been reported by others[11].

TABLE 1 Restoration of appressorium formation in *M. grisea* on the hydrophobic surface of a GelBond-sheet, under treatment with glisoprenin C (A: 6,6 μM Glisoprenin C; B: 13,2 μM Glisoprenin C). In the controls without glisoprenin 95,6 % ± 2,4 % of germinating conidia formed an appressorium.

Compound formation [%]	Concentration	Appressorium	
	[μg/ml]	A	B
none		8.3 ± 2.5	0.7 ± 0.6
sn-1,2-dioctanoylglycerol	20	88.3 ± 5.5	46.7 ± 13.8
	50	93.2 ± 2.7	76.0 ± 4.0
sn-1-oleoyl-2-acetyl-glycerol	100	1.5 ± 0.7	n.t.
1,16-hexadecanediol	1	6.3 ± 1.0	n.t.
chlorophenylthio-cAMP	200	7.7 ± 3.7	n.t.

n.t.: not tested

Experimental

Organisms used. All *Magnaporthe grisea* strains used in this study were obtained from Dr. B. Speakman, BASF AG, Ludwigshafen. The fungi were cultivated as previously described[5].

Test systems for appressorium formation. Appressorium formation was followed as reported before. Compounds were added dissolved in methanol, the solvents final concentration was <2%. No solvent effects on germ tube elongation or appressorium formation were observed at this concentration. Experiments were performed as triplicates and 3 repeats of 100 conidia each were evaluated in each test. Concentrations of the compounds tested were:0.5, 1, 2, 4, 6, 8, 10, 15, 20, 25, 40, 50, 80 and 100 mg litre[-1].

Acknowledgements
This work was supported by the Bundesministerium für Bildung, Wissenschaft, Forschung and Technologie and the BASF AG, Ludwigshafen. We thank S. Mensch and R. Reiss for expert technical assistance

REFERENCES
Dean, R. A., (1997) Signal pathways and appressorium morphogenesis. *Annu. Rev. Phytopathol.*, **35**, 211-34.

Gilbert, R.D., Dean, R.A., (1996) Chemical signals responsible for appressorium formation in the rice blast fungus *Magnaporthe grisea*. *Physiological and Molecular Plant Pathology*, **48**, 335-46.

Lee, Y.-H., & Dean R. A., (1993) cAMP regulates infection structure formation in the plant pathogenic fungus *Magnaporthe grisea*. *Plant Cell*, **5**, 693-700.

Lee, Y.-H., & Dean, R. A., (1994) Hydrophobicity of contact surface induces appressorium formation in *Magnaporthe grisea*. *FEMS Microbiol. Lett.*, **115**, 71-76.

Thines, E., Eilbert, F., Anke, H. & Sterner, O., (1998) Glisoprenins C, D and E, new inhibitors of appressorium formation in *Magnaporthe grisea*, from cultures of *Gliocladium roseum*. 1. Production and biological activities. *J. Antibiotics*, **51**, 117-22. produced by *Gliocladium* sp. FO-1513. 1. Production, isolation and physico-chemical and biological properties. *J. Antibiotics*, **45**, 1202-06.

Sterner, O., Thines, E., Eilbert, F. & Anke, H., (1998) Glisoprenins C, D and E, new inhibitors of appressorium formation in *Magnaporthe grisea*, from cultures of *Gliocladium roseum*. 2. Structural elucidation. *J. Antibiotics*, **51**, 228-31.

Thines, E., Eilbert, F., Sterner, O. & Anke, H., (1997) Signal transduction leading to appressorium formation in germinating conidia of *Magnaporthe grisea*: effects of second messengers diacylglycerols, ceramides and sphingomyelin. *FEMS Microbiol. Lett.*, **156**, 91-94.

Quest, A. F. G., Raben, D. M. & Bell, R. M., (1996) Diacylglycerols biosynthetic intermediates and lipid second messengers. In *Lipid Second Messengers*, ed. R. M. Bell, J. H. Exton and S. M. Prescott, Plenum Press, New York and London, pp.1-58

Thines, E., Eilbert, F., Sterner, O. & Anke, H., (1997) Glisoprenin A, an inhibitor of the signal transduction pathway leading to appressorium formation in germinating conidia of *Magnaporthe grisea* on hydrophobic surfaces. *FEMS Microbiol. Lett.*, **151**, 219-24.

Tomoda, H., Huang, X.-H., Nishida, H., Masuma, R., Kim, Y. K. & Omura, S., (1992) Glisoprenins, new inhibitors of acyl-CoA: cholesterol acyltransferase.

Hamer, J. E. & Holden, D. W., (1997) Linking approaches in the study of fungal pathogenesis: a commentary. *Fungal Gen. Biol.*, **21**, 11-16.

THE ROLE OF CARBOHYDRATES IN THE PATHOGENICITY OF THE RICE BLAST FUNGUS *MAGNAPORTHE GRISEA*

A.J. FOSTER AND N.J. TALBOT
School of Biological Sciences, University of Exeter,
Washington Singer Laboratories,
Perry Road, Exeter, EX4 4QG, UK

1. Introduction

To understand, and ultimately control, the ability of *Magnaporthe grisea* to infect rice it will be necessary to dissect the underlying genetic and biochemical adaptations associated. Using molecular genetic, cytological and biochemical approaches there has been significant progress in unravelling the basis of appressorium formation and function (Dean, 1997). In contrast relatively little is known of the characteristic biology associated with growth *in planta*. Such growth will depend on the ability of the fungus to assimilate photosynthetically derived sugars and to convert these into forms useful for its own nutrition, or for storage and subsequent mobilisation. Most plant sugars are unavailable to the fungus prior to penetration such that the forms of nutrient available will change qualitatively and quantitatively through the stages of pathogenic development. Different enzymes for carbohydrate storage and mobilisation might therefore be expected act at each stage and these enzymes, and the proteins which control their activity, may be required for successful plant colonisation. Identification of proteins involved in nutrition during pathogenesis will indicate which metabolic pathways are involved and may identify novel fungicide targets. The first clues as to what these factors might be involved have come with the recent identification, via insertional mutagenesis, of the *PTH1* and *PTH9* genes (Sweigard *et al.*, 1998). These have homology to *GRR1*, a yeast gene involved glucose regulation of transcription, (catabolite repression), and coupling the nutritional status of the cell to the progression of the cell cycle (Flick and Johnston, 1991; Li and Johnston, 1997), and *NTH1*, a neutral trehalase encoding gene (Kopp *et al.*, 1993; Kopp *et al.*, 1994) respectively. Mutant strains of *M. grisea* which carry an insertion within the respective homologues of these genes show reduced pathogenicity toward susceptible hosts suggesting that regulation of carbohydrate metabolism, possibly involving catabolite repression, and trehalose mobilisation both play a role in pathogenesis. As a first step toward understanding the carbohydrate fluxes which occur during pathogenesis we are characterising the *PTH9* neutral trehalase encoding gene in greater detail. In this review we consider the storage

D. Tharreau et al. (eds.), Advances in Rice Blast Research, 271–280.

carbohydrates present in fungi generally and reflect on which of these may have particular significance for pathogenesis of the rice blast fungus as well as the possible regulation of pathogenicity associated metabolic processes.

2. Storage carbohydrates in fungi: a general perspective

A simple classification of carbohydrates recognises three main groups: mono-saccharides, disaccharides and polysaccharides. The derivatives of certain soluble sugars, especially the sugar alcohols (or polyols, for example mannitol), fall among the first group and are common constituents of fungi. Polyols are the most extensively studied fungal secondary metabolites (Rast and Pfyffer, 1989) and can form a significant proportion of biomass: up to 20% of the dry weight of mycelium and up to 40% of the dry weight of certain spore types (Pfyffer and Rast, 1980). Among the disaccharides one especially - trehalose - is noted for its wide distribution in fungi (Lewis and Smith, 1967). Polysaccharides are also important as constituents of cell walls (e.g. glucans and mannans) (reviewed by Peberdy, 1989), for storage (glucans and glycogen) and also for reactions which occur at the cell surface, such as self non-self recognition (glyco-proteins).

3. Potential carbohydrate fluxes during the disease cycle of *M. grisea*

In simple nutritional terms what the rice blast fungus must achieve in its pathogenic life style is to break into plants and consume any ready source of nutrients, reproduce itself many times, and then distribute itself to new uninfected hosts. Clearly adjustments of central carbohydrate metabolism must occur during these processes. This discussion will consider what these adjustments might be and where and when they might occur.

Considerable progress has been made in understanding the disease cycle of the rice blast fungus. Asexual spores (conidia) are believed to be responsible for the dissemination of the disease in the field. These are produced at sites of lesions on infected rice plants and can dispersed onto uninfected plants. As soon as the spore is hydrated a strong conidial attachment to the leaf is formed via release of spore tip mucilage from the conidial apex (Hamer *et al.*, 1988). Spore tip mucilage has a carbohydrate component which binds concanavalinA, suggesting it contains a-linked glucosyl or mannosyl residues (Howard and Valent, 1996). Germination then occurs immediately upon attachment (Howard, 1994).

M. grisea conidia germinate in distilled water independently of the surface type or of exogenous nutrients (Xiao *et al.*, 1994; Howard, 1994). In contrast to other fungal species germination occurs very rapidly in *Magnaporthe grisea* and the reserves present in the conidia are able to support growth of a short germ tube and formation of a fully functional appressorium without exogenous nutrients. Clearly endogenous nutrients stored within the fungus play a critical role during the early pre-penetrative phase of the disease cycle and yet very little is known about the nutrients used or even what forms of reserve are available. The fungus is known to accumulate mannitol within mycelium (Yamada *et al.*, 1961) and both mannitol dehydrogenase and hexokinase activities, necessary for conversion of mannitol to fructose-6-phosphate, have been detected (Hult *et al.*, 1980). Mannitol levels do not change significantly during the growth of the fungus in culture (Yamada *et al.*, 1961) and the existence of a mannitol cycle in the fungus has been suggested to be a mechanism for the generation of NADPH which is required, in particular, for synthesis of lipids (Hult *et al.*, 1980). Investigations in our laboratory have detected mannitol within conidia, along with the disaccharide trehalose, suggesting that these carbohydrates could act as possible candidate endogenous carbon sources for initial stages of development. Mannitol, for example, is known to decrease during the germination of spores of various fungal species (see Gottlieb, 1978, and references therein). In a number of fungi trehalose degradation during spore germination has also been observed (reviewed by Thevelein, 1984). In contrast to *M. grisea*, in the model fungi *Aspergillus nidulans* and *Neurospora crassa* spore germination requires an external carbon source; this carbon source may trigger germination but is not necessarily consumed during the process since non-metabolisable carbon sources also allow the process to proceed (d'Enfert, 1997). As *M. grisea* conidia germinate readily in distilled water there is a clear need to identify the endogenous nutrients used during conidial germination.

4. Trehalose metabolism and the pre-penetration growth phase

The non-reducing disaccharide trehalose, (α-D-glucopyranosyl-α-D-glucopyranoside), is a major component of spores and stationary phase cells of yeast and other fungi and has been considered to be a potential carbohydrate reservoir (Thevelein, 1994). Research in recent years however, suggests a second role for trehalose as a protectant against forms of environmental stress (Van Laere, 1989; Wiemken, 1990). In budding yeast, for example, trehalose accumulates during glucose or nitrogen starvation, or in cells subjected to heat shock (Thevelein, 1994). Consistent with this, increased trehalose content has been correlated with enhanced resistance to stress (reviewed in Jorge *et al.*, 1997). Accumulated trehalose is metabolised upon return to conditions favourable for active growth (Thevelein, 1994), or recovery form heat shock (Hottiger *et al.*, 1987; Nwaka *et al.*, 1994). Trehalose can protect enzymatic activity *in vitro* (Hottiger *et al.*, 1994; De Virgilio *et al.*, 1994) suggesting accumulation *in vivo* may fulfil a similar role. Trehalase (α,α-trehalose glucohydrolase, EC 3.2.1.28) hydrolyses trehalose yielding two molecules of glucose. Trehalase activity has been detected across a broad range of fungal species. Trehalase assays undertaken to date have detected activity with either an acidic and/or a neutral pH optimum (Thevelein, 1984). Neutral trehalase activity tends to be cytosolic while acidic activity is often associated with vacuoles or the periplasm. Trehalase enzymes have been studied extensively in *S. cerevisiae* where both acidic and neutral activities have been recorded (Londesborough and Varimo, 1984). One acidic trehalase and two neutral trehalase encoding genes have since been identified and characterised (Kopp *et al.*, 1993; Wolfe and Lohan, 1994; Nwaka *et al.*, 1995).

The fact that trehalose degradation during spore germination is a feature of a number of fungal species, where the process has been analysed, suggests the disaccharide may be an important endogenous nutrient during the early stages of the rice blast disease cycle. The *PTH9* gene is however dispensable for germination suggesting that, if trehalose is used for germination, other trehalase activities must exist.

Upon germination on an appropriate surface, the *M. grisea* germ tube grows a short distance before growth becomes depolarised and apical hooking and swelling occurs. Later a septum is formed separating the terminal cell which differentiates to become an appressorium. The high turgor generated by this cell, which allows mechanical penetration of the plant cuticle, appears to result principally from the accumulation of glycerol as a compatible solute (de Jong *et al.*, 1997). In *A. nidulans* degradation of trehalose within the spore is concomitant with accumulation of glycerol, and is apparently independent of exogenous carbon sources (d'Enfert and Fontaine, 1997). Here glycerol accumulation appears to be a prerequisite for anisotropic growth and germ tube elongation. Given that glycerol has been implicated in appressorium turgor generation in *M. grisea* we are currently testing whether trehalose is a potential source (via glucose and glycolysis) for biosynthesis of glycerol. An alternative source for production of glycerol may, however, be glycogen, another widely distributed carbon source among fungi. Consistent with this alternative biosynthetic route, loss of putative glycogen bodies within the appressoria with appressorium maturation has been detected using electron microscopy (Bourett and Howard, 1990). Determining the pathway used for glycerol formation within appressoria is therefore an important goal for research in our laboratory. In *S. cerevisiae* glycerol, the sole compatible solute produced during osmotic stress, is formed by a NADH-dependent glycerol-3-phosphate dehydrogenase catalysing reduction of dihydroxyacetone phosphate to glycerol-3-phosphate which is then converted to glycerol by two specific phosphatases, GPP1 and GPP2 (Norbeck *et al.*, 1996). It is important to establish whether *M. grisea* has a similar pathway for glycerol formation within the appressorium.

5. Potential nutrient fluxes during the *in planta* growth phase

Once *M. grisea* has gained entry to the plant, exogenous nutrients will become available and it may be worth speculating that carbohydrate metabolism changes radically at this point. Little detail is known however. Once within the host *M. grisea* grows within the first 2-3 epidermal cells with characteristic swollen hyphae. (Bourett and Howard, 1990). Infection hyphae, with a more typical growth design, then ramify throughout the epidermal and mesophyll cells as colonisation proceeds. Sequestration of plant nutrients by the fungus must rapidly follow entry into the plant and could then act as a reserve in *M. grisea* hyphae for subsequent sporulation. The most abundant nutrient within the plant will be photosynthetically derived sucrose. During the initial, essentially biotrophic, growth phase it would clearly be advantageous for *M. grisea* to convert this sucrose into a form not readily used by the plant. Such conversions have been reported in a number of biotrophic interactions and may act as a 'biochemical valve', that is a device to maintain nutrient flow from plant to fungus, but not from fungus to plant (see Harley and Smith, 1983). The carbohydrate components of blast-infected leaves suggests that mannitol, arabitol and trehalose accumulate in hyphae during the interaction (Hwang *et al.*, 1989). Consistent with this idea, mannitol is not found in healthy plants generally and can not be utilised by most plants (Lewis and Smith, 1967). Mannitol and arabitol have been independently detected in heavily blast infect rice leaves and the proportion of arabitol among total polyols was found to increase when compared to the polyol composition of cultured *M. grisea* (Pfyffer *et al.*, 1990). Trehalose has not been detected within healthy plants generally and fungal trehalose may not be metabolically available to the plant, although a number of plant species have trehalase enzymes (Muller *et al.*, 1995). Since rice plants have not been found to possess arabitol or mannitol, these both may act as a *M. grisea* carbohydrate store which can not be metabolised by the rice plant. Fungal nutrient stores may therefore be biochemically (and physically) inaccessible to the rice plant. Conversion of sucrose by the fungus would first require inversion to give fructose and glucose by an invertase. Plants possess invertases themselves and they may have a protective role in the production of toxic reducing sugars such as melibiose (Cochrane, 1958). In *M. grisea*-infected rice plants, acid invertase shows a 2-3 fold increase in activity in plants at 5 days post-inoculation and a 3-5 fold increase in plants 7 days post-inoculation (Hwang *et al.*, 1989), but it was not possible to show whether the increase was plant or fungus derived. *M. grisea* may, however, use the fructose produced as a result of this activity for polyol formation because high levels of sucrose and glucose, but not fructose, have been detected in infected plants (Hwang *et al.*, 1989). Some glucose meanwhile could obviously be channelled into the formation of trehalose or glycogen.

Polyols and trehalose may also play a role in osmo-regulation of *M. grisea* within the rice plant. We have shown that arabitol concentration increases within the soluble carbohydrates of *M. grisea* with application of hyper-osmotic stress (K.P.Dixon, unpublished results). Due to its non-reducing nature, trehalose may accumulate to high concentrations without toxicity while glycerol is well known as a compatible solute in yeast. Trehalose has been implicated in the response of yeast to environmental stress and accumulation of the disaccharide during various stresses is correlated with increased resistance (Van Laere, 1989; Wiemken, 1990; see also Nwaka *et al.*, 1998). It is possible to envisage a requirement for similar stress responses during growth of *M. grisea in planta*, in particular in overcoming plant defence mechanisms.

In a compatible *M. grisea*-rice interaction, lesions become visible on the surface of the plant leaf approximately 72 hours after the initial germination event. At this point up to 10% of the total leaf biomass may be fungal tissue (Talbot *et al.*, 1993). Conidiophores, bearing large numbers of conidia, become apparent in the lesions between five to seven days after infection (Ou, 1985), and this sporulation event is undoubtedly preceded by mobilisation of stored carbohydrates by the fungus.

6. Regulation of pathogenesis and carbon metabolism

A number of regulatory proteins have been identified and implicated in pathogenesis. Consideration should therefore be given as to whether these may act on the metabolic processes discussed. cAMP-dependent signalling, for example, is known to regulate appressorium function; the cAMP-dependent protein kinase encoding gene, *CPKA*, gene was isolated following the observation that appressorium formation was cAMP dependent under certain conditions (Mitchell and Dean, 1995). *cpka-* null mutants are non-pathogenic because they have a penetration deficiency, suggesting that they may not generate the high turgor levels of wild type appressoria and that *CPKA* may somehow regulate turgor accumulation (Xu *et al.*, 1997). cAMP has also been shown to act as a second messenger during the activation of trehalases in response to spore germination and stress (Thevelein, 1988). One possibility is that the assumed reduced turgor in *cpka* - strains results from a lack of induction of trehalase activity by the kinase depriving the appressorium of metabolic precursors for glycerol formation. We have shown that, in common with the predicted protein products all other fungal neutral trehalase encoding genes (Nwaka *et al.*, 1998), *NTH1*p has a consensus site for phosphorylation by PKA suggesting that *NTH1*p is a target for PKA regulation of during pathogenesis. This hypothesis is currently being tested.

It has also been demonstrated that a MAP kinase pathway is required in *M. grisea* for both appressorium formation and for growth *in planta*. The *PMK1 M. grisea* gene is a homologue of the *S. cerevisiae FUS3/KSS1*, MAP kinase encoding genes. No appressoria are formed in *PMK1⁻* mutants of *M. grisea* and *PMK1⁻* mutants do not grow invasively in rice plants even if injected into the leaf sheaf. This suggests that the action of *PMK1* may be required at least two time points. Interestingly mutants still respond to cAMP during the initial stages of the infection process suggesting that the gene may act after the cAMP signal for appressorium formation.

It has been postulated that starvation stress *in planta* may act as a trigger for pathogenesis (Talbot *et al.*, 1993; Talbot *et al.*, 1997). The *MPG1* gene, which is highly expressed during pathogenesis, is induced by carbon or nitrogen starvation stress (Talbot *et al.*, 1993; Lau and Hamer, 1996; Beckerman and Ebbole, 1996). The *NPR1* and *NPR2* genes which regulate nitrogen uptake and *MPG1* expression of, are also required for full pathogenicity (Lau and Hamer, 1996). If starvation stress acts to trigger pathogenesis then trehalose metabolism, which has an established role in the stress response of yeast, may be required to maintain fungal viability during the stress. In common with the promoter of *S. cerevisiae NTH1*, the promoter of *M. grisea NTH1* has closely spaced stress response elements and is expressed in response to hyper osmotic stress. We are currently exploring the genetic basis for this regulation of neutral trehalase and would like to know what significance it has for the pathogenic development of *M. grisea*. It is clear from the above that regulation of pathogenesis is complex, multifaceted and co-operative. It seems likely that elements of carbohydrate metabolism will emerge as targets for this regulation.

7. Future Perspectives

With initiatives aimed at sequencing the entire *M. grisea* genome there is unprecedented potential for progress in the identification of genes in the near future. Distinguishing those genes which have a role in pathogenesis will obviously follow more slowly. What is clear from progress to date is that there may be only a small set of pathogenicity specific genes and that the pathogenic character of *M. grisea* may depend instead on the unique expression pattern and regulation of genes homologous to those present in other, non-pathogenic fungi. Among these, some genes involved in carbohydrate metabolism may be required for pathogenesis and we can expect the identification of more genes such as *NTH1* and *PTH1*. It is however the determination of the unique and integrated pattern of gene expression and its regulation which accompanies successful fungal colonisation of rice that is the challenge for future research.

8. References

Beckerman, J. L., Ebbole, D. J. (1996) *MPG1*, a gene encoding a fungal hydrophobin of *Magnaporthe grisea*, is involved in surface recognition. *Mol. Plant Microbe Interact.*, 9, 6, 450-6

Bourett, T. M. and Howard, R. J. (1990) *in vitro* development of penetration structures in the rice blast fungus *Magnaporthe grisea. Can. J. Bot.*, 68, 329-342.

Carrillo, D., Vicente-Soler, J. and Gacto, M. (1994) Cyclic AMP signalling pathway and trehalase activation in the fission yeast *Schizosaccharomyces pombe. Microbiol*, 140, 1467-1472

Cansado, J., Soto T., Fernandez, J., Vicente-Soler, J., Gacto, M. (1998) Characterisation of mutants devoid of neutral trehalase activity in the fission yeast *Schizosaccharomyces pombe*: partial protection from heat shock and high-salt stress. *J. Bacteriol.* 180, 5 1342-1345

Cochrane, V. W. (1958) Physiology of fungi. John Wiley, New York

Dean, R. A. (1997) Signal pathways and appressorium morphogenesis. *Annu. Rev. Phytopath.*, 35, 211-234

De Jong, J. McCormack, B. Smirnoff, N. and Talbot, N. J. (1997) Glycerol generates turgor in rice blast. *Nature*, 389, 244

d'Enfert, C. (1997) Fungal spore germination: Insights from the molecular genetics of *Aspergillus nidulans* and *Neurospora crassa*. *Fung. Genet. Biol.* 21, 2,163-172

d'Enfert, C. and Fontaine, T. (1997) Molecular characterisation of the *Aspergillus nidulans treA* gene encoding an acid trehalase required for growth on trehalose. *Mol. Microbiol*, 24, 1, 203-216

Flick, J. S. and Johnston, M. (1991) *GRR1* of *Saccharomyces cerevisiae* is required for glucose repression and encodes a protein with leucine-rich repeats. *Mol. Cell Biol.*, 11, 5101-5112

Gottlieb, D. (1978) *The germination of fungus spores.* (Patterns of progress : microbiology series). Meadowfield Press Ltd, Shildon, U.K.

Hamer, J. E., Howard, R. J., Chumley, F. G. and Valent, B. (1988) A mechanism for surface attachment of spores of a plant pathogenic fungus. *Science*, 239, 288-290.

Harley, J. L. and Smith, S. E. (1983) *Mycorrhizal Symbiosis.* Academic Press: London, U.K

Hottiger, T. Schmutz, P. and Wiemken, A. (1987) Heat-Induced accumulation and futile cycling of trehalose in *Saccharomyces cerevisiae*. *J. Bacteriol*, 169, 12, 5518-5522.

Hottiger, T. De Virgilio, C. Hall, M: Boller, T. and Wiemken, A. (1994) The role of trehalose synthesis for the acquisition of thermotolerance in yeast. II. Physiological concentrations of trehalose increase the thermal stability of proteins in vitro. *Eur. J. Biochem*, 219, 187-193.

Howard, R. J. (1994) Cell biology of pathogenesis. In *Rice Blast Disease*. Ed. Zeigler R.S., Leong S.A. and Teng, P.S. Wallingford, U.K.: CAB International 3-22

Howard, R. J., Valent, B. (1996) Breaking and entering: host penetration by the fungal rice blast pathogen *Magnaporthe grisea*. *Annu. Rev. Microbiol.*, 50, 491-512

Hult, K., Veide, A. and Gatenbeck, S. (1980) The distribution of the NADPH regenerating mannitol cycle among fungal species. *Arch. Microbiol.*, 128, 2 253-255

Hwang, B. K., Kim, D. K. and Kim, Y. B. (1989) Carbohydrate-composition and acid invertase activity in rice leaves infected with Pyricularia-oryzae. *J. Phytopath-Phytopath. Zeitschrift*, 125, 2, 124-132

Jorge, J. A., Polizeli, Md. L. T. M., Thevelein, J. M. and Terenzi, H. F. (1997) Trehalases and trehalose hydrolysis in fungi. *FEMS Micrbiol. Lett.*, 154, 165-171

Kopp, M., Muller, H. and Holzer, H. (1993) Molecular analysis of the neutral trehalase gene from *Saccharomyces cerevisiae*. *J. Biol. Chem*, 268, 7, 4766-4774.

Kopp, M., Nwaka, S. and Holzer, H. (1994) Corrected sequence of the yeast neutral trehalase-encoding gene (*NTH1*): biological implications. *Gene* 150, 2, 403-4

Lau, G. and Hamer, J. E. (1996) Regulatory Genes Controlling *MPG1* Expression and Pathogenicity in the Rice Blast Fungus *Magnaporthe grisea*. *Plant Cell*, 8, 771-781

Lee, Y- H. and Dean R. A. (1993) cAMP Regulates Infection Structure Formation in the Plant Pathogenic Fungus *Magnaporthe grisea*. *Plant Cell*, 5, 693-700.

Lee, Y- H. and Dean, R.A. (1994) Hydrophobicity of contact surfaces induces appressorium formation of *Magnaporthe grisea*. *FEMS Microbiol Lett.*, 115, 71-76

Lewis, D. H. and Smith, D. C. (1967) Sugar alcohols in fungi and green plants. *New Phytol.*, 66, 143-184.

Li, F. N., Johnston, M. (1997) *GRR1* of *Saccharomyces cerevisiae* is connected to the ubiquitin proteolysis machinery through Skp1: coupling glucose sensing to gene expression and the cell cycle *EMBO J.*, 1997, 16, 18, 5629-5638

Londesborough , J. and Varimo, K. (1984) Characterisation of two trehalases in bakers yeast. *Biochem J.*, 219, 511-518.

Mitchell, T. K. and Dean R. A. (1995) The cAMP-dependent protein kinase catalytic subunit is required for appressorium formation and pathogenesis by the rice blast pathogen *Magnaporthe grisea*. *Plant Cell*, 7, 1869-1978.

Mittenbuhler, K. and Holzer, H. (1988) Purification and characterisation of acid trehalase from the yeast suc2 mutant. *J. Biol. Chem.*, 263, 17, 8537-8543.

Muller, J., Boller, T. and Wiemken, A. (1995) Trehalose and trehalase in plants: Recent developments. *Plant Sci.*, 112, 1, 1-9

Norbeck, J., Pahlman, A. K., Akhtar, N., Blomberg, A. and Adler, L. (1996) Purification and characterization of two isoenzymes of DL- glycerol-3-phosphatase from *Saccharomyces cerevisiae* - Identification of the corresponding GPP1 and GPP2 genes and evidence for osmotic regulation of Gpp2p expression by the osmosensing mitogen-activated protein kinase signal transduction pathway. *J. Biol. Chem.*, 271, 23, 13875-13881

Nwaka, S., Kopp, M. and Holzer, H. (1995) Expression and function of the trehalase genes *NTH1* and YBR0106 in *Saccharomyces cerevisiae*. *J. Bio. Chem.* 270, 17, 10193-10198.

Nwaka, S, and Holzer, H. (1998) Molecular biology of trehalose and the trehalases in the yeast *Saccharomyces cerevisiae*. *Prog. Nucleic. Acid Res. Mol. Biol.* 58: 197-237

Ou, S. H. (1985) *Rice Diseases* (Kew, Surrey: Commonwealth Mycological Institute, C.A.B.) pp. 109-201.

Peberdy, J. F. (1989) Fungal cell walls - a review. In Biochemistry of Cell Walls and Membranes in Fungi (ed. P.J. Kuhn, A.P.J. Trinci, M.J. Jung and M.W. Goosey). Springer-Verlag, Berlin, Germany

Pfyffer, G. E. Boraschi-Gaia, C., Weber, B., Hoesch, L., Orpin, C. G. and Rast, D. M. (1990) A further report on the occurrence of acyclic sugar alcohols in fungi. *Mycol. Res.*, 94, 219-222

Rast, D. M. and Pfyffer, G. E. (1989) Acyclic polyols and higher taxa of fungi. *Bot. J. Linn. Soc.* 99, 39-57

Sweigard , J. A., Carrol, A. M., Farrall, L. Chumley, F. G. and Valent, B. (1998) *Magnaporthe grisea* pathogenicity genes obtained through insertional mutagenesis. *Mol. Plant Microbe Interact.*, 11, 404-412

Talbot, N. J., Ebbole, D. J. and Hamer, J. E. (1993) Identification and characterisation of *MPG1*, a gene involved in pathogenicity from the rice blast fungus *Magnaporthe grisea*. *Plant Cell*, 5, 1575-1590.

Talbot, N. J., McCafferty, H. R. K., Ma, M., Moore K. and Hamer J. E. (1997) Nitrogen starvation of the rice blast fungus *Magnaporthe grisea* may act as an environmental cue for disease symptom expression. *Physiol. Molec. Plant Path.* 50, 3, 179-195

Thevelein, J. M. (1984) Regulation of Trehalose Mobilisation in Fungi. *Microbiol Rev*, 48 (1), 42-59

Thevelein , J. M. (1988) Regulation of trehalase activity by phosphorylation-dephosphorylation during developmental transitions in fungi. *Exp. Mycol*, 12, 1-7.

Thevelein, J. M. (1994) Signal Tranduction in Yeast. *Yeast*, 10, 1753-1790

Valent, B., Crawford, M. S., Weaver, C. G. and Chumley, F. G. (1986) Genetic studies of fertility and pathogenicity in *Magnaporthe grisea*. *Iowa Stat. J. Res.*, 60, 569- 594

Van Laere, A. (1989) Trehalose, reserve and/or stress metabolite. *FEMS Microbiol. Rev.*, 63, 201-210.

Wiemken, A. (1990) Trehalose in yeast, stress protectant rather than reserve carbohydrate. *Antonie Leeuwenhoek*, 58, 209-217.

Wolfe, K. H. and Lohan A. J. (1994) Sequence around the centromere of *Saccharomyces cerevisiae* chromosome II: similarity of CEN2 to CEN4. *Yeast* , Suppl. A: S41-6

Xu, J. R. and Hamer, J. E. (1996) MAP kinase and cAMP signalling regulate infection structure formation and pathogenic growth in the rice blast fungus *Magnaporthe grisea*. *Genes Dev.*, 10, 21, 2696-2706

Xu, J. R. Urban, M., Sweigard, J. A. and Hamer, J. E. (1997) The *CPKA* gene of *Magnaporthe grisea* is essential for appressorial penetration. *Molec. Plant Microbe Interact.* 10, 187-194

Yamada, H., Okamoto K., Kodama, K., Noguchi, F. and Tanaka, S. (1961) Enzymic studies on mannitol formation in *Piricularia oryzae*. *J. Biochem.* 49, 5, 404-410

POSSIBLE ROLE OF CONCANAVALIN A-BINDING GLYCOPROTEIN(S) SECRETED FROM GERMINATING CONIDIA OF *MAGNAPORTHE GRISEA*

T. TERAOKA, M. NAGAOKA, K. HIRANO, H. TAKAHASHI and
D. HOSOKAWA
*Departement of Applied Biological Science, Tokyo University of
Agriculture and Technology, Fuchu, Tokyo 183-8509, Japan*

1. Introduction

Many plant species contain carbohydrate-binding proteins, which are commonly referred to as lectins or agglutinins. But their physiological role in plants has not yet been elucidated. One of the possible roles is the idea that lectins play a role in the plant's defense mechanism as does the immunoglobulin in the animal owing to their binding specificity.

We found the novel rice lectin in seedlings through the studies between rice plant and its pathogens, such as *Xanthomonas campestris* pv. *oryzae* and *Magnaporthe grisea* [1]. In rice plant, only the seed lectin such as rice bran lectin has been reported until our discovery. The seed lectin are specific to N-acetyl-glucosamine and its β-1,4-chitinoligomers and naturally distrubutes in seeds. On the other hand, our novel rice lectin is specific to glucose and mannose residues in much the same way as Concanavalin A (ConA) is. So the lectin is named the mannose-binding rice lectin (MRL). The MRL has a potential acitivity to agglutinate the intact cell of *X. campestris* and also conidia and protoplasts of *M. grisea*, and distributes in all growth parts, mainly in shoot and root. Moreover, it is composed of some isolectins and thought to belong to a family of stress-inductible proteins, because one of the isolectins s almost identical to salt and drought stress-inducible *salT* gene [2]. Then we focus our attention on the hypothesis that MRL may function as a plant antibody or a receptor in host-parasite interaction. If our hypothesis is correct, the counterpart, *M. grisea* should have molecule(s) interacting with MRL in the infection process, naturally in fluid of the germinating conidia. One of the essential approach to verify our hypothesis was to clarify the existence of the molecule(s) interacting with MRL / Con A at infection of *M. grisea* and, if there were, their characterization and possible role in infection.

D. Tharreau et al. (eds.), Advances in Rice Blast Research, 281–285.

2. Existence of the molecule(s) interacting with MRL / ConA at infection of *M. grisea*

Already some reports have been presented to support the existence of the molecule(s) we expect; Hamer *et al.* [3] reported that germinating conidia of *M. grisea* secretes the ConA-binding molecule associated with spore adhesion on surface. Xiao *et al.* [4] also proposed that extracellular glycoprotein(s) bound to ConA has an important role in appressorium formation. From these reports, probably the molecule(s) interacting with MRL / ConA exists at infection site in or out *M. grisea*. So we set up aplan using commercially available ConA instead of MRL as shown in Fig.1.

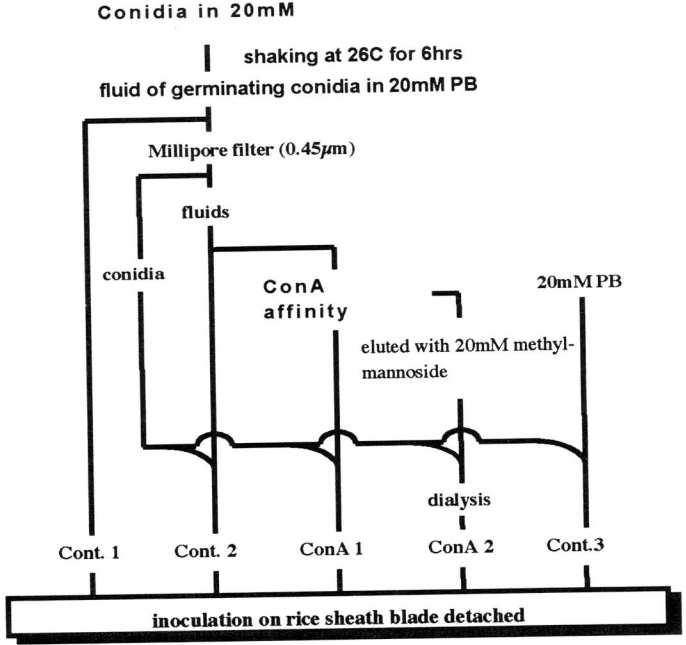

Fig. 1. Plan for verification of evidence of ConA-binding molecule(s) secreted from germinating conidia of *M. grisea* and the role in the infection.

Firstly, conidia of *M. grisea* suspended in 20mM phosphate buffer (PB, pH 7) was shaken slowly to promote germination and the release of molecules secreted. Then the fluid was separated with by Millipore filter (0.45μM) and passed through the ConA affinity column to trap the molecules interacting with MRL / ConA. When conidia were mixed again with the solution passed through the ConA column and inoculated on rice sheath blade detached, the growth of infection hypha and the appressorium formation were severely suppressed 48 hours after inoculation (Fig. 2). In another line, the ConA column was eluted with 20mM methyl-mannoside in PB to recover the molecules trapped. When conidia were mixed with the eluate after dialysis to PB and inoculated similarly, the inhibitory effect on the growth of infection hypha was restored to some degree (Fig. 2). Furthermore, when the fluid of germinating conidia was previously treated with periodic acid, heating at 120° for 5 min, or protease K respectively, similar inhibitory effect on growth of infection hypha was observed as described above. These effects did not depend on races and rice cultivars tested, although the inhibitory effect appeared more severely in the compatible than in the incompatible ones. These results suggest that some glycoproteins or proteoglycans, which can interact with MRL, are actually secreted in the infection process and may play important roles in agressiveness of infection hypha, and also in appressorium formation. We must remark that the

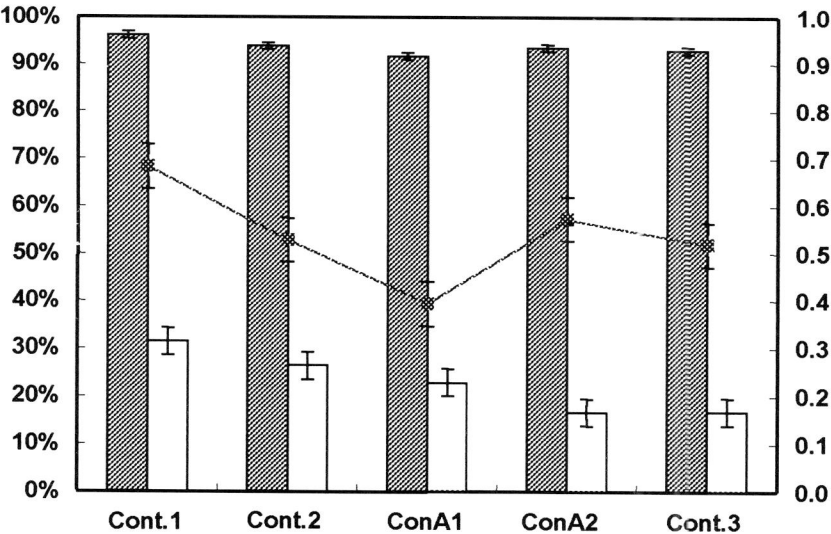

Fig. 2 Effect of Con A-binding molecule(s) supposed to be secreted from the germinating conidia of *Magnaporthe grisea* (race; 007) on their infection to compatible rice cultivar (cv. Aichiasahi). ▨ ; % of spore germination, ▢ ; % of appressorium formation, ⊶ ; growth degree of infection hypha.

secretion may not be dependent on the races of *Magnaporthe grisea* and also the inhibitory effect may not be specific to race-cultivar interaction.

Well, how many molecules were associated with the inhibitory effect? In Western blot analysis of the fluid with MRL and MRL-specific monoclonal antibody, some molecules were detected. And similar bands were detected in the eluate from ConA affinity column with 20mM methyl-mannoside.

3. Interaction between MRL and ConA-binding molecules in infection

Assuming that MRL can funtion as a plant antibody or a receptor in host-parasite interaction, the treatment with MRL-specific antibody before inoculation or simultaneously with inoculation should encourage the inhibitory effect of the ConA-binding molecule(s) secreted in the fluid by blocking MRL. Conversely, the treatment with ConA before inoculation or simultaneously with inoculation should suppress the effect of the ConA-binding molecule(s) by cooperating with MRL. In both cases their treatments after inoculation when *M. grisea* have infected should have no effect, assuming that MRL and ConA-binding molecule(s) can function by interacting with one another infection site.

As we expected, the MRL-specific antibody encouraged the growth of infection hypha when treated at 10 hours before inoculation and simultaneously with inoculation, but not at 10 hours after inoculation. This effect of the treatment with antibody was stronger in the incompatible than in the compatible. On the contrary, ConA (50μg/ml) strongly inhibited the growth of infection hypha when treated at 10 hours before inoculation and simultaneously with inoculation, but at 10 hours after inoculation. This effect of the treatment with ConA was stronger in the compatible than in the incompatible, especially when using conidia pretreated by shaking as inoculum since there was decrease in the amount of the ConA-binding molecules. In other race-cultivar combination tested, almost the same effects were observed. Interestingly, in these treatments the effect on appressorium formation was not parallel to that on the growth of infection hypha.

4. Conclusion

All our results suggest that *M. grisea* secrets some glycoproteins in infection process which interact with ConA / MRL, and that one or some of them play important roles in appressorium formation and aggressiveness of infection hypha. Probably the glycopreoteins(s) associated with appressorium formation may be different from the glycoprotein(s) associated with aggressiveness of infection hypha, besause the effect of treatments with MRL-specific antibody and ConA on appressorium formation was not always parallel to that on aggressiveness of infection hypha. Hereafter we must try to fractionate the fluid and identify the glycoprotein(s) associated with each infection step, resulting in verification of general pathogenic factors in *M. grisea*. Anyway, all our results also supports strongly our hypothesis that MRL can function as a plant antibody or a receptor in host-parasite interaction, and that MRL may not be associated with race-specific resistance but with general resistance.

5. References

1. Teraoka, T., Sakakibara, T., Den, E., Hosokawa, D., Watanabe, M. (1990) A novel rice lectin specific to mannoside / glucoside residues in rice seedlings. *Agric. Biol. Chem.* **54**, 3053-3056

2. Hirano, K., Teraoka, T., Takahashi, H., and Hosokawa, D. (1998) Characterization and Gene Cloning of the Novel Mannose-Binding Rice Lectin. Abstract, section 1.9.12, Seventh Fifth International Congress of Plant Pathology, Edinburgh, Scotland

3. Kanoh, D., Toyoki, T., Haga, M., and Sekiwaza, T. (1988) A kinetic analysis on the enhancement of superoxide generation from rice leaf disks stimulated with proteoglucomannan after treatment with concanavalin A. *Agric. Biol. Chem.* **52**, 3163

4. Hamer, J.E., Howard R.J., Chumley, F.G., and Valent, B. (1988) A mechanism for surface attachment in spores of a plant pathogenic fungus. *Science* **239**, 288-290.

5. Xiao, J.-Z, Ohshima, A., Kamakura, T., Ishiyama, T., and Yamagushi, I. (1994) Extracellular glycoprotein(s) associated with cellular differentiation of *Magnaporthe grisea. Molecular Plant-Microbe Interaction* **7**, 639-644

MONOCLONAL ANTIBODIES INTERFERRING WITH APPRESSORIUM FORMATION AND PATHOGENESIS OF MAGNAPORTHE GRISEA

FUCHENG LIN   JIANXIANG WU   DEBAO LI *

Biotechnology Institute, Zhejiang   University at Huajia Pool, Hangzhou 310029, People's Republic of China

1. Introduction

Magnaporthe grisea (T. T. Hebert) Yaegashi & Udagawa [anamorph: *Pyricularia grisea* (Cooke) Sacc.] is a typical hermaphroditic ascomycete and best known as the causal agent of rice blast disease (Ou, 1985). Like many other fungal pathogens, *M. grisea* develops a specific infection structure, an appressorium, to adhere tightly to and then penetrate the host surfaces. Hamer, et al (1988) found that the mucilage of conidia of the fungus is important for them to attach the rice surface and form appressorium. Dean (1997) suggested that the fungus commonly relies on thigmotropic sensing mechanisms to decipher host surfaces, especially hydrophobic surfaces, appropriate for appressorium formation, then cAMP accumulates to cause the cascade of protein phosphorylated reaction, regulating the expression of genes for appressorium formation. Talbot, et al (1996) isolated MPG1 encoding a hydrophobin product which is important in the development of appressorium formation and for fungal pathogenesis. We have developed a series of monoclonal antibodies to this fungus, which could interfere with the appressorium formation. We hope to learn what extracellular sensing mechanisms are involved in initiation and signaling of appressorium.

* To whom correspondence should be addressed:
Tel. 0086-571-6971184. Fax. 0086-571-6961525. E-mail. lidb@public.hz.zj.cn

D. Tharreau et al. (eds.), Advances in Rice Blast Research, 286–302.
© 2000 *Kluwer Academic Publishers. Printed in the Netherlands.*

2. Materials and Methods

2.1. FUNGAL STRAIN AND CULTURAL CONDITION

M. grisea 91-11b1 was kindly given by Mr. Rongyac Cai, Rice Blast Disease Research Lab, Zhejiang Agricultural Acacemy, Hangzhou 310021, China, used throughout the experiments and routinely maintained on complete medium. Fungal cultures were grown on oatmeal agar (50g of oatmeal plus 20ml tomato juice per liter) at 23C under fluorescent light to improve conidiation. Conid.a were collected from 7-day-old cultures and washed three times with cisti_led Millipore water.

2.2. CONIDIAL GERMINATION AND APPRESSORIUM FORMATION

Fifty-μl drops of conidial suspension (10^4 conidia per ml) was placed on slides with cellophone film, sealed in moistened Petri dishes, and incubated at room temperature for 20 h. The percentages of germinated and germinating conidia induced to form appressoria were determined by microscope, 100 conidia per replicate in five experiments, with three replicates per treatment. Then, the germinating conidia with appressoria were pooled and centrifuged as the immuno-antigens.

2.3. PRODUCTION OF MONOCLONAL ANTIBODIES

According to the protocol of Pain, et al (1992), two female Balb-c mice (6- to 8-week-old) were given three intraperitoneal injections (1~2 mg of the germinating conidia with appressoria) at intervals of 2 to 3 weeks, before a final tail vein injection 3 days prior to fusion. The spleen cells from immunized mice were fused with NS1 myeloma cells. The resulting hybridomas were screened for antibody production 10 days after the fusion by indirect ELISA on plates coated with the germinating conidia with appressoria , on which the conidia had been incubated at 21C for 24h and attached tightly to the surfaces of the plates. Selected cell lines were cloned by limiting dilution and re-screened by ELISA. Antibodies were used in the form of tissue culture supernatant (TCS), ascite and purified IgG from ascites by the procedure described by Chen, *et al* (1990).

2.4. INDIRECT IMMUNOFLUORESCENCE

The fungal conidia from 7-day-old oatmeal cultures of *M. grisea* 91-11B1 were washed three times by centrifugation and resuspension in 0.02 M PBS (pH 7.0) and then diluted to give a concentration of 10^4 conidia / ml. Droplets of conidium suspension (20 µl) were placed in the wells of cleaned multiwell microscope slides. After 24h at 21C in the dark, conidia germinated to form appressorium. Fungal cell were immunolabelled, without prior chemical fixation, in the following steps: (1) washed with PBS, 5 min; (2) treated with 10% (v/v) normal mouse serum (Sigma) in PBS, 30 min; (3) treated with undiluted TCS, 3h; (4) washed three times with PBS, 5 min each; (5) treated with FITC conjugated sheep anti-mouse IgG (whole molecule), affinity isolated antibody (Sigma, Saint Louis, USA) diluted 1: 40 in PBS, 2 h; (6) washed three times with PBS 5 min each. Glycerol containing 0.1% (w/v) *p*-phenylenediamine and 10% (v/v) PBS was added to the slides and the cells observed with a light microscope with UV-epifluorescence. FITC fluorescence was viewed with excitation filter BP 450-490. Photomicrographs were recorded on Kodak film .

2.5. INTERFERING WITH APPRESSORIUM FORMATION BY MONOCLONAL ANTIBODIES

Droplets of conidium suspension (10 µl , 2×10^4 conidia ml^{-1}) were placed in the wells of cleaned multiwell microscope slides. Meanwhile, IgGs from mice ascites which could specificlly recognize the fungus in indirect immunofluorescence were diluted, then , added 10µ to the conidium suspension to make the mixed drops containing 10^4 conidia ml^{-1} with different concentrations of IgGs. After that, the slides were kept in moistened incubator at 23C for 24h. Appressorium formation was checked for 8 times at intervals of 3 h.

2.6. EFFECT OF cAMP ON INTERFERING WITH APPRESSORIUM FORMATION BY MONOCLONAL ANTIBODIES

Droplets of conidium suspension (10 μl , 2×10^4 conidia ml^{-1}) with 10 mM cAMP or 2.5 mM 3-isobutyl-1-methylxanthine (IBMX) (Sigma, Saint Louis, USA) were placed in the wells of cleaned multiwell microscope slides. Meanwhile, IgGs from mice ascites which could strongly interfere with the appressorium formation were diluted, then , added 10μl to the conidium suspension to make the mixed drops containing 10^4 conidia ml^{-1} with different concentrations of IgGs. After that, the slides were kept in moistened incubator at 23C for 24h. Conidia germinated to form appressorium were checked for 8 times at intervals of 3 h.

2.7. EFFECT OF MONOCLONAL ANTIBODIES ON THE FUNGAL PATHOGENSIS

Droplets of conidium suspension (5 μl , 2×10^4 conidia / ml) were placed on the surface of 5-leaf-time rice (Yufongzhao, indica) leaf fragments (5.0cm long) carefully, which would keep in Petri dishes with wet autoclaved cotton in a growth chamber at 25C with a 13-hour-light and a 11-hour-dark cycle. Meanwhile, IgGs from mice ascites which could strongly interfere with the appressorium formation were diluted, then , added 5μ to the conidium suspension to make the mixed drops containing 10^4 conidia / ml with different concentrations of IgGs. Sizes of disease leaf lesions were measured for five days at intervals of one day.

2.8.   WESTERN BLOTTING

One gram of conidia and conidia with germ tubes were collected , and added with 1 ml 1% SDS with 1 mM PMSF and 40 mM β-mercaptoethanol in an Eppendorf tube, respectively, then, tubes were boiled 30 min. After that, supernatant was removed by centrifugation at 12000 rpm . Samples were analyzed by SDS-PAGE (using 10% acrylamide gels; reducing sample buffer containing 0.125 M Tris-HCl pH 6.8, 205 (v/v) glycerol, 40 mg/ml SDS; 25 µg/ml bromophenol blue and 10% (v/v) β-mercaptoethanol) and Western blotting (Pain et al, 1992).

3.   Results

3.1. SCREENING AND SELECTION OF HYBRIDOMAS

Tissue culture supernatants (TCS) from growing hybridomas were screened by ELISA on plates coated with the germinating conidia with appressoria of *M. grisea*. A total of 850 wells contained hydridoma clones. TCS from 230 of these showed binding to them, but 23 showed preferential binding in the initial screening assays. Eleven of these hybridomas were selected for further study and were cloned to produce MAb-secreting cell lines (See Table 1).

TABLE 1. Characteristics of monoclonal antibodies(MAbs) to *Magnaporthe grisea* [a]

Mabs	Titre of TCS [b]	Localization antigens in fungal infection structures [c]
4A1	1:64000	conidia
2B4	1:64000	conidia , germ tubes and appressorium
2H4	1:32000	germ tubes and appressorium
1D1	1:32000	germ tubes
2B11	1:16000	conidia
3H8	1:16000	germ tubes
2A3	1:16000	germ tubes
4G9	1:16000	conidia. germ tubes
3G4	1:8000	germ tubes
5H10	1:8000	conidia
1F8	1:8000	conidia, germ tubes

a. Conidia suspension of *Magnaporthe grisea* 91-11B1 were collected from 7-day-old cultures at the concentration of 10^4 conidia ml^{-1}, and placed on multiwell plates for ELASA and immunoflurescence assays;

b. The germinating conidia with appressoria on the surfaces of multiwell plates were incubated with Mabs, horseradish   peroxidase conjugated (HRP) anti-mouse IgG and TMB (3,3',5,5'-tetramethyl-benzidene) stabilized substrate (Promega). Titres were determined in a reader.

c. The localization of antigens were viewed by indirect immmunofluorescence as described in Materials and Method.

3.2. INDIRECT IMMUNOFLUORESCENCE

Antigens recognized by 11 Mabs had different distributions (see table 1). Four Mabs, such as 4A1, 2B4, 2H4 and 1D1, strongly react with the cell wall surfaces of the pathogen. 4A1 labeled conidia strongly and uniformly, but much weaker labeling of germ tubes and appressoria; 2B4 labeled conidia, germ tubes and appressoria uniformly; 2H4 labeled appressoria mostly strongly, with almost no labeling conidia and germ tube; 1D1 labeled germ tubes strongly with almost no labeling conidia and appressoria (see Figure 1).

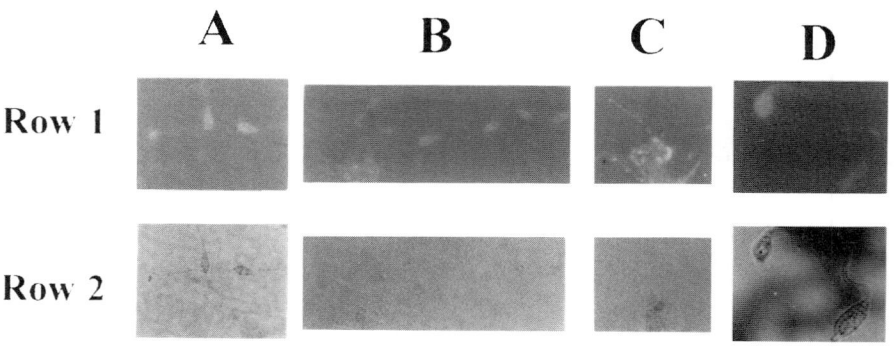

Figure 1. Indirect immunoflorescence indicating the antigen localication in the infection structures of Magnaporthe grisea by monoclonal antiboodies.

The Indirect immunoflorescence steps described as in Material and Methods. FITC fluorescence was viewed with exitation filter BP 450-490. Row 1 presents the pictures under UV-epifluorescence and Row 2 as a check under light.
(A) 2B4 labeledd conidia, germ tubes and appressoria uniformly.
(B) 4A1 labeled conidia strongly and uniformly.
(C) 1D1 labeled germ tubes.
(D) 2E4 labeled apppressoria most strongly.

3.3. INTERFERING WITH APPRESSORIUM FORMATION BY MONOCLONAL ANTIBODIES AND EFFECT OF cAMP

Results showed that 2B4, 4A1, 1D1 and 2H4 could interfere with the appressorium formation effectively (see table 2). 2B4 was the strongest one to inhibit the formation even at the concentration of 5 µg/ml (see Figure 2). The weakest was 2H4 which need 20 µg/ml to interfere with the appressorium formation. 10mM cAMP and 2.5 mM IBMX "removed" the inhibition to the appressorium formation by MAbs effects and restored normal percentages of appressorium formation as CK (see table 2 and Figure 3).

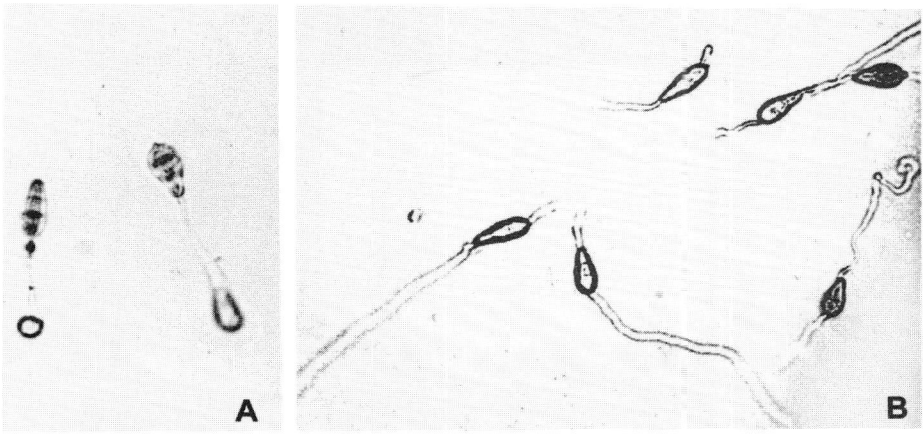

Fugure 3. Effect of cAMP on interfering with appresso.ium formation by monoclonal antibody 2B4.

Conidium suspension (10^4 conidia / ml) with (A) or without (B, as a check) 10 mM cAMP containing 20 µg / ml IgG of 2B4 was added on polyenthyene multiwell microscope slide, and kept in moistened incubator at 23 C for 12.0 h..
(A) 10 mM cAMP restored appressorium formation containing IgG of 2B4.
(B) 2B4 effectively interfered with the appressorium formation.

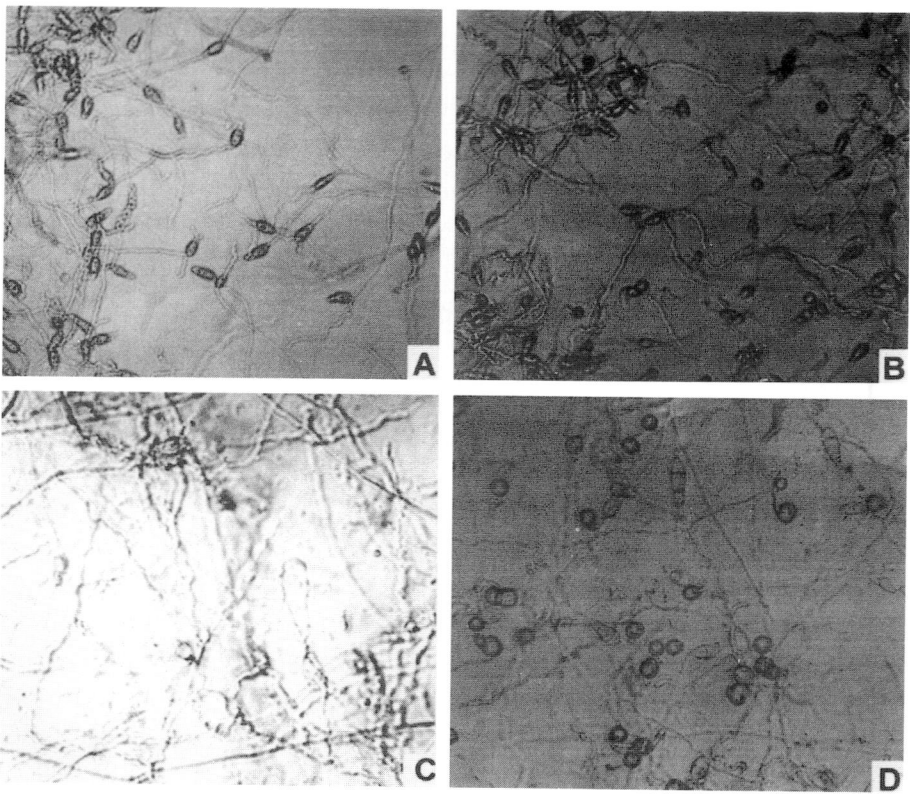

Figure 2. Interfering with appressorum formation by monoclonal antibody 2B4.

Conidium suspension (10^4 conidia / ml) with 5 µg / ml IgG of 2B4 was added on polyenthyene multiwell microscope slide, and kept in moistened incubator at 23 C.
(A) and (B , as a check) kept for 9 h..
(B) and (D, as a check) kept for 21 h..

TABLE 2   Interfering with appressorium formation of *Magnaporthe grisea* and its pathogenesis by monoclonal antibodies [a]

MAbs	Concentrations of IgG	Conidial germination	Appressorium formation [b]	Length Leaf lesion of blast disease [c]
	(μg/ml)	(% ± SD)	(% ± SD)	(mm)
2B4	20	95.3 ± 2.5	1.2 ± 1.2	0.0
+ 10mM cAMP		96.4 ± 2.4	90.3 ± 3.6	
+ 2.5mM IBMX		95.4 ± 2.7	89.7 ± 3.4	
	10	96.1 ± 3.1	2.3 ± 1.2	0.0
	5	95.2 ± 2.2	5.4 ± 1.3	0.3
4A1	20	97.0 ± 2.1	3.1 ± 1.0	0.0
+ 10mM cAMP		95.4 ± 2.5	91.3 ± 3.8	
+ 2.5mM IBMX		97.4 ± 2.3	88.3 ± 3.7	
	10	95.0 ± 1.8	6.2 ± 1.7	0.2
1D1	20	95.8 ± 2.4	2.9 ± 1.4	0.0
+ 10mM cAMP		97.4 ± 2.0	88.3 ± 3.8	
+ 2.5mM IBMX		95.8 ± 2.5	89.1 ± 3.9	
	10	94.9 ± 1.6	5.9 ± 1.3	0.3
2H4	20	97.2 ± 2.1	6.3 ± 1.4	0.3
+ 10mM cAMP		95.4 ± 2.3	91.3 ± 2.8	
+ 2.5mM IBMX		95.7 ± 2.5	90.3 ± 3.8	
10mM cAMP		96.4 ± 2.3	94.3 ± 3.3	
2.5mM IBMX		97.1 ± 2.5	93.3 ± 3.1	
CK		96.3 ± 2.1	89.4 ± 3.5	10.3

a, A conidial suspension (10^4 conidia   per milliliter) was placed on polyethylene multiwell plates for assays; b, Appressorium formation was determined as described in Material and Methods; c, Length of leaf lesions were measured after 5days; CK ,distilled water as a check.

3.4. EFFECT OF MONOCLONAL ANTIBODIES ON THE FUNGAL PATHOGENESIS

All of the four MAbs could effectively inhibit the disease leaf lesions development (see table 2 and Figure 4). The strongest one is clone 2B4 which could inhibit the disease leaf lesions development at the concentration of 5 μg / ml of IgG.

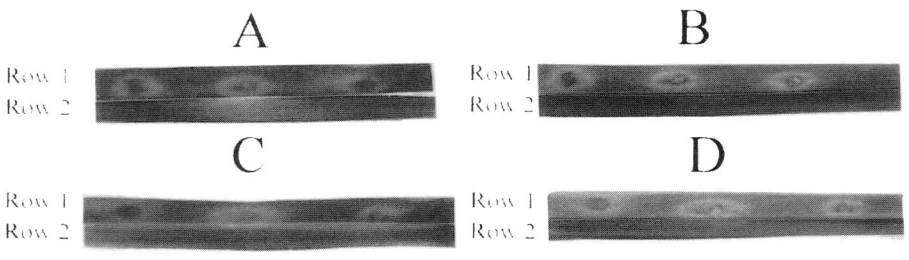

Figure 4. Effect of monoclonal antibodies on the fungal pathogenesis.

Check and treatment shared one leaf fragment as described in Materials and Methods. Row 1. check and Row 2 treatment. All of the four Mabs could effectively inhibited the disease leaf development after 5 days.

(A) 2B4, 5 μg / ml.

(B) 4A1, 10 μg / ml.

(C) 1D1, 10 μg / ml.

(D) 2H4, 20 μg / ml.

3.5. WESTERN BLOT ANALYSIS

The result indicated that 2B4 recognized 15KD protein antigen from conidia and germ tubes, 4A1 could recognized 31 KD protein antigen from conidia and carbonic anhydrase from the component of midrange molecular marker (Promega, USA), 1D1 recognized 60KD protein antigen from germ tubes and 2H4 almost did not recognize any protein bands in NC membranes both from germ tubes and conidia (see Table 3 and Figure 5)

TABLE 3   Results of western blot analysis of monoclonal antibodies

Monoclonal antibodies	sample sources	antigen proteins
2B4	conidia	15KD
	germ tubes	15KD
4A1	conidia	31KD
	midrange molecular marker	31KD
1D1	germ tubes	60 KD
2H4	--	--

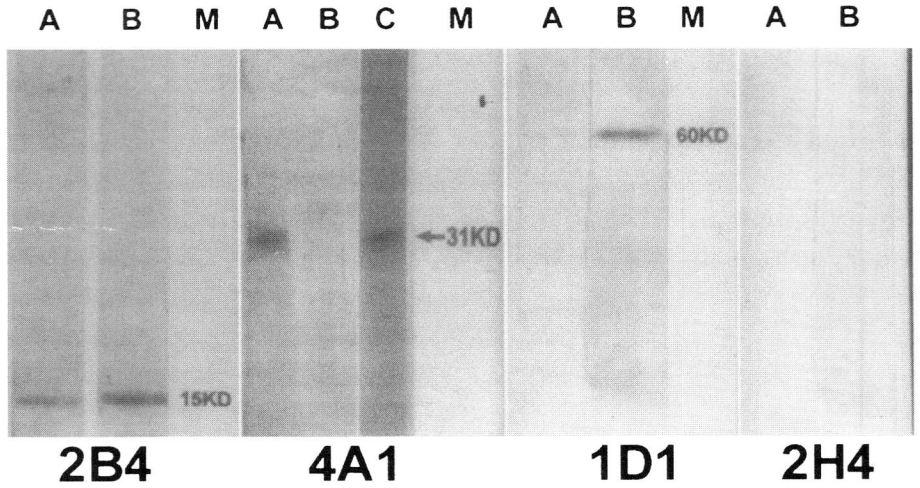

Figure 5. Western blot analysis

Column A indicates the sample from conidia, B germ tubes , C carbonic anhydrase and M marker.

2B4 recognized 15 KD protein antigen both from conidia and germ tubes.

4A1 recognized 31 KD protein antigen from conidia and the component of midrange molecular marker (Promega, USA).

1D1 recognized 60 KD protein antigen from germ tubes.

2H4 almost did not recognize any protein bands in NC membranes both from conidia and germ tubes.

4. Discussion

Appressorium formation of *Magnaporthe grisea* proceeds by a complex morphogenetic sequence. The initial steps in appressorial development are regulated by multiple external signals. Recent researches provide that hardness and hydrophobicity, especially plant cutin monomers *cis*-9,10-epoxy-18-hydroxyoctadecanoic acid and *cis*-9-octadecen-1-ol (Gilbert, *et al*, 1996) are important environmental factors in stimulating appressorium formation. These external signals are received by receptors, such as a hydrophobin MPG1 (Talbot, et al, 1996) or glycoproteins (Dean, 1997), on cell surfaces of the pathogen. Different antigens on

(Dean, 1997), on cell surfaces of the pathogen. Different antigens on the surfaces of conidia, germ tubes and appressoria could be recognized by four of eleven monoclonal antibodies, respectively, according to the indirect immunofluoresence evidence. All of the four monoclonal antibodies could interfere with the appressorium formation of *Magnaporthe grisea* and inhibit the development of leaf lesion of rice blast disease. It means the antibodies could block certain receptors on the cell surfaces of the pathogen, which could receive signals and switch on appressorium formation.

Since its initial researches (Hiatt, et al, 1989), the expression of functional antibodies in transgenic plants has been considered highly promising for plant disease control. Recently, van Engelen, *et al* (1994) have created an efficient system for high-level expression of monoclonal antibody 21CS in tobacco roots, which was raised against a cutinase produced by fungus *Botrytis cinerea*, and had the potential to interfere with pathogenesis. It is also potentially significant to transfer genes of monoclonal antibodies interfering with appressorium formation of *M. grisea* into rice plant and thus control the destructive disease.

Earlier experiments indicted the existence of a cAMP- dependent signaling in surface recognition and appressorium formation (Lee and Dean, 1993). Our researches also showed that exogerous cAMP could restore appressorium formation of germination conidia even interfered with relatively high concentration of IgGs of monoclonal antibodies. It means appressoria could still effectively form in   high concentration of exogenous cAMP without successful attachment of the pathogen on hydrophobic surfaces. These evidence confirmed that cAMP accumulation is a key step for appressorial morphogenesis and downstream to the interaction between the external signals and the receptors on the cell surfaces of the pathogens in signal transduction pathway of appressorium formation.

Western blot analysis indicted three different monoclonal antibodies 2B4, 4A1 and 1D1 could recognize the 15, 31 and 60 KD protein antigens from conidia or germ tubes in bands of NC membrane   blotted from SDS-PAGE. 2B4 recognized 15 KD proteins both from conidia and germ tubes, which appeared to be a hydrophobin MPG1 protein. In fact, 2B4 had strongest inhibitory effect on appressorium formation amongst these 11 monoclonal antibodies obtained. It would be significant to isolate the antigen protein and study its characteristics further. 4A1 recognized 31 KD protein only from conidia, which is consistent with the result of indirect immunofluoresence. It was firstly found that 4A1 also recognized carbonic anhydrase (CA) which is a central enzyme to both transport and metabolic processes at the cellular level (Henry, 1996). Membrane-associated CA, with an extracellular orientation, also appears to be important in acidifying the outer boundary layer through the catalyzed of excreted CO_2, which facilitates cellular ammonia transport by providing H^+ ions for the protonation of NH_3, thus maintaining the trans-membrane NH3 gradient according to recent researches. But, the function of CA in the initial process of apppressorium formation by *M. grisea* leaves to be unclear. Thus, studies on role of CA on the appressorium formation appear to be necessary and important. 1D1 labeled germ tubes without labeling of conidia and appressoria in indirect immunofluoresence assay, recognized 60 KD protein from germ tubes, which might be a glycopretein on germ tubes. Next, isolation, purification and characteristics of this protein are also necessary. Indirect immunofluoresence also showed 2H4 only specifically binding to appressoria with almost no labeling of conidia and germ tubes. It was also found that 2H4 did not recognize any proteins from conidia and germ tubes. Whether 2H4 could recognize antigen proteins from appressia is still unknown.

It would be also meaningful to screen conidia cDNA expression libraries by these four monoclonal antibodies to get positive clones regarding with antigens for receiving environmental stimuli to initiate appressorium formation on the cell surfaces of the pathogen.

Acknowledgments

This work was funded by National Nature Science Foundation of China

University of Exeter, UK and our staff for their advice and help.

References

Chen, B, Wu, M. and Ye, Q. (1990) Comparative studies of different purification methods for monoclonal antibody (McAb), Chinese Journal of Virology 6, 122-126.

Dean, R.A. (1997) Signal pathways and appressorium morphogenesis, *Annual Review of Phytopathology* 35, 211-234.

van Engelen, F.A., Schouten A., Molthoff, J.W., Roosien J., Salinas J., Dirkse, W.G., Schots, A., Bakker, J., Gommers, F.J., Jongsma, M.A., Bosch, D. and Stiekema, W.J. (1994) Coordinate expression of antibody subunit genes yields high levels of functional antibodies in roots of transgenic tobacco, *Plant Molecular Biology* 26, 1701-1710.

Gilbert, R.D., Johnson, A.M. and Dean, R.A. (1996) Chemical .signals responsible for appressorium formation in the rice blast fungus *Magnaporthe grisea, Physiological and Molecular Plant Pathology* 48, 335-346.

Haitt, A., Cafferkey, R., Bowdish, K., (1989) Production of antibodies in transgenic plants, *Nature* 342,469-470.

Hamer, J.E., Howard, R.J., Chumley, F.G. and Valent, B. (1988) A mechanism for surface attachment in spores of a plant pathogenic fungus, *Science* 239, 288-290.

Henry, R.P. (1996) Multiple roles of carbonic anhydrase in cellular transport and metabolism, *Annual Review Physiology* 58, 523-538.

Lee, Y.H., Dean, R.A. (1993) cAMP regulates infection structure formation in the plant pathogenic fungus *Magnaporthe grisea, Plant Cell* 5, 693-700.

Ou, S.H. (1985) *Rice Diseases*, 2nd ed, Kew, UK, Commonwealth Mycological Institute.

Pain, N.A., O'Connell, R.J.O., Bailey, J.A. and Green, J.R. (1992) Monoclonal antibodies which show restricted binding to four *Colletotrichum* species: *C. lindemuthionum, C. malvarum, C. orbiculare* and *C. trifolii, Physiology and Molecular Plant*

Pathology 40, 111-126.

Talbot, N.J., Kershaw, M.J., Wakley, G.E., de Vries, O.M.H., Wessels, J.G.H. and Hamer, J.E. (1996) MPG1 encodes a fungal hydrophobin involved in surface interactions during infection-related development of *Magnaporthe grisea, Plant Cell* 8, 985-999.

IDENTIFICATION OF AVIRULENCE GENES OF *MAGNAPORTHE GRISEA* AND SELECTION OF MOLECULAR MARKERS

N. YASUDA, Y. FUJITA, M. TSUJIMOTO
Laboratory of Plant Disease Control, Hokuriku National Agricultural Experiment Station
Inada1-2-1, Jyouetsu, 943-0193 Niigata, Japan

1. Introduction

The molecular basis of cultivar specificity in rice blast disease has been studied in certain race-cultivar combinationsSweigard *et al.*,1993; Smith and Leong,1994; Leong *et al.*, 1994; Valent and Chumly 1994; Farman and Leong, 1996; Mandel *et al.*,1997. Among these, some specificity are unstable (Valent and Chumley 1994) and others are stable (Mandel *et al.*, 1997). And the we can see various degree of resistance reaction can be seen:.from no visible symptom to big brown lesions. It s considered that the molecular mechanisms behind race-cultivar interactions varies with the host-pathogen combinations. Thus we started studies toward cloning of other avirulence genes.

2. Identification of avirulence genes

We performed crosses between two isolates pathogenic to rice in order to identify genes that control specificity toward certain cultivars. We conducted three pairs of crosses.

Two isolates, Y90-71 and 3514-R-2, are sexually compatible but differ in their pathogenicity on rice cultivar 'Hattan3'. Y90-71 was avirulent on 'Hattan3', but 3514-R-2 was virulent. We crossed these isolates and scored the segregation for avirulence/virulence among the progenies. The distinction between avirulent and virulent progenies on 'Hattan3' is usually clear as the parents. The ratio of avirulent to virulent progeny toward 'Hattan3' was 95:90. A χ^2 value of 0.14 indicated that segregation was consistent with a 1:1 ratio. This data suggested that avirulence or virulence toward 'Hattan3' is determined by a single gene difference between Y90-71 and 3514-R-2. Therefore, the major gene identified here is "avirulence gene" which determines virulence or avirulence on rice cultivar 'Hattan3'.

Another genetic cross was conducted between field isolates, Y93-154a-1 and Y93-165g-1. These two isolates differ in their pathogenicity on rice cultivars 'Aichiasahi' and 'Ishikarishiroke'. In the similar way, we scored the segregation for avirulence/virulence

D. Tharreau et al. (eds.), Advances in Rice Blast Research, 303–307.

among the progenies. The ratio of avirulent to virulent progeny toward 'Aichiasahi' was 70:67. A value of 0.0657 indicated that segregation was consistent with a 1:1 ratio. The ratio of avirulent to virulent progeny toward 'Ishikarishiroke' this time was 68:69. The data also fitted a 1:1 ratio. Here we identified two different avirulence genes. One is for avirulence toward 'Aichiasahi' and the other is for avirulence toward 'Ishikarishiroke'.

Last cross was conducted between two isolates, Y93-164a-1 and Y93-245c-2. These two isolates differ in their pathogenicity on rice cultivar 'K59'. That's why we could do a genetic analysis of avirulence on 'K59' using these isolates. The result of this test revealed the ratio of avirulent to virulent progeny to be 34:38. This data also fitted a 1:1 ratio. This means that the avirulence toward 'K59' is under single gene control.

Here we identified four different avirulence genes. Isolate Y90-71 carries the avirulence gene toward 'Hattan3',Y93-165g-1 carries the avirulence gene toward 'Aichiasahi', Y93-164a-1 carries the avirulence gene toward 'Ishikarishiroke' and Y93-245c-2 carries the avirulence gene toward 'K59'.

3. Identification of resistance gene corresponding to the avirulence gene

We considered the resistance gene of rice cultivar which corresponds to the avirulence gene. Rice cultivar 'Ishikarishiroke' is known to possess the resistance gene, *Pi-i*. *Pi-i* was identified using 7 fungus isolates (Yamasaki and Kiyosawa, 1966). Among these 7 isolates, one isolate, Ken 54-20, was postulated to have an avirulence gene named *Av-i* which corresponds to *Pi-i*. And the incompatibility between 'Ishikarishiroke' and Ken54-20 was explained by the interaction of *Pi-i* and Av-i(Yamasaki and Kiyosawa, 1966). But, whether the avirulence gene toward 'Ishikarishiroke' of Y93-164a-1 is the same as Av-i remains undetermined. To answer this question, rice cultivar 'Norin3' was used as the representative parent susceptible to Ken54-20 and Y93-164a-1. And we analyzed the F3 lines from the cross between 'Ishikarishiroke' and 'Norin3'.

TABLE 1. Segregation for resistance to two blast fungus isolates:Y93-164a-1 and Ken54-20 in the F_3 family from the cross Norin3 x Ishikarishiroke

Reaction to Ken54-20	Number of F_3 lines classified				Expected ratio
	by reaction to Y93-164a-1				
	Res.	Seg.	Sus.	Total	
Resistant	28	0	0	28	
Segregating	0	46	0	46	
Susceptible	0	0	26	26	
Total	28	46	26	100	1:2:1

$$chi^2=0.03$$

For the gene analysis, the F_3 lines were inoculated with Ken54-20 and Y93-164a-1 (Table). To the fungus strain, Ken54-20, a segregation ratio of F_3 lines for resistant segregating: susceptible was 28:46:26. A χ^2 value of 0.03 indicated that segregation was consistent with a 1:2:1 ratio. This data confirmed single gene control for resistance of 'Ishikarishiroke' to Ken54-20. As for the reaction of F_3 lines to Y93-164a-1, all of the F_3 lines showed the same reaction as to Ken54-20. The reaction of F_3 lines to Y93-164a-1 coincided exactly with the reaction to Ken54-20. That is, all of the resistant lines to Ken54-20 also showed resistance to Y93-164a-1, all of the segregating lines to Ken54-20 also showed segregating reaction to Y93-164a-1, and all of the susceptible lines to Ken54-20 also showed susceptibility to Y93-164a-1. This means that the resistance gene of 'Ishikarishiroke' affecting resistance to Ken54-20 also controls resistance to Y93-164a-1. Therefore, we conclude that the avirulence gene of Y93-164a-1 toward 'Ishikarishiroke' corresponds to *Av-i*.

We are preparing for the identification of resistance genes corresponding to the other avirulence genes.

4. Selection of molecular markers toward map based cloning

As we could identify each avirulence gene as a single locus, we are attempting to clone these genes by physical mapping and chromosome walking. Then we started with the selection of molecular markers linked to the avirulence genes.

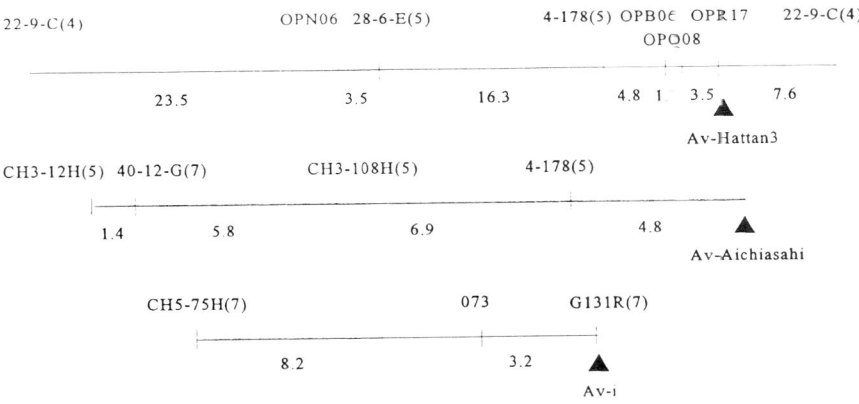

Figure 1. Genetic map location of the avirulence genes.

For the analysis of the avirulence gene toward 'Hattan3', we used 22 RFLP markers provided by S.A.Leong's group at the University of Wisconsin (Nitta *et al.*, 1997) and 800 RAPD markers purchased from Operon Technologies. About 45% of RFLP markers and less than 10% of RAPD markers detected polymorphism between parental isolates. For the analysis of the avirulence gene toward 'Aichiasahi' and Av-i, 26 RFLP markers from Leong's group and 46 RFLP markers provided by Nagao Hayashi at Aichi-ken Agriculture Research Center were used. About 45% of the markers detected polymorphism between parental isolates. With these markers, we analyzed progenies from the cross and detected genetic linkage between the molecular markers and the avirulence gene. Figure shows the map location of avirulence genes constructed by MAPMAKER/EXP ver3.0. The avirulence gene toward 'Hattan3' was linked to four RFLP markers and four RAPD markers by the analysis of 46 progenies. Marker *22-9-C* detected two different loci, and both of them were found to be linked to the avirulence gene. Two markers, *28-6-E* and *4-178* were located on chromosome 5 of the map constructed by S.A.Leong's group. RAPD marker, *OPR17*, was still cosegregated with the avirulence gene.

By means of a segregation analysis of 52 progenies, four markers within linkages to the avirulence gene toward 'Aichiasahi' were found. Three markers out of four were located on chromosome 5 in Leong's map. As for the map location of *Av-i*, Three markers were found to be linked to the gene. Two markers were located on chromosome 7 in Leong's map. Marker *073* is the marker provided by N.Hayashi, and it was an independent marker in his map. Marker *G131R* was cosegregated with the avirulence gene.

Toward map-based cloning, we will increase the number of progenies and find starting points on either side of the locus for a chromosome walk.

5. Acknowledgments

We thank Sally Ann Leong and Nagao Hayashi for generously providing RFLP makers.

6. References

Farman, M.L. and Leong, S.A.(1996) Genetic analysis and mapping of avirulence genes in *Magnaporthe grisea*. In: Bos, C..J. (ed.) *Fungal genetics: principles and practice*. Marcel Dekker Inc., New York, Basel, Hong Kong, pp.295-315.

Leong, S.A., Farman, M., Smith, J., Budde, A., Tosa, Y., Nitta, N. (1994) Molecular genetic approach to the study of cultivar specificity in the rice blast fungus. In: Zeigler, R., Leong, S.A., Tang, P. (ed.) *Rice Blast Disease*. CABI, London, pp 88-110.

Mandel, M.A., Crouch, V.W., Gunawardena, U.P., Harper, T.M. and Orbach, M.J.(1997) Physical mapping of the *Magnaporthe grisea AVR1-MARA* locus reveals the virulent allele contains two deletions. *Molecular Plant Microbe Interactions* 10, 1102-1105.

Nitta, N., Farman, M.L. and Leong, S.A. (1997) Genome organization of *Magnaporthe grisea*: integration of genetic maps, clustering of transposable elements and identification of genome duplications and rearrangements. *Theoretical and Applied Genetics* 95, 20-32.

Smith, J.R. and Leong, S.A.(1994) Mapping of a *Magnaporthe grisea* locus affecting rice (*Oryza sativa*) cultivar specificity. *Theoretical and Applied Genetics* 88, 901-908.

Sweigard, J.A., Valent, B., Orbach, M.J., Walter, A.M., Rafalski, A. and Chumley F. (1993) Genetic map of the rice blast fungus *Magnaporthe grisea* (n=7). In: O'Brien, S.J. (ed.) *Genetic Maps, Vol. 6*.Cold Spring Harbor Laboratory, New York, pp. 3.113-3.115.

Valent, B. and Chumley, F.G. (1994) Avirulence genes and mechanisms of genetic instability in the rice blast fungus. In: Zeigler, R., Leong, S.A., Tang, P. (ed.) *Rice Blast Disease*. CABI, London, pp 111-134.

Yamasaki, Y. and Kiyosawa, S. (1966) Studies on inheritance of resistance of rice varieties to blast. 1. Inheritance of Japanese varieties to several strains of the fungus. *Bull. Natl. Inst. Agr. Sci.* D14, 39-69.

PROGRESS ON UNDERSTANDING THE MOLECULAR BASIS OF CULTIVAR
SPECIFICITY IN THE INTERACTION OF THE *MAGNAPORTHE GRISEA AVR1-
CO39* WITH RICE VARIETY CO39

S.A. LEONG,[1] M.L. FARMAN,[3] R.S. CHAUHAN,[2] N. PUNEKAR,[2] S.
MAYAMA,4 H. NAKAYASHI,[4] Y. ETO,[4] Y. TOSA,[4] P. RONALD[5] and
H.-B. ZHANG.[6]
[1]USDA-ARS, Plant Disease Resistance Research Unit
[2]Department of Plant Pathology, University of Wisconsin
1630 Linden Dr.
Madison, WI 53506 U.S.A.

[3]Department of Plant Pathology
University of Kentucky
Lexington, KT U.S.A

[4]Laboratory of Plant Pathology
Kobe University
Rokkadai-Cho, Kobe 657, Japan

[5]Department of Plant Pathology
University of California at Davis
Davis, CA 95616, USA

[6]Texas A & M BAC Center
College Station, TX 77843-2123, USA

1.   Cloning of *M. grisea* Cultivar Specificity Gene *AVR1-CO39*.

A gene conferring cultivar-specific interactions with rice cultivar CO39 was isolated
from *Magnaporthe grisea* strain 2539 using a map-based cloning approach (Farman and
Leong, 1998). *AVR1-CO39* was previously found to map on the largest chromosome *of
M. grisea* between markers 5-10-F and CH5-120H (Nitta *et al.*, 1997; Smith and Leong,
1994) (Figure 1A). To define the physical extent of the chromosome walk to be
undertaken, RecA-mediated Achilles' cleavage at these markers was employed. The
cleavages released a 610 kb genomic DNA fragment containing *AVR1-CO39* (Figure
1B). A chromosome walk across the 610 kb fragment was completed in about 20 steps
from both flanking markers and revealed that the relationship of genetic to physical
distance in this chromosomal region varied by up to fourteen-fold. This ratio was highly
nonaverage on the CH5-120H proximal side of the locus (1 cM/~150 kb). Thus the
physical map position of the *AVR1-CO39* gene was not correlated with its genetically
defined map position. Several gaps in the cosmid genomic DNA library were also

D. Tharreau et al. (eds.), Advances in Rice Blast Research, 308–315.
© 2000 *Kluwer Academic Publishers. Printed in the Netherlands.*

encountered in the walk necessitating the use of Achillles' cleavage to release a 310 kb fragment spanning the gaps and containing *AVR1-CO39*. Subclones of this fragment provided new entry points for chromosome walking in the region proximal to *AVR1-CO39*. Interestingly, the cosmids immediately flanking the *AVR1-CO39* locus identified muliple crossover events. Moreoever, a region that was unclonable was found adjacent to the avirulence gene.

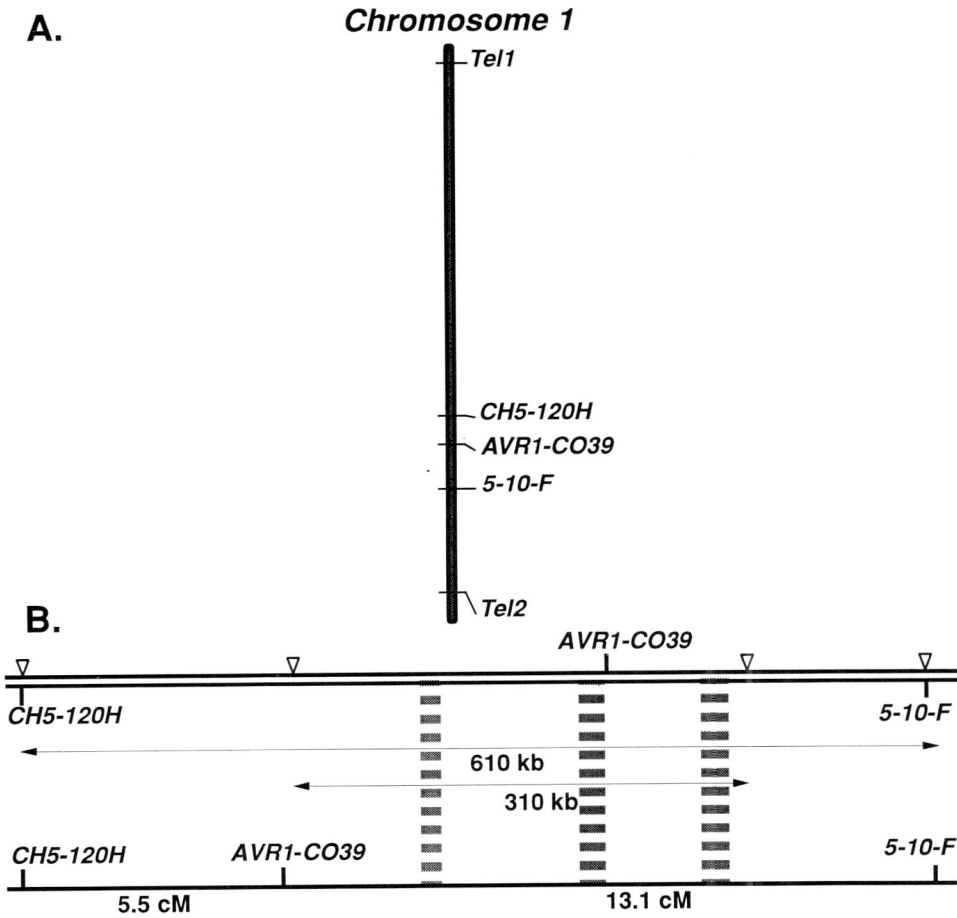

Figure 1. A. Chromosome 1 of *M. grisea*. B. Physical and genetic map positions of *AVR1-CO39* on chromosome 1 of *M. grisea*. Triangles indicate postions of cleavage by Achilles' cleavage. Fragments released by Achilles' cleavage are shown with their relative size. Regions with bars indicate locations of gaps found in the cosmid library of total genomic DNA of strain 2539.

The *AVR1-CO39* locus was finally delimited to a 1.05 kb region by subcloning and transformation of Guy11, a strain virulent on CO39. DNA sequence analysis revealed several small open reading frames (orfs) (Figure 2). The sequence surrounding the ATG of one open reading frame of 89 amino acids (orf 3) matched four out of five of the consensus bases found in fungal translational start sites. This open reading frame also contains a putative signal peptide consisting of a hydrophobic amino terminus punctuated by a lysine in position two and putative cleavage sites. Interestingly, the sequence TTATTTAT, similar to that found in inflammatory mediator genes such as TNF (Caput *et al.*, 1986), was found in the 3' flanking region of open reading frame 3. Site-directed mutations of the open reading frames on one strand were created in order to assess their role in avirulence. The translational start codon of each open reading frame was converted to TTT; such mutations in open reading frames 1 and 3 led to a loss of avirulence as did frameshift mutations in these open reading frames while mutation of open reading frame 2 did not affect avirulence. The absence of a splice site and a lariat sequence as well as the presence of a putative TATA element immediately upstream of the ATG of open reading frame 1 may indicate that open reading frame 1 overlaps sequences critical to the promotion of transcription of *AVR1-CO39*. Transcript mapping is underway to investigate this possibility. RT-PCR using various primers across the open-reading frame-containing region with RNA isolated from vegetative cells and susceptible plants infected with 2539 generated a product that is regulated by carbon source limitation and which, based on the primers used, overlaps the open reading frame 3 region.

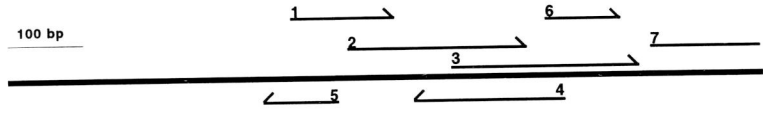

Figure 2. Open reading frame analysis of the 1.05 kb fragment containing *AVR1-CO39*.

The distribution of *AVR1-CO39* in *M. grisea* was investigated by probing genomic DNA from a large sample of host-specific forms of *M. grisea* with the 1.05 kb probe. The data to date indicate that isolates infecting rice, *Digitaria* and wheat lack *AVR1-CO39* while about half of the isolates surveyed from *Eleucine* and nearly all isolates from *Setaria* contain putative homologs of the gene. A definitive conclusion must await additional probings with the cDNA for *AVR1-CO39* and isolation and sequence analysis of the hybridizing DNA sequences.

A detailed analysis of the *AVR1-CO39* locus from rice isolate Guy11 indicated that at least 20 kb of DNA corresponding to and containing the *AVR1-CO39* locus of 2539 was absent. The endpoints of the putative deletion in other rice isolates lacking *AVR1-CO39* homology were examined using a PCR-based method. The organization of the DNA in this region of the genome appeared to be conserved in these isolates suggesting that *AVR1-CO39* may have been deleted before the appearance of rice-infecting isolates. Pedigree analysis based on RFLP patterns obtained by Southern hybridization with the 1.05 kb *AVR1-CO39* probe suggested that *AVR1-CO39* was inherited from strain 4091-5-8 , the weeping love grass-infecting parent of 2539 (Leung *et al.*, 1988).

2. Mapping of the Corresponding Disease Resistance Gene in Indica Rice Variety CO39.

Indica rice line CO39, originally bred for blast resistance in India, has been used either as a universal susceptible or a recurrent parent in the development of near-isogenic lines carrying single genes for blast resistance (Mackill and Bonman, 1992; Wang *et al.*, 1995. However, CO39 shows resistance to blast pathogen races/lineages from different parts of the world (Smith and Leong, 1994; Wang *et al.*, 1995; Zeilger *et al.*, 1995). Preliminary genetic analysis of resistance to isolates carrying *AVR1-CO39* in two F2 families derived from the cross CO39 (resistant) X 51583 (susceptible) revealed that resistance in CO39 is controlled by a single dominant gene (Table 1). This study has been extended in order to confirm this interpretation and to determine the map position of the resistance gene.

TABLE 1. Segregation of resistance and susceptibility in F2 families derived from a cross of CO39 X 51583

Family	No. Plants	Resistant	Susceptible	X^2	P
#262	253	199	54	1.8	0.17
#263	199	150	49	0.105	0.9
Total	452	349	103	1.18	0.3

Fifteen mirosatellite (simple sequence repeats) markers (Chen *et al.*, 1997), mapping to four rice chromosomes 1, 6, 11 & 12, known to carry disease resistance genes (McCouch, 1994), were tested for polymorphism between CO39(resistant) and 51583(susceptible). Most of the tested RM (microsatellite) markers showed polymorphism between the resistant and susceptible lines. Cosegregation of polymorphic RM markers with resistance/susceptibility was tested using bulked segregant analysis as well as individual F3 families. Microsatellite marker, RM202, located on chromosome 11, co-segregated with blast resistance ir resistant/susceptible F3 families as well as bulks (Figure 3). Linkage relationship between RM202 and resistance in CO39 is being established by testing the response of segregating F2 populations.). The linkage relationship between RM202 as well as RFLP markers bracketing RM202 on Chromosome 11 and resistance gene in CO39 is being established by testing the resistance response of segregating F2 populations. The centromere proximal marker RG247 on chromosome 11 (Figure 4) was shown to cosegregate with

resistance in F3 families and bulks. A BAC library of nuclear DNA of CO39 was prepared at the Texas A&M BAC Center. The library contains 25,000 clones with an average insert size of 110 kb to give a 5X coverage of the genome.
We plan to employ the gene golfing method of Zhang and Wing (1997) to identify BAC clones containing the resistance gene of CO39.

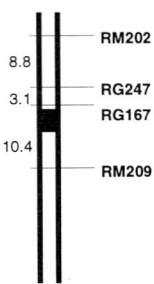

Figure 3. Centromere region of chromosome 11 of rice.

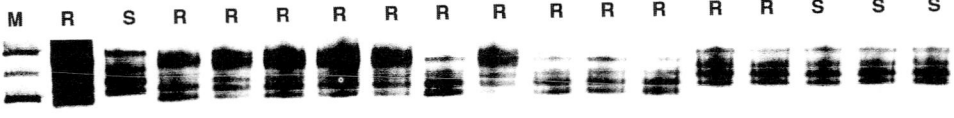

M R   S   R   R   R   R   R   R   R   R   R   R   R   S   S   S

Figure 4. Segregation of marker RM202 with resistance and susceptibility to AVRI-CO39 in resistant/susceptible F3 families and bulks. Genomic DNA from F3 families derived from F2 individuals which had been scored for resistance (R) or susceptibility (S) was used as a template for the polymerase chain reaction with RM202 primers. Products were electrophoresed on a 4.5 % acrylamide-urea sequencing gel and visualized by staining with silver.

Structural identities between R genes from different plant species have facilitated the cloning of R genes homologs from monocots using degenerate oligonucleotides derived from dicot R genes (Leister *et al.*, 1998). This strategy is also being used for cloning the R gene in CO39. Degenerate oligonucleotide primers from the conserved regions of both the leucine-rich repeats (LRR) and serine-threonine kinase domains of the bacterial blight resistance gene *Xa21* from rice (Song *et al.*, 1995) amplified a 850bp fragment in CO39 and 51583. Digestion of the amplicons with *Sau*3A or *Msp*I revealed novel RFLPs associated with each amplicon (Figure 5). DNA sequence analysis of clones obtained from the amplicon of CO39 revealed many novel R-gene-like homologs (Table 2). RFLP analysis of CO39 and 51583 with cloned fragments have indicated that these are single copy loci with allelic variation between CO39 and 51583. These R-gene-like homologs are being mapped by testing their segregation behavior in F2 populations of crosses between CO39x51583.

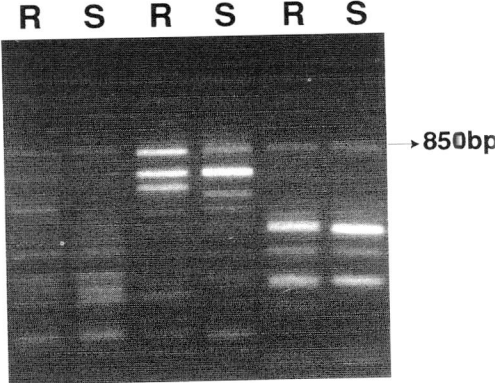

Figure 5. Restriction fragment length polymorphisms revealed after digestion of primary amplicons from CO39 (Resistant; R) and 51582 (Susceptible; S) with *Sau*3A, *Msp*I, and *Hae*III Digests were analyzed by electrophoresis on 1.0 % agarose and visualized by staining with ethidium bromide. The left two lanes contain amplicon DNA digested with *Sau*3A, the middle two lanes contain amplicon DNA digested with *Msp*I, and the right two lanes contain amplicon DNA digested with *Hae*II.

TABLE 2. DNA sequence analysis of putative resistance gene homologs.

Amplicon	Size (bp)	Sequence Identity	Probability
CO39 (R10)	775	Protein Kinase *Xa21*	$3.5e^{-51}$
CO39 (R16)	950	Serine-Threonine-specific Kinase	$1.4e^{-21}$
CO39 (R17)	600	Receptor Kinase-like Protein	$4.5e^{-27}$
CO39 (R22)	250	Disease Resistance Protein	$3.9e^{-19}$
CO39 (R23)	800	Protein Kinase *Xa21*	$1.5e^{-26}$

Amplicons were cloned in the T:A cloning site of vector pCRII from InVitrogen (Madison, Wisconsin). DNA sequencing was conducted from both ends of the insert using the SP6 and T7 promoters, respectively. Approximately 400 bases of each strand was sequenced from each ~800 bp insert DNA. Sequence data was analyzed using the Blastx program of the NCBI.

3.   References

Caput, D., Beutler, B., Hartog, K., Thayer, R., Brown-Shimer, S., and Cerami, A (1986) Identification of a common nucleotide sequence in the 3'-untranslated region of mRNA molecules specifying inflammatory mediators, *Proc. Natl. Acad. Sci. USA* 83, 1670-1674.

Chen, X., Temnykh, S., Xu, Y., Cho, Y.G., and McCouch, S.R. (1997) Development of a microsatellite framework map providing genome-wide coverage in rice (*Oryza sativa* l.), *Theor. Appl. Genet*. 95, 553-567.

Farman, M.L. and Leong, S.A. (1998) Chromosome Walking to the *AVR1-CO39* Avirulence Gene of *Magnaporthe grisea*: Discrepancy between the Physical and Genetic Maps, *Genetics* 150, 1049-1058.

Leister, D., Kurth, J., Laurie, D.A., Yano, M, Sasaki, T., Devos, K., Graner, A. and Schulze-Lefert, P. (1998) Rapid reorganization of resistance gene homologues in cereal genomes, *Proc. Natl. Acad. Sci. USA* 95, 370-375.

Leung, H., Borromeo, E.S., Bernardo, M.A., and Notteghem, J.L. (1988) Genetic analysis of virulence in the rice blast fungus *Magnaporthe grisea*, *Phytopathology* 78, 1227-1233.

McCouch, S.R., Nelson, R.J., Tohme, J., and Zeigler, R.S. (1994) Mapping of blast resistance genes in rice, in R.S. Zeigler, S.A. Leong, P.S. Teng (eds.), *Rice Blast Disease*, CAB International, Wallingford, pp. 167-186.

Nitta, N., Farman, M. and Leong, S.A. (1997) Genome organization of *Magnaporthe grisea*: Integration of genetic maps, clustering of transposable elements, and identification of genome duplications and rearrangements, *Theor. Appl. Genet*. 95, 20-32.

Smith, J.R. and Leong, S.A. (1994) Mapping of a *Magnaporthe grisea* locus affecting rice (*Oryza sativa*) cultivar specificity, *Theor. Appl. Genet*. 88, 901-908.

Song, W.-Y., Wang, G.-L., Chen, L.-L., Kim, H.-S., Pi, L.-Y., Holsten, T., Gardner, J., Wang, B., Zhai, W.-X., Zhu, L.-H., Fauquet, C., and Ronald, P. (1995)   A receptor kinase-like protein encoded by the rice disease resistance gene, *Xa21*, *Science* 270, 1804-1806.

Wang, G.L., Mackill, D.J., Bonman, J. M., McCouch, S.R., Champoux, M.C., and Nelson, R.J. (1995) RFLP mapping of genes conferring complete and partial resistance to blast in a durably resistant rice cultivar, *Genetics* 136, 1421-1434

Zeigler, R.S., Cuoc, L.X., Scott, R. P., Bernado, M.A., Chen, D.H., Valent, B., and Nelson, R.J. (1995) The relationship between lineage and virulence in *Pyricularia grisea* in the Philippines, *Phytopathology* 85, 443-451.

Zhang, H.-B. and Wing, R.A. (1997) Physical mapping of the rice genome with bacs, *Plant Molecular Biology* 35, 115-127.

ANALYSIS OF TRANSPOSABLE-ELEMENT CLUSTERING PATTERNS IN *MAGNAPORTHE GRISEA* GENOME USING BAC LIBRARY

M. Nishimura, S. Nakamura, N. Hayashi[1], M. Masuya[2], S. Asakawa[3], N. Shimizu[3], H. Kaku, A. Hasebe, and S. Kawasaki

National Institute of Agrobiological Resources, 2-1-2, Kan-nondai, Tsukuba, Ibaraki 305-8602, Japan, [1]National Agricultural Research Center, 3-1-1, Kan-nondai, Tsukuba, Ibaraki 305-8666, Japan, [2]Dept. of Information and Computer Science, Kagoshima University, 1-21-40, Korimoto, Kagoshima 890-0065, Japan, [3]Dept. of Molecular Biology, Keio University, School of Medicine, 35, Shinanomachi, Shinjuku, Tokyo 160-0016, Japan

1. Introduction

Transposable elements have been found in a wide range of organisms from bacteria to higher eucaryotes [1]. They are thought to be responsible for the generation of genomic diversity. The study on the distribution of some transposable elements in genomes of *E.coli* [2, 3], *S. cerevisiae* [4], allium [5], maize [6] and mouse [7] showed that these transposable elements were not randomly distributed in the genome. These results suggest that "transposon islands" are present in the genomes. In rice-pathogenic *M.grisea*, seven transposons have been reported: retrotransposons, MGSR1 [8], MGR583 [9, 10], MAGGY [11], *fosbury* [12], and Mg-SINE [13], and two DNA transposons, MGR586 [9, 14] and Pot2 [15]. The analysis using some cosmid clones also suggested that some of them co-occurred in the fungal genome [12, 16]. However, clustering patterns of these transposable elements in the fungal genome was difficult to analyze with cosmids because of their small insert size.

We have recently constructed a BAC (bacterial artificial chromosome) library of the rice-pathogenic *M. grisea* [17]. The library consisted of 5760 clones with an average insert size of 120 kbp, therefor it covered about 18 genomes based on the estimated genome size of 38 Mbp [9]. BAC system was developed as a vector for construction of large insert-sized libraries [18]. It has several advantages; low rate of chimerism, high efficiency of cloning or recovery of long insert DNA, stability of the insert DNA, and so on. It could be applicable not only to the study of genome analysis and positional cloning, but also to study of gene distribution.

In this chapter, as an application of the BAC library, we will discuss the co-occurring patterns of transposons in *M. grisea* genome.

D. Tharreau et al. (eds.), Advances in Rice Blast Research, 316–322.
© 2000 *Kluwer Academic Publishers. Printed in the Netherlands.*

2. Construction of *M. grisea* BAC library

A BAC library of *M. grisea* was constructed from rice-pathogenic Chinese field isolate, CHNOS 60-2-3, which had strong and stable fertility towards other rice-pathogenic *M.grisea* strains. Partially digested high molecular weight DNA of *M. grisea* (200-500 kbp) was ligated to dephosphorylated BAC vector. The ligation mixture was electroplated into *E. coli* cell with the cloning efficiency of 1×10^6 clones/μg vector. This library consisted of 5760 clones. All the clones were replicated onto two nylon membranes (2880 clones/membrane) for efficient screening.

The estimated insert size of the BAC clones ranged from 35 to 175 kbp, with an average of 120 kbp. This library was estimated to be about 13-genome equivalent, therefore, it covered more than 99.999 % of the genome. The representability of the library with regards to the original genome was tested by hybridization of the BAC membranes using a single-copy RFLP marker and DNA transposon MGR586 as a probe. These results were consistent with the estimated coverage of the library. A contig map covered a 240 kbp region around the RFLP marker indicated that there were no chimerism or deletion [17].

3. Distribution of *M. grisea* transposons in the BAC library

As a model of gene distribution study using BAC membrane, we analyzed the transposon co-occurring patterns in the fungal genome. Five out of seven transposons in the rice pathogen, namely MGSR1, MGR586, MGR583, Pot2, MAGGY were used in this study. They were hybridized with the BAC membranes for the transposon distribution analysis. *Fosbury* was not used in this study, as it was almost identical to MAGGY. The other transposon, Mg-SINE, was not present in our isolate. The percentage of the clones containing each transposon is shown in Table 1. The highest figure in the percentage of positive clones was obtained in Pot2, followed by MAGGY, MGR583 and MGR586. MGSR1 showed the lowest value. The percentage of positive clones of each transposon seemed to correspond to their copy numbers in *M. grisea* genome.

Table 1. The percentage of the BAC clones containing each transposon

Transposon	Positive clone (%)	Copy number
MGSR1	10	40 [9]
MGR586	12	55
MGR583	13	60
MAGGY	18	100 [11]
Pot2	23	100 [15]

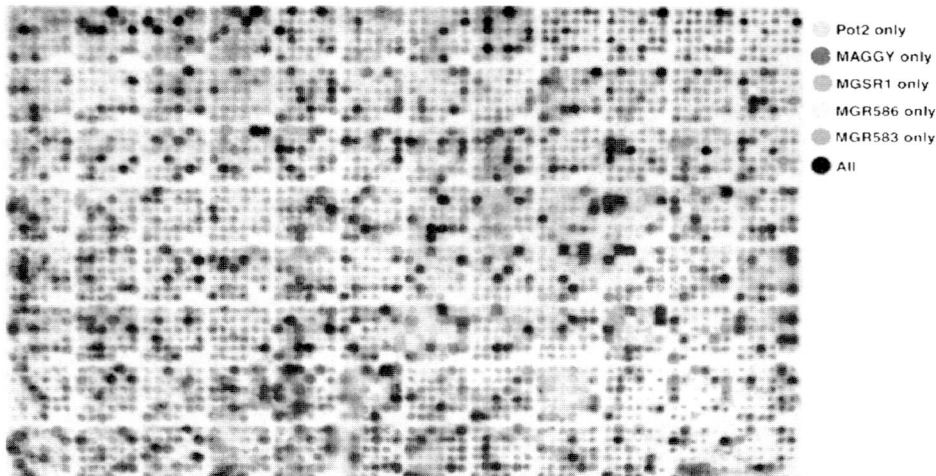

Pot2 only
MAGGY only
MGSR1 only
MGR586 only
MGR583 only
All

Figure 1. Transposon co-occurring pattern in the *M. grisea* BAC library. Five BAC membranes hybridized
to each transposons, namely MGSR1, MGR586, MGR583, Pot2, and MAGGY were overlapped.
Dark-colored dots showed the clones contained several transposons simultaneously.

To analyze the co-occurring pattern of transposons, these five BAC membranes which
were hybridized to each transposon, were overlapped (*Figure 1*). The result that some of
the clones contained several transposons simultaneously suggested that transposons
were clustering in the fungal genome.

We determined the percentage of clones that hybridized with different transposon
combination among 3000 clones to study the clustering pattern among these
transposons. The compatibility among the five transposons is shown in Table 2. The
percentage of positive clones with each transposon combination obtained by experiment
is shown in column A, obtained by calculation on the assumption that random
distribution of transposons is shown in column B. The figure on column C is the
maximum probability estimated from the percentage of the BAC clones containing each
transposon in Table 1. The value of (A-B)/(C-B) shows the compatibility of each set of
transposon combination. When these transposons are distributed randomly in the
genome, (A-B)/(C-B) value would be zero, while transposons co-occurred, that will
give larger values. In Table 2, all the (A-B)/(C-B) values were much greater than zero.
These results showed that transposons were clustering in the genome.

Table 2A shows the compatibility between two transposons. Two combinations of
Pot2 with MGSR1 or MGR586 showed higher compatibility, beside, combination of
MGR583 and MAGGY, and MGSR1 with MGR586, or MGR583, lower compatibility.
Table 2B and 2C is the compatibility of three-, four- and five- transposon combinations.
The results gave the same tendency as in Table 2A. As shown in Table 2C, the
percentage of BAC clones having five transposons was 1.4 %. Considering 10 % of
BAC clones were positive to MGSR1 (Table 1), 14% of MGSR1containing clones was

Table 2. Compatibility among several transposons.

A. The compatibility between two transposons.

Two-transposon combination	A: Positive Clone (%)	B: Calculated Probability (%)	C: Maximum probability (%)	(A-B)/(C-B): Compatibility
MGSR1 & Pot2	7.6	2.3	10	0.69
MGR586 & Pot2	8.8	2.8	12	0.65
MGR583 & Pot2	9.0	3.0	13	0.60
MGSR1 & MAGGY	6.5	1.8	10	0.57
MAGGY & Pot2	10.9	4.1	18	0.49
MGR586 & MGR583	6.4	1.6	12	0.46
MGR586 & MAGGY	6.5	2.2	12	0.44
MGSR1 & MGR583	4.8	1.3	10	0.40
MGSR1 & MGR586	4.7	1.2	10	0.40
MGR583 & MAGGY	6.0	2.3	13	0.35

B. The Compatibility among three transposons.

Three-transposon combination	A: Positive clone (%)	B: Calculated Probability (%)	C: Maximum probability (%)	(A-B)/(C-B): Compatibility
MGSR1 & MAGGY & Pot2	5.2	0.41	10	0.50
MGRS1 & MGR583 & Pot2	4.3	0.30	10	0.41
MGSR1 & MGR586 & Pot2	4.0	0.28	10	0.38
MGSR1 & MGR583 & MAGGY	3.5	0.23	10	0.33
MGR586 & MAGGY & Pot2	4.2	0.50	12	0.321
MGR586 & MGR583 & Pot2	4.1	0.36	12	0.32
MGSR1 & MGR586 & MAGGY	3.2	0.22	10	0.30
MGR583 & MAGGY & Pot2	4.0	0.54	13	0.28
MGSR1 & MGR586 & MGR583	2.3	0.16	10	0.22
MGR586 & MGR583 & MAGGY	2.3	0.28	12	0.17

C. The compatibility among four or five transposons.

Four- or Five- transposon combination	A: Positive clone (%)	B: Calculated probability (%)	C: Maximum probability (%)	(A-B)/(C-B): Compatibility
MGSR1 & MGR586 & MGR583 & MAGGY	1.6	0.028	10	0.16
MGSR1 & MGR586 & MAGGY & Pot2	3.1	0.054	10	0.31
MGSR1 & MGR586 & MGR583 & Pot2	2.0	0.036	10	0.20
MGSR1 & MGR586 & MAGGY & Pot2	2.6	0.050	10	0.26
MGR586 & MGR583 & MAGGY & Pot2	1.9	0.065	12	0.15
MGSR1 & MGR586 & MGR583 & MAGGY & Pot2	1.4	0.0065	10	0.14

also carried other four transposons simultaneously. These results strongly indicated that the five transposons clustered to form, namely, "transposon islands" in the fungal genome. Based on these results, the estimated number of the clusters with more than four transposons in one genome was 17, in the other words, there will be two to three cluster regions per chromosome.

The co-occurring tendency of the five transposons in the fungal genome was schematically shown in *Figure 2* based on the data represented in Table 2A. Pot2 was the most compatible with other four transposons. MGSR1 showed lower compatibility with MGR586 or MGR583. MAGGY showed the least compatible with MGR583.

Since some transposons of *M. grisea* were found in tandem arrays or nested to each other [9, 10, 11, 12, 13, 15], these results confirmed our observation.

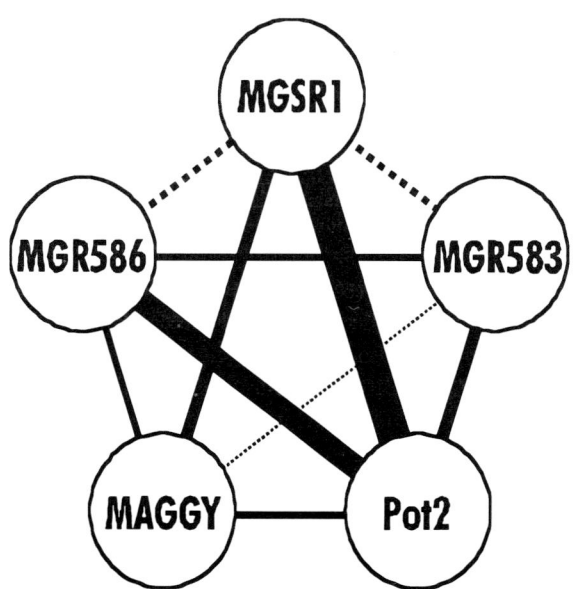

Figure 2. Schematic representation of the co-occurring tendency among the five transposons in *M. genome.* Thickness of the lines indicate strength of compatibility among five transposons, MGSR1, MGR586, MGR583, Pot2 and MAGGY. Bold lines indicate that they are highly clustered in the genome, while dotted lines, lower clustered.

4. Conclusion

Using BAC library, it was easy to give a clear indication on transposon clustering pattern in *M. grisea* genome. Further analysis of the BAC clones with several transposons simultaneously will give detailed information of regions in the genome where transposon clusters are.

So far, the applications of BAC library have been only focused on genome analysis and positional cloning. As shown in this study, BACs are also helpful for the study of gene distribution. As BACs are capable to represent its original genome, it will also greatly facilitate the study of genome structure and genome dynamcs.

5. Acknowledgements

The *M. grisea* strain CHNOS 60-2-3 was kindly provided from Dr. H. Naito of National Agriculture Research Center. We thank Dr. Y. Tosa of Kobe University for providing MAGGY probes. We are grateful to Dr. T. Kamakura of RIKEN for his useful advice on preparing intact genetic DNA.

6. References

1. Berg, D. E. and Howe, M. M.: *Mobile DNA*, American Society for Microbiology, Washington, DC, 1989.

2. Boyd, E. F. and Hartl, D. L.: Nonrandom location of IS1 elements in the genomes of natural isolates of *Escherichia coli.*, *Mol. Biol. Evol.* **14** (1997), 725–73.

3. Lewis, L. A., Gopaul, S., and Marsh, C.: The non-random pattern of insertion of IS2 into the *hemB* gene of *Escherichia coli.*, *Microbiol. Immunol.* **38** (1994), 461–465.

4. Lochmuller, H., Stucka, R., and Feldmann, H.: A hot-spot for transposition of various Ty elements on chromosome V in *Saccharomyces cerevisiae*, *Curr Genet.* **16** (1989), 247–252.

5. Pearce, S. R., Pich, U., Harrison, G., Flavell, A. J., Heslop-Harrison, J. S., Schubert, I., and Kumar, A.: The Ty1-copia group retrotransposons of *Allium cepa* are distributed throughout the chromosomes but are enriched in the terminal heterochromatin, *Chromosome Res.* **4** (1996), 357–364.

6. Boyle, A. L., Ballard, S. G., and Ward, D. C.: Differential distribution of long and short interspersed element sequences in the mouse genome: chromosome karyotyping by fluorescence in situ hybridization, *Proc. Natl. Acad. Sci. USA* **87** (1990), 7757–7761.

7. SanMiguel, P., Tikhonov, A., Jin, Y., Motchoulskaia, N., Zakharov, D., Melake-Berhan, A., Springer, P. S., Edwards, K. J.,Lee, M., Avramova, Z., and Bennetzen, J. L.: Nested retrotransposons in the intergenic regions of the maize genome, *Science.* **274** (1996), 765–768.

8. Sone, T., Suto, M., and Tomita, F.: Host species-specific repetitive DNA sequence in the genome of *Magnaporthe grisea*, the rice blast fungus, *Biosci. Biotech. Biochem.* **57** (1993). 1228–1230.

9. Hamer, J. E., Farrall, L., Orbach, M., Valent, B., and Chumley, G.: Host species–specific conservation of a family of repeated DNA sequences in the genome of a fungal plant pathogen, *Proc. Natl. Acad. Sci. USA.* **86** (1989), 9981–9985.

10. Kachroo, P., Ahuja, M., Leong, S., and Chattoo, B. B.: Organisation and molecular analysis of repeated DNA sequences in the rice blast fungus *Magnaporthe grisea*, Curr Genet. **31** (1997), 361–369.

11. Farman, M. L., Tosa, Y., Nitta, N., and Leong, S. A.: MAGGY, a retrotransposon in the genome of the rice blast fungus *Magnaporthe grisea*, *Mol. Gen. Genet.* **251** (1996), 665–674.

12. Shull, V., and Hamer, J. E.: Genetic differentiation in the rice blast fungus revealed by the distribution of the *fosbury* retrotransposon, *Fungal Genetic and Biology* **20** (1996), 59–69.

13. Kachroo, P., Leong, S. A., and Chatto, B. B.: Mg-SINE: a short interspersed nuclear element from the rice blast fungus, *Magnaporthe grisea*, *Proc. Natl. Acad. Sci. USA.* **92** (1995), 11125–11129.

14. Farman, M. L., Taura, T., and Leong, S. A.: The *Magnaporthe grisea* DNA fingerprinting probe MGR586 contains the 3' end of an inverted repeat transposon, *Mol. Gen. Genet.* **251** (1996), 675–681.

15. Kachroo, P., Leong, S. A., and Chatto, B. B.: Pot2, an inverted repeat transposon from the rice blast fungus *Magnaporthe grisea*, *Mol. Gen. Genet.* **245** (1994), 339–348.

16. Nitta, N., Farman, M. L., and Leong, S. A.: Genome organization of *Magnaporthe grisea*: integration of genetic maps, clustering of transposable elements and identification of genome duplication and rearrangements, *Theor. Appl. Genet.* **95** (1997), 20–32.

17. Nishimura, M., Nakamura, S., Hayashi, N., Asakawa, S., Shimizu, N., Kaku, H., Hasebe, A., and Kawasaki, S.: Construction of a BAC library of the rice blast fungus *Magnaporthe grisea* and finding specific genome regions in which transposons tend to cluster, *Biosci. Biotech. Biochem.* **62** (1998), 1515–1521.

18. Shizuya, H., Birren, B., Kim, U., Mancino, V., Slepak, T., Tachiiri, Y., and Simon, M.: Cloning and stable maintenance of 300-kilobase-pair fragment of human DNA in *Escherichia coli* using an F-factor-based vector, *Proc. Natl. Acad. Sci. USA.* **89** (1992), 8794–8797.

RETROTRANSPOSITION OF MAGGY IN *PYRICULARIA* *SPP*.

HITOSHI NAKAYASHIKI, KANAKO KIYOTOMI, YUKIO TOSA AND
SHIGEYUKI MAYAMA
*Laboratory of Plant Pathology, Faculty of Agriculture, Kobe University,
Kobe 657-8501, JAPAN*

1. Abstract

MAGGY is a gypsy-like LTR retrotransposon found in rice isolates of the blast fungus,
Pyricularia grisea (teleomorph, *Magnaporthe grisea*). In this study, we examined
transposition of MAGGY in *Pyricularia* isolates that did not originally possess a
MAGGY element. Genomic Southern analysis of MAGGY-transformants suggested that
transposition of MAGGY occurred in all isolates tested. In contrast, no transposition was
observed in all transformants with a modified MAGGY containing a 513-bp deletion in
the reverse transcriptase domain. When a MAGGY derivative carrying an artificial intron
was introduced into the wheat isolate of *P. grisea*, loss of the intron was observed. These
results showed that MAGGY can undergo autonomous RNA-mediated transposition in all
Pyricularia isolates. This supports the hypothesis that MAGGY was acquired
horizontally after the differentiation of MAGGY-carriers and its non-carriers.

2. Introduction

Eukaryotic genomes contain long-terminal repeat (LTR) retrotransposons, transposable
elements that use an RNA intermediate during duplication (Boeke *et al.*, 1985). Several
LTR retrotransposons have been identified recently in various species of fungi, especially
in pathogenic filamentous fungi (Daboussi, 1997). MAGGY is an LTR retrotransposon
isolated from a rice isolate of the blast fungus, *Pyricularia grisea* (teleomorph,
Magnaporthe grisea), comprising two ORFs and 253-bp LTRs (Farman *et al.*, 1996).
ORF1 encodes a predicted peptide of 457 amino acids corresponding to the *gag* protein of
retroelements. ORF2 has a typical character of a *pol* gene encoding a polypeptide with
protease, reverse transcriptase, RNaseH and endonuclease domains.

    Pyricularia spp. consist of many subgroups that cause the blast disease on diverse
plant species of *Gramineae, Cannaceae, Cyperaceae,* and *Zingiberaceae* (Kozaka and Kato,
1980). Genomic Southern analysis revealed that MAGGY was not ubiquitous in genomes
of *Pyricularia* subgroups (Tosa *et al.*, 1995). MAGGY was present in multiple copies in
Pyricularia isolates from rice, foxtail millet and some grasses, but absent in those from

D. Tharreau et al. (eds.), Advances in Rice Blast Research, 323–329.

wheat, finger millet, crabgrass, mioga etc. The sporadic distribution of MAGGY in *Pyricularia* subgroups raises questions regarding the ability of the element to transpose in non-carriers of MAGGY. In this study we show that MAGGY is an active element which transposes autonomously via an RNA intermediate in *Pyricularia* genomes that do not originally possess this element.

3. Results

3.1. CONSTRUCTION OF pMGY70 AND ITS DERIVATIVES

Plasmids used in this study are shown in Figure 1. Genomic subclone plasmids, pMGY-RF and pMGY-LF2 containing partial MAGGY sequences were kindly provided by Dr. Sally Leong. The plasmid pMGY70 is approximately 10 kb long and contains a full length MAGGY copy, constructed by ligating a *Dra*I-*Bam*HI fragment of pMGY-RF with the *Bam*HI - *Xba*I fragment of pMGY-LF2. pMGY70ΔEV is a plasmid with a deletion of a 513-bp *Eco*RV fragment (nt. 2241 - 2754) corresponding to the protease and reverse transcriptase domains of MAGGY. Therefore, it cannot provide proteins required for retrotransposition. Plasmid pMGY70-INT was constructed by inserting an artificial intron to *Pst*I site (nt. 5497) in 3' LTR of pMGY70 through several subclonings. The artificial intron was synthesized referring to intron 2 of a *THR1* reductase gene isolated from *Colletotrichum lagenarium* (Perpetua *et al.*, 1996).

3.2. TRANSPOSITION OF MAGGY IN *PYRICULARIA GRISEA* ISOLATES

Three *P. grisea* isolates (the wheat, finger millet and crabgrass isolates), and a *P. zingiberi* isolate were examined for new hosts of MAGGY. Several studies suggested that the finger

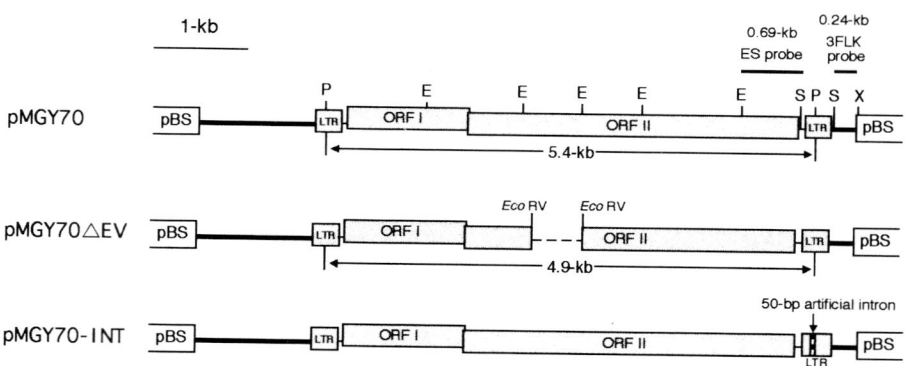

Figure 1. Structure of a plasmid containing MAGGY and its derivatives. Thick line between LTR and pBS sequence represents the flanking genomic sequence of *P. grisea*. Segments used as probes are indicated by lines. pBS, pBluescript SK+ II; E, *Eco*RI; P, *Pst*I; S, *Sma*I, X, *Xba*I.

millet and wheat isolates of *P. grisea* were much closer to the rice isolate, an original host of MAGGY, than was the crabgrass isolate (Shull and Hamer, 1994; Kusaba *et al.*, 1988). *P. zingiberi* has some characteristic features, exemplified by the formation of a sclerotium, thus is classified into a different species from *P. grisea*. These fungal isolates were transformed with pMGY70 or pMGY70ΔEV by means of PEG-mediated co-transformation with plasmid pSH75 (Kimura and Tsuge, 1993) that carries a hygromycin B phosphotransferase gene. Genomic DNA of two independent pMGY70-transformants and one pMGY70ΔEV-transformant per fungus were analyzed by Southern blots using a restriction enzyme, *Pst*I, and two probes, ES and 3FLK. The ES probe was a 0.69-kb *Eco*RI-*Sma*I fragment of pMGY70 corresponding to the 3' region of ORF 2, and the 3FLK probe was a 0.22-kb *Sma*I-*Xba*I fragment of pMGY70 corresponding to genomic sequence downstream of the 3' LTR (Figure 1). *Pst*I digests hybridized with the ES probe represents a signal with a single molecular weight, a signal of 5.4-kb in case of complete MAGGY and that of 4.9-kb in case of MAGGY with the deletion, independent of the genomic position in which MAGGY inserted. The intensity of the band reflects the total copy number of MAGGY in all cells, thus is expected to correlate with total frequency of MAGGY transposition. In contrast, the intensity of band detected with the 3FLK probe is thought to be constant irrespective of MAGGY transposition. Therefore, the blot probed with the 3FLK fragment was used as a quantitative control for the transposition assay. In Figure 2, a 0.54-kb signal was observed in all pMGY70-transformants, and its intensity

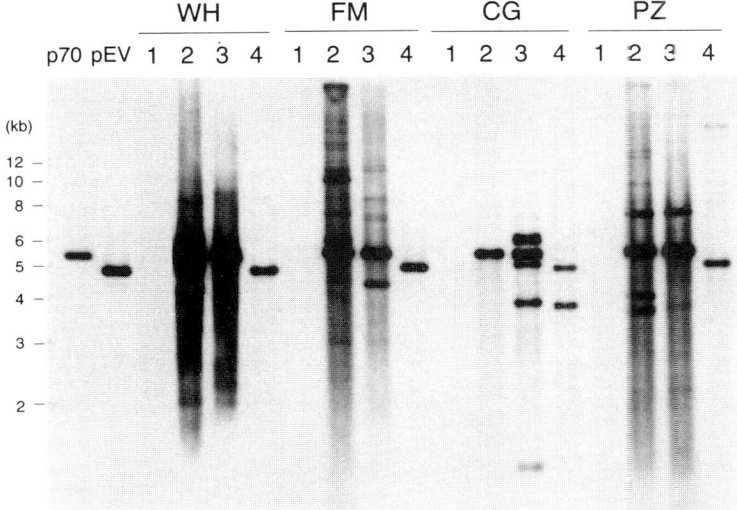

Figure 2. Southern blot analysis of *Pyricularia* isolates transformed with pMGY70 or pMGY70ΔEV. Genomic DNA was digested with *Pst*I, fractionated on a 0.7% TAE agarose gel, and probed with the ES fragment of MAGGY. Control plasmids were used as size markers. Lane 1, non-transformant; Lane 2-3, transformants with pMGY70; Lane 4, transformants with pMGY70ΔEV; p70, pMGY70; pEV, pMGY70ΔEV; WH, wheat isolate of *Pyricularia grisea*; FM, finger millet isolate of *Pyricularia grisea*; CG, crab grass isolate of *Pyricularia grisea*; PZ, *Pyricularia zingiberi*.

was higher compared with that of 4.9-kb in a corresponding pMGY70ΔEV-transformant (Figure 2). The results described above suggest that this MAGGY copy can transpose in the all genomes of *Pyricularia* spp., but that a derivative carrying a deletion in the *pol* gene cannot. Transposition of MAGGY seems to be autonomous, since it is not compensated by other endogenous retrotransposons. The frequency of MAGGY transposition appeared to be different among isolates tested. Active transposition was suggested in the wheat isolates of *P. grisea* and the *P. zingiberi* isolate, whereas transposition of MAGGY in the crabgrass isolate was relative low. Based on these results, transposition of MAGGY in these fungal isolates occurred in the order: the wheat isolate or the *P. zingiberi* isolate, the finger millet isolate, and the crabgrass isolate. As a control, the membrane was stripped and reprobed with the 3FLK fragment. No significant difference was observed in the intensity of signals between pMGY70-transformant and pMGY70ΔEV-transformant in all isolates (data not shown).

3.3. TRANSPOSITION OF MAGGY IN THE WHEAT ISOLATE OF *PYRICULARIA GRISEA* OCCURS VIA AN RNA INTERMEDIATE

In order to clarify whether MAGGY transposes through an RNA intermediate, pMGY70-INT, which carries a 50-bp artificial intron in the 3' LTR region (Figure 1), was introduced into the wheat isolate of *P. grisea*. If MAGGY transposes through reverse

(A)   **(B)**

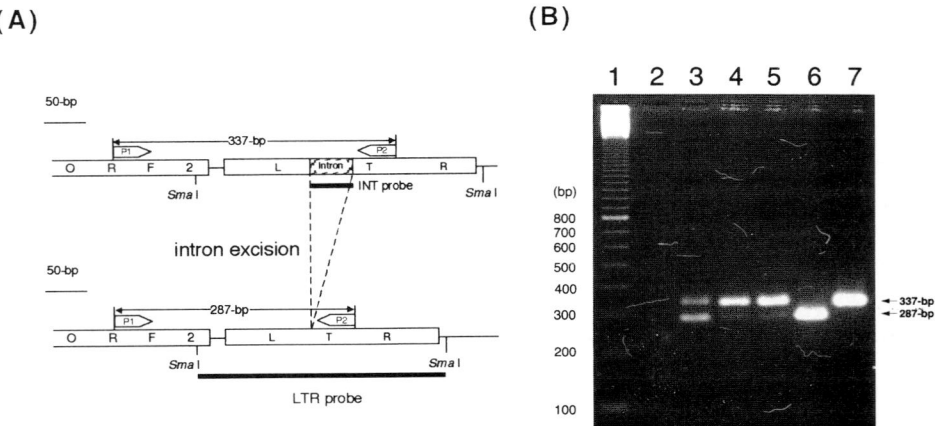

Figure 3. PCR analysis for the loss of an artificial intron during MAGGY transposition in the wheat isolate of *P. grisea*.
(A) Map of 3' LTR region of pMGY70-INT and its expected intron-less derivative, showing the positions of primers (P1 and P2) used in the transposition assay by PCR, and segments used for probing blots. Sizes of the intron-less and intron-containing PCR products are indicated.
(B) PCR transposition assay of pMGY70-INT transformants of the wheat isolate. Genomic DNA of three individual transformants, one non-transformant, and control plasmids, pMGY70 and pMGY70-INT were subjected to PCR amplification using Primers, P1 and P2. Amplified fragments were fractionated on a 3.5% TAE agarose gel, and visualized by ethidium bromide staining. Lane 1, 100-bp ladder; Lane 2, non-transformant; Lane 3-5, transformants with pMGY70-INT; Lane 6, pMGY70; Lane 7, pMGY70-INT.

transcription, we expect the appearance of intron-deleted elements in addition to the master intron-containing element. Loss of the intron in transformants was assayed by PCR using primers, P1 and P2 shown in Figure 3A. Expected sizes of the amplified intron-less and intron-containing fragments of 337-bp and 287-bp, respectively, were observed in lanes 6 and 7 of Figure 3B using pMGY70 and pMGY70-INT as templates. In all transformants, the fragment corresponding to the intron-less MAGGY was amplified in addition to the fragment corresponding to the intron-containing MAGGY, whereas no fragment was observed from non-transformed *P. grisea*. Following PCR-Southern analysis with probes corresponding to the LTR and the intron regions revealed that both the upper and the lower amplified fragments in the transformants are derived from the LTR domain, but that the upper contains the intron sequence whereas the lower does not (data not shown). These results suggest that the intron sequence was excised during the transposition of MAGGY, thus MAGGY undergoes an RNA-mediated transposition.

4. Discussion

In *Pyricularia* species, the distribution of retrotransposons, for example MAGGY, *fosbury*, and *grh*, are sporadic (Dobinson *et al.*, 1993; Tosa *et al.*, 1995; Farman *et al.*, 1996; Shull and Hamer, 1996). MAGGY and *fosbury* are present in rice and *Setaria* pathogens of *Pyricularia*, whereas the distribution of *grh* is localized in *Eluesine* pathogen only. Based on a dendrogram constructed by rDNA RFLP (Kusaba *et al.*, 1998), *Pyricularia* isolates of MAGGY-carriers were classified into the same cluster and all *grh*-carriers were classified into a single rDNA type which stood apart from the cluster of MAGGY-carriers. Several phylogenic studies of retrotransposons based on its conservative reverse transcriptase domain revealed that the distribution of closely related retrotransposons does not always follow the phylogenic relationship of their host species (Flavell *et al.*, 1995). Therefore, many researchers have pointed out the possibility of horizontal transfer of retrotransposons (Mizrokhi and Mazo, 1990). However, no direct evidence has been obtained up to now. Our results presented here support the concept that MAGGY was acquired horizontally since the results rule out the alternative explanation that MAGGY was lost in non-MAGGY-carriers.

MAGGY seemed to transpose more actively in *P. zingiberi* than in the crabgrass isolates of *P. grisea* suggesting that the activity of MAGGY in *Pyricularia* species did not solely depend on the general genetical similarity of the new host to the original one. Some factors other than compatibility to cellular machinery might be involved in the different activity of MAGGY in *Pyricularia* species. A specific host gene was reported to have the ability to control a retrotransposon in *Drosophila*. A single allele called *flamenco* in *D. melanogaster* was repressed the transposition of *gypsy* (Prud'homme *et al.*, 1994). Even though its gene and product have not been isolated so far, it was shown that its restrictive effects on transposition was occurred at the transcriptional level (Pelisson *et al.*, 1994).

Plant DNA transposons, exemplified by the maize *En/Spm* and *Ac/Ds* elements, are used as gene tagging tools in a wide range of heterologous plant species in addition to their original hosts (Sundaresan, 1996). In other eukaryotic organisms such as *Saccharomyces cerevisiae* and *Drosophila melanogaster*, several genes have been tagged by their own LTR retrotransposons Ty1 and *copia*, and successfully isolated (Bingham *et al.*, 1981; Garfinkel *et al.*, 1988). Our results demonstrated that MAGGY is a candidate for a gene tagging tool in filamentous fungi, especially in the *Ascomycotina*. We have found that RNA-mediated transposition of MAGGY occurred in C. *lagenarium* and *Alternaria longipes* by the PCR assay (Nakayashiki *et al.*, submitted). Analysis on the regulation of MAGGY transposition and target site specificity will be important points to establish MAGGY as a tagging tool in filamentous fungi.

Pathogenic fungi isolated from the field often show changes in their phenotype in culture, for example, morphological changes such as colony color or shape, loss of pathogenicity, reduction of spore production etc. These observations lead us to hypothesize that fungal transposable elements in field isolates may be active. Indeed, most active fungal transposable elements are derived from field isolates. Transposable elements could contribute to such genetic variation in nature, and it might result in the survival of the fittest. On the other hand, fungal transposable elements found in laboratory strains are commonly inactive. Laboratory strains are usually established as a result of continuous selection for phenotypic stability. In such fungal strains, transposable elements might be neutralized by host suppression mechanisms.

Little is known about the molecular mechanism of MAGGY transposition. Studies on retrotransposons are sometimes complicated by their high copy number and relic copies in host genomes. Introduction of MAGGY into a new host should facilitate the elucidation of mechanisms of retrotransposition and the interaction between a retrotransposon and a host at the molecular level.

5. Acknowledgments

We thank Dr. Yasuyuki Kubo for technical advice on fungal transformation, Dr. Takashi Tsuge for providing plasmid vector pSH75, and Dr. Sally Leong for providing original MAGGY subclones, pMGY-RF and pMGY-LF2. We are much indebted to Dr. Hajime Kato, the former professor of Kobe university, for valuable suggestions.

6. References

Bingham, P. M., Levis, R. and Rubin, G. (1981) Cloning of DNA sequences form the *white* locus of *D. melanogaster* by a novel and general method, Cell **25**, 693-704.
Boeke, J. D., Garfinkel, D. J., Styles, C. A. and Fink, G. R. (1985) Ty elements transpose through an RNA intermediate, Cell **40**, 491-500.
Daboussi, M. J. (1997) Fungal transposable elements and genome evolution, Genetica **100**, 253-260.
Dobinson, K. F., Harris, R. E. and Hamer, J. E. (1993) *Grasshopper*, a long terminal repeat (LTR) retroelement in the phytopathogenic fungus *Magnaporthe grisea*, Mol. Plant Microbe Interact. **6**, 114-26.
Farman, M. L., Tosa, Y., Nitta, N. and Leong, S. A. (1996) MAGGY, a retrotransposon in the genome of the rice blast fungus *Magnaporthe grisea*, Mol. Gen. Genet. **251**, 665-74.
Flavell, A. J., Jackson, V., Iqbal, M. P., Riach, I. and Waddell, S. (1995) Ty1-*copia* group retrotransposon sequences in amphibia and reptilia, Mol. Gen. Genet. **246**, 65-71.

Garfinkel, D. J., Mastrangelo, M. F., Sanders, N. J., Shafer, B. K. and Strathern, J. N. (1988) Transposon tagging using Ty elements in yeast, Genetics **120**, 95-108.

Kimura, N. and Tsuge, T. (1993) Gene cluster involved in melanin biosynthesis of the filamentous fungus *Alternaria alternata*, J. Bacteriol. **175**, 4427-35.

Kozaka, T. and Kato, H. (1980) Morphology, host range, and classification of the blast fungus, in Y. Yamasaki and T. Kozaka (eds.), Rice Blast Disease and Breeding of Resistant Cultivars, Hakuyusha, Tokyo, pp. 47-80.

Kusaba, M., Eto, E., Tosa, Y., Nakayashiki, H. and Mayama, M. Genetic diversity in *Pyricularia* isolates from various hosts revealed by RFLP of nuclear ribosomal DNA and the distribution of the MAGGY retrotransposon, (submitted).

Mizrokhi, L. J. and Mazo, A. M. (1990) Evidence for horizontal transmission of the mobile element *jockey* between distant *Drosophila* species, Proc. Natl. Acad. Sci. USA **87**, 9216-20.

Nakayashiki, H., Kiyotomi, K., Tosa, Y., and Mayama, S. Transposition of the retrotransposon MAGGY in heterologous species of filamentous fungi. (submitted).

Pelisson, A., Song, S. U., Prud'homme, N., Smith, P. A., Bucheton, A. and Corces, V. G. (1994) *Gypsy* transposition correlates with the production of a retroviral envelope-like protein under the tissue-specific control of the *Drosophila flamenco* gene, EMBO J. **13**, 4401-1.

Perpetua, N. S., Kubo, Y., Yasuda, N., Takano, Y. and Furusawa, I. (1996) Cloning and characterization of a melanin biosynthetic *THR1* reductase gene essential for appressorial penetration of *Colletotrichum lagenarium*, Mol. Plant Microbe Interact. **9**, 323-9.

Prud'homme, N., Gans, M., Masson, M., Terzian, C. and Bucheton, A. (1995) *Famenco*, a gene controlling the *gypsy* retrovirus of *Drosophila melanogaster*, Genetics **139**, 697-711.

Shull, V. and Hamer, J. E. (1994) Genomic structure and variability in *Pyricularia grisea.*, in R.S. Zeigler, S. A. Leong, P.S. Teng (eds.), Rice Blast Disease, CAB international, Wallingford, UK. pp.65-86.

Shull, V. and Hamer, J. E. (1996) Genetic differentiation in the rice blast fungus revealed by the distribution of the *Fosbury* retrotransposon, Fungal Genet. Biol. **20**, 59-69.

Sundaresan, V. (1996) Horizontal spread of transposon mutagenesis: new uses for old elements, Trend in Plant Science **1**, 184-190.

Tosa, Y., Nakayashiki, H., Hyodo, H., Mayama, S., Kato, H. and S, A. L. (1995) Distribution of retrotransposon MAGGY in *Pyricularia* species, Ann. Phytopathol. Soc. Jpn. **61**, 549-554.

DEVELOPMENTS IN RICE BLAST RESEARCH OVER THE PAST FIVE YEARS: WE ENTER THE GENOMICS ERA

B. Valent

DuPont Agricultural Products, P.O. Box 80402, Wilmington, DE 19880-0402, USA

Research that will impact our ability to provide rice farmers with more durable resistance to blast disease includes: understanding the molecular mechanisms behind host-pathogen interactions; understanding the population biology and potential for evolution of the rice blast pathogen; and understanding the evolution and behavior of rice resistance genes. Since the last International Rice Blast Conference, significant progress has been made in identifying pathogen genes required for the fungus to infect rice, and in understanding the population biology of the pathogen. There have also been major breakthroughs in the cloning and molecular characterization of resistance genes from diverse plants. New molecular details on the structure of the *Pi-ta* rice blast resistance gene will be described and compared with characterized resistance genes from other host/pathogen systems. The recent rapid progress in genomic sciences, including high through-put DNA sequencing, is also impacting rice blast research. The first fungal genome to be sequenced, that of the yeast *Saccharomyces cerevisiae*, contains nearly 6400 genes. Rice genome sciences are at an advanced state due to major genome initiatives such as in Japan, China and Korea, and a new international project to sequence the entire rice genome has set goals for completion of 40% of the rice genome sequence by the year 2003. Genomic research at DuPont focused on identifying the estimated 9000 genes of the rice blast fungus will be described, along with discussion of the future of rice blast research.

MOLECULAR ANALYSIS OF DISEASE RESISTANCE IN RICE

Pamela Ronald , Wen-Yuan Song, Dahu Chen, Yuwei Shen, Guo-Liang Wang, Li Ya Pi, Randy Ruan, Lili Chen, S. Zhang, C. Fauquet.

Department of Plant Pathology and Center for Engineering Plants for Resistance Against Pathogens, University of California, Davis, 95616

International Laboratory for Tropical Agricultural Biotechnology, Scripps, San Diego

The rice Xa21 gene, which confers resistance to Xanthomonas oryzae pv. oryzae, was isolated by positional cloning. Transgenic plants containing the Xa21 gene showed resistance to 31 isolates of Xoo. The sequence of the predicted protein, which carries both a leucine-rich repeat motif and a serine-threonine kinase, suggests a role in cell surface recognition of a pathogen ligand and subsequent activation of an intracellular defense response. In support for this model, we have found that Xa21 encodes an active kinase with serine threonine specificity. The kinase domain interacts with a rice catalase and a Xoo GTP binding protein in the yeast two hybrid system suggesting the role of both rice and Xoo proteins in transducing the resistance response. Confirmation of the biological relevance of these interactions is underway. We have found that one member of the Xa21 gene family, Xa21D, encodes a truncated protein due to the presence of a retrotransposon-like sequence in the LRR domain. Interestingly, although Xa21D lacks a kinase, it confers partial resistance to Xoo in transgenic plants, indicating that the presence of the Xa21D LRR domain is sufficient for pathogen recognition. We are also studying the molecular mechanism durable resistance to rice blast in the West African upland rice variety, Moroberekan. Two major resistance loci, Pi5(t) and Pi7(t), were identified in Moroberekan through RFLP analysis of a recombinant inbred (RI) (Wang et al. 1994). We are now fine mapping these loci to facilitate the cloning of Pi5(t) and Pi7(t). Bulk segregant AFLP analysis was applied to identify DNA markers linked to the respective resistance loci. Out of 1700 primer combinations screened in a Pi5(t) segregating population, thirty-three markers were identified to be linked with Pi5(t). Out of 1500 primer combinations screened in the Pi7(t) populations, only three linked markers were identified. The chromosomal locations of the linked markers were determined using a previously reported F2 mapping population developed from a cross between Labelle and Black Gora (Rodona and Mackill, 1997). Pi5(t) mapped to rice chromosome 9 rather than chromosome 4 as previously reported (Wang et al. 1994),and Pi7(t) mapped to rice chromosome 11. High-resolution mapping of the Pi5(t) locus using an F4 segregating population, consisting of 3,000 individual derived from the cross of RIL260 and CO39, is in progress.

MOLECULAR GENETIC ANALYSIS OF BLAST RESISTANCE IN CO39 CORRESPONDING TO AVIRULENCE GENE *AVR1-CO39* OF *MAGNAPORTHE GRISEA*

R.S. CHAUHAN[1], M.L. FARMAN[2], P. RONALD[3] and S.A. LEONG. [1,4]

[1]Department of Plant Pathology, University of Wisconsin, Madison, WI 53706, USA; [2]Department of Plant Pathology, University of Kentucky, Lexington, KY 40546; [3]Department of Plant Pathology, University of California at Davis, Davis, CA 95616, USA; [4]Plant Disease Resistance Unit, USDA Agricultural Research Service, 1630 Linden Dr., Madison, WI 53706, USA

Indica rice line CO39- originally bred for blast resistance in India- has been used either as a universal susceptible or a recurrent parent in the development of near-isogenic lines carrying single genes for blast resistance. However, CO39 is still showing resistance to blast pathogen races/lineages from different parts of the world. Genetic analysis of avirulence in *M. grisea* strain 2539, avirulent on CO39, has lead to the identification and eventual cloning of a stable cultivar specificity gene *AVR1-CO39*. These studies have also indicated that resistance in CO39 is controlled by a single dominant gene. The present study has been carried out with the objective of mapping and cloning blast resistance gene in CO39.

Fifteen mirosat (simple sequence repeats) markers, mapping to four rice chromosomes 1, 6, 11 & 12, known to carry disease resistance genes, were tested for polymorphism between CO39(resistant) and 51583(susceptible). Most of the test RM markers showed polymorphism between the resistant and susceptible lines. Co-segregation of polymorphic RM markers with resistance/susceptibility was tested using bulked segregant analysis as well as individual F3 families. Microsat marker, RM202, located on chromosome 11, co-segregated with blast resistance in resistant/susceptible F3 families as well as bulks. Linkage relationship between RM202 and resistance gene in CO39 is being established by testing the response of segregating F2 populations. RFLP markers bracketing RM202 on chromosome 11 are also being tested for cosegregation with the target resistance gene.

Structural homologies betwen R genes from different plant species have facilitated the cloning of R gene from monocots using degenerate oligos from dicot R genes. This strategy is also being used for cloning the R gene in CO39. Degenerate oligonucleotide primers from the conserved regions of both the leucine-rich repeats (LRR) and serine-threonine kinase domains of Xa21, have amplified a 850bp fragment in CO39 and 51583. DNA sequence analysis of the cloned amplicon has revealed novel R-gene-like homologs.

ANALYSIS OF ACQUIRED RESISTANCE RELATED PROTEIN RIR1 IN TRANSGENIC RICE PLANTS

U. Schaffrath[1] and R. Dudler[2]

[1] Institut für Pflanzenphysiologie (Bio III), RWTH Aachen, D-52056 Aachen
[2] Institute of Plant biology, University of Zurich, CH-8008 Zurich

In rice local acquired resistance against *Pyricularia oryzae* (Cav.) can be induced by a preinoculation with the non-host pathogen *Pseudomonas syringae* pv. *syringae*. Several cDNAs corresponding to transcripts that accumulate concomitantly with resistance induction have been cloned and characterized.

Our goal was to verify the causality between gene activation and pathogen resistance. For that study we have choosen the RIR1 cDNA which is a member of a new gene family and which has no homology to previous described genes.

In a first step, we have investigated the structure and localisation of the RIR1 protein by biochemical means. Using antibodys raised against a RIR1-fusion protein we show that RIR1 is secreted from rice protoplasts transiently expressing 35S::RIR1 construct and that the protein accumulates in the cell wall compartment of rice leaves upon inoculation with *P. syringae* pv. *syringae*.

In parallel, we have constructed transgenic rice plants overexpressing the RIR1 cDNA either in sense and antisence direction. Both constructs are driven by the cauliflower mosaic virus 35S promotor and thus plants should either overexpress or suppress the gene in all tissue. We succeeded in regeneration of healthy plants that in fact producing seeds. Southern and northern analyses are presented to confirm the genetics of our transformants.

MAJOR AND MINOR GENES FOR RICE BLAST RESISTANCE FROM 'LEMONT' AND 'TEQING' TAGGED USING RFLP

R.E.Tabien[1], Z. Li[2], M.A. Marchetti[3], S.R.M. Pinson[3]

[1] Plant Breeding and Biotechnology Division, Philippine Rice Research Institute, Muñoz, Nueva Ecija, Philippines
[2] Plant Breeding, Genetics and Biochemistry Division, International Rice Research Institute, P.O. Box 933 1099 Manila, Philippines
[3] USDA-ARS, Route 7, Box 999, Immes Road, Beaumont, Texas 77713, USA

Spray inoculation using five races of rice blast prevalent in Southern U.S. (IC-17, IB-49, IB-54, IG-1, IE-1) and field nursery evaluation of a recombinant inbred (RI) population of 'Lemont' and 'Teqing' were done to map major genes and quantitative trait loci (QTLs) for resistance. Using RFLP markers as tag for each gene, major genes and their combinations were evaluated at the blast nursery and the temporal expression of QTLs was studied. Greenhouse phenotype of 219 F_8 RILs correlated with 167 RFLP markers revealed four major genes. Three genes from Teqing, *Pi-q5*, *Pi-q1*, and *Pi-q6* were present in *chromosomes 2, 6*, and *12* and bracketed with RG520-RZ446a, C236-RZ508, and RG869-L102, respectively. One Lemont gene, *Pi-b2*, was mapped in *chromosome 11* between L457b and RZ536a. QTL mapping identified nine loci for area under disease progress curve (AUDPC), six for percent diseased leaf area (%DLA), and eight for Standard Evaluation System for blast (SES rating). Analyses of the major genes and their combination detected both direct effect and gene x gene interaction. QTLs identified were temporally expressed, and were generally expressed 26-40 days after seeding (DAS).

References
Inukai T, Zeigler RS, Sarkarung S, Bronson M, Dung LV, Kinoshita T, Nelson RJ. 1996. Theor Appl Genet 93:560-567
McCouch SR, Nelson RJ, Tohme J, Zeigler RS 1994. In: Zeigler RS, Leong SA and Teng PS (eds). Rice blast disease. CAB Inter, Wallingford, England, pp167-186
Naqvi NI, Chattoo BB. 1996. Genome 39:26-30
Pan Q, Wang L, Ikehashi H, Tanisaka T. 1996. Phytopathology 86:1071-1075
Wang GL, Mackill DJ, Bonman JM, McCouch SR, Champoux MC, Nelson RJ. 1994. Genetics 136:1421-1434

PEDIGREE ANALYSIS OF SELECTED RICE BREEDING LINES FOR BLAST RESISTANCE

R.C. Sharma[1], S.M. Shrestha[1], S. Sarkarung[2], R. J. Nelson[3], and H. Leung[2]

[1] Institute of Agriculture and Animal Science, Rampur, Chitwan, Nepal
[2] International Rice Research Institute, P.O. Box 933, 1099, Manila, Philippines
[3] Centro Internacional de la Papa (CIP), Lima, Peru

Blast, caused by *Pyricularia grisea* Sacc., is a major disease of rice (*Oryza sativa* L.). Knowledge of the blast resistance genes carried by rice varieties can be useful for the management of existing genotypes, and for the production of improved ones. Substantial research effort has been invested in the mapping of blast resistance genes. The availability of molecular markers linked to blast resistance genes, together with pedigree data, pathotype data and selective genetic analysis, can provide information about the genotypes and phenotypes of rice varieties. This study was conducted to correlate molecular marker data with pathotype data for tracing blast resistance in selected pedigrees of advanced breeding lines of rice. Molecular analysis of advanced rice breeding lines and their parents was done using RFLP markers at the International Rice Research (IRRI), Philippines. Selected rice breeding lines and their parents were inoculated with well characterized blast isolate at IRRI, Philippines, and field evaluated under natural blast infection at Rampur, Nepal. Molecular and pathotype analyses revealed association between blast reaction and molecular markers in several pedigrees. Single donor parent for molecular markers linked to blast resistance could be traced in certain pedigrees.

EVALUATION OF BLAST RESISTANCE SPECTRUM IN RICE VARIETIES

Nguyen Trinh Toan, La Tuan Nghia, Nguyen Thi Ninh Thuan, Nguyen Thanh Thuy, Ngo Vinh Vien[*], Vu Duc Quang, Tran Duy Quy
Agricultural Genetics Institute, Tuliem, Hanoi, Vietnam.
[*]National Institute of Plant Protection, Tuliem, Hanoi, Vietnam.

For the deployment of blast resistant varieties, the evaluation of blast resistance/susceptibility spectrum of rice is an important part of rice breeding and improvement programs. The purpose of this study is to describe the disease reaction patterns of rice varieties infected with representative blast isolates and to identify potential parents for blast resistance breeding.

37 rice varieties including 20 from IRRI (10 NILs carrying single, and 10 elite lines carrying multiple resistance genes), plus 17 improved and Vietnam local varieties were choosen for the research. 20 representative isolates of blast fungus population in the Northern and Central Vietnam that had been identified by using MGR-DNA fingerprinting were used to infect these rice varieties. CO39 and LTH, the highly susceptible cultivars, were used as susceptible checks. Disease reaction patterns of the varieties tested were analyzed using NTSYS-UPGMA program.

The two susceptible checks were tightly clustered together (>90%), indicating that either of them represents good virulence checks. F98-7 carrying the gene $Pi\text{-}k^m$ was observed to be not effective for the Vietnamese blast population since it forms tight cluster with the susceptible checks. On the other hand, gene combinations effective for the existing blast population are $Pi\text{-}1$ (C101LAC), $Pi\text{-}2$ (C101A51) and $Pi\text{-}ta^2$ (F128-1). The differential disease reaction patterns observed among the tested varieties provided the needed information in the identification of potential parents for blast resistance breeding. Of the tested varieties, CR203 was susceptible to 80 % of the representative blast isolates. Since CR203 is one of the most cultivated varieties in North Vietnam, pathogenicity test showed that this variety is a good recipient for crosses with other resistant donors. On the other hand, some local cultivars were found to be highly resistant. Varieties falling into tight cluster with the NILs carrying the effective genes were identified to be the potential donors of resistance genes for later resistance breeding programs. These varieties/lines are: A3052, A3049, A3046, Tamthom, Chiembac, Tetep, and C70.

IDENTIFICATION OF AFLP MARKERS LINKED TO BLAST RESISTANCE GENES IN RICE

Vu Duc Quang, Luu Thi Ngoc Huyen, Nguyen Thi Kim Dung, Doan Thi Hoa, Peter Breyne*, Ann van Gysel*
Agricultural Genetics Institute, Hanoi, Vietnam
*State University Ghent, VIB, Belgium

Rice blast is the most dangerous disease of rice in all rice producing countries. The pathogen may cause rice yield losses of more than 15%, occasionally to 20-30%. Breeding for resistance to blast disease is one of the priorities of rice breeding programs in Vietnam. This research aims to identify, by means of AFLP techniques, molecular markers closely linked to blast resistance genes in rice for the use in rice breeding.

Two F2 populations generated from crosses between CR203 (blast sensitive) with lines 188 and 299 (blast resistant) of 132 and 145 individual plants, respectively, were used in the experiments. Phenotyping was carried out on the F3 populations, based on which F2 individual plants were judged for resistance/susceptibility. Segregation analysis showed that the blast resistance traits in the lines 188 and 299 were dominant monogenic. Genomic DNA was extracted from lyophilized leaf tissues of the F2 individuals. After that the DNA was digested with EcoRI and MseI, followed by ligation with EcoRI and MseI adaptors. The selective amplification was performed using about 30 different primer combinations, each of the primers being designed to contain two additional nucleotides. As the results, 301 and 188 markers were detected within the two populations, respectively. The data was analyzed by MapMaker program. It was shown that at min. LOD = 4.00 and max. distance = 35.00 the markers of populations 1 and 2 were separated into 14 and 15 linkage groups, respectively. For population 1, the Blast1 locus was flanked by markers CGCA14 and ATAG2 at distance of 16.5 and 12.7 cM respectively. For population2, the Blast2 lied between markers CGTC1 and CCAG5 at distance of 8.9 and 16.4 cM respectively. Searches for more tightly linked markers are under way.

EVALUATION OF NEAR-ISOGENIC LINES AGAINST BLAST PATHOGEN UNDER NURSERY CONDITIONS IN BANGLADESH

H.U. Ahmed, M. M. Rahman, and M.A.T Mia
Plant Pathology Division, Bangladesh Rice Research Institute (BRRI)
Joydebpur, Gazipur 1701, Bangladesh

Ten near-isogenic lines in CO39 or LTH backgrounds were tested under BRRI blast nursery conditions for two consecutive years. Single line of around one hundred plants constituted a plot for each entry. The treatments were laid out in randomized complete block design with three replications. Plants were exposed to natural inoculum supplemented with chopped blast infected leaves collected from previous years' blast nursery. Disease severity data was recorded following standard evaluation system, and percent diseased leaf area was noted per plot basis for six times at 3-5 day intervals. Area under disease progress curves were calculated and, analyzed. Most of the R-genes were ineffective to the present blast population(s) in the nursery. However, the patterns of blast epidemiology on different R-genes were variable. Results indicated that the exotic R-genes may not be useful for the development of blast resistant rice variety for cultivation in Bangladesh. Future research plan on blast disease of rice will be discussed.

Reference:
1. Ahmed, H. U., Haque, M. A., Shahjahan, A. K. M. and Miah, S. A. 1985. Blast resistance in rice germplasm in Bangladesh. *Bangladesh J. Plant Pathology* 1(1): 73-74.
2. Mackill, D.J. and Bonman, J.M. 1992. Inheritance of blast resistance in near-isogenic lines of rice. *Phytopathology* 82: 746-749.
3. Shahjahan, A. K. M. 1994. Practical approaches to rice blast management in tropical monsoon ecosystems, with special reference to Bangladesh, p465-488. In R.S. Zeigler, S. A. Leong and P.S. Teng (ed.), Rice Blast Disease. IRRI & CABI Publication.
4. Zhongzhuan Ling, Mew, T.W., Jiulin Wang and Cailin Lei. 1995 Development of near-isogenic lines as international differentials of the blast pathogen. *IRRN* 20:1 13-14.

IDENTIFICATION OF A NEW RESISTANCE GENE TO A CHINESE BLAST FUNGUS ISOLATE IN JAPANESE RICE CULTIVAR AICHI ASAHI

N. Hayashi[1], I. Ando[2], T. Imbe[3]
[1] Aichi-ken Agriculture Research Center, Mountainous Region Agricultural Research Institute, Inabu, Aichi 441-2513, Japan.
[2] AFFRC, MAFF, 1-2-1 Kasumigaseki, Chiyoda-ku, Tokyo, Japan
[3] International Rice Research Institute, P. O. Box 933, Manila, Philippines.

Gene analysis for the high resistance of the rice cultivar 'Aichi Asahi' and some other Japanese cultivars to a blast fungus isolate CHNOS58-3-1 from China was performed. All the Japanese differential cultivars were resistant to the isolate except for 'Pi No. 4' which showed a moderate resistance. Analysis of the F_2 population of a cross with the susceptible cultivar 'Reiho' and the resistant cultivar Aichi Asahi to the isolate indicated that the resistance of Aichi Asahi to the isolate was conferred by one dominant gene. To identify the gene in other Japanese differential cultivars, 'AK lines' which were derived from a cross of 'Aichi Asahi' X 'K59' and assumed to harbor no known genes except for the new one, were used for the allelism tests. The new completely dominant resistance gene was detected in 14 differential cultivars but not in 'Pi No. 4', 'Yashiro-mochi', and 'K1' and was designated as *Pi19(t)*. *Pi19(t)* was allelic or closely linked to *Pita²* on chromosome 12. *Pi19(t)* was extensively distributed among Japanese traditional local cultivars.

DIFFERENCE I N INDUCED RESISTANCE IN SASANISHIKI NEAR ISOGENIC LINES BY INCOMPATIBLE RICE BLAST FUNGUS

T. Ashizawa, K. Zenbayashi and S. Koizumi
Department of Lowland Farming, Tohoku National Agricultural Experiment Station, Yotsuya, Omagari, Akita 014-0102, Japan

In Japan, nine near isogenic lines (NILs) with different complete resistance genes to blast were breeded from the rice cv. Sasanishiki and some of the NILs are now cultivated as a multiline. We studied induced resistance in the Sasanishiki NILs by incompatible rice blast fungus for the basis of effective control of blast with multilines. In rice leaves inoculated with incompatible *Pyricularia grisea* of the tested eight NILs, epidermal cell reactions at the penetration sites of the fungus were grouped into three types, i. e. no reaction, cytoplasmic granulation and deep browning. The percentage of no reaction at the penetration sites (nearly 80%) was almost the same among the seven near isogenic lines, and cytoplasmic glanulation was observed in about 20% of the epidermal cells of the seven NILs including *Pi-z* line (the NIL possessing the complete resistance gene *Pi-z*). On the other hand, deep browning was seen in 13% of the penetration sites of *Pi-i* line (the NIL having the complete resistance gene *Pi-i*), although a few deep browning epidermal cells were found in the other NILs. These reactions affected visual lesions appeared on leaf blades on the NILs such as brown spots on *Pi-i* lines and no symptoms on the other NILs. Enlargement of blast lesions on leaf blades was reduced by pre-spray-inoculation of incompatible *P. grisea*, and the reduction on *Pi-i line* (50%) was higher than that on *Pi-z* line(10%). The reduction in the lesion enlargement was also observed on uninoculated parts adjacent to the points inoculated with the incompatible *P. grisea*, and the reduction rate was 43% for *Pi-i* line and 30% for *Pi-z* line. Pre-inoculation of the incompatible fungus also suppressed development of invaded hyphae of the compatible isolate in epidermal cells. Deep browned epidermal cells by the incompatible fungus inhibited penetration of compatible hyphae into them, although cytoplasmic glanulation did not have no effect on the penetration.

GENETIC ANALYSIS OF PARTIAL RESISTANCE TO LEAF BLAST IN THE RICE CULTIVAR CHUBU 32

K. Zenbayashi1, T. Ashizawa1, T. Tani2 and S. Koizumi1
1 Department of Lowland Farming, Tohoku National Agricultural Experiment Station
Yotsuya, Omagari, Akita 014-0102, Japan
2 Mountainous Region Agricultural Research Institute, Aichi Agricultural Research Center, Inabu, Kitashitara, Auchi, 441-2513, Japan

The rice cultivar Chubu 32 exhibits a high level of partial resistance to leaf blast in Japan. We analyzed the inheritance of this partial resistance using 149 F3 lines from a single cross between the rice cultivar Norin 29 showing a low level of partial resistance to the disease and Chubu 32. The two cultivars have no complete resistance genes to pathogenic races of Pyricularia grisea distributed in Japan. Chubu 32 was used as a pollen parent in the cross with Norin 29. Partial resistance to leaf blast in the F3 lines was evaluated in an upland nursery trial under natural conditions at Omagari, Akita, Japan. Disease scoring of their parent cultivars and each F3 line was based on a visual estimate of percentage of diseased leaf area. The F3 lines segregated between their parents for leaf blast severity and the histogram of the disease severity in them showed two peaks, indicating the presence of an incomplete dominant gene. We also conducted inoculation tests in greenhouse. From the 149 F3 lines, six lines were selected and their eighth leaf stage seedlings were inoculated with two compatible isolates of P. grisea Ina85-182 and IBOS8-1-1, belonging to the same pathogenic race. The degree of disease severity of the six lines inoculated with Ina85-182 corresponded to the result of nursery trial. However, IBOS8-1-1 caused severe leaf blast on the three F3 lines exhibiting high level of partial resistance in the nursery trial. This result suggests that IBOS8-1- has specific pathogenicity to the partial resistance in Chubu 32.

THE USE OF THE RESISTANT VARIETIES FOR BLAST CONTROL IN VIETNAM

Ngo Vinh Vien[1], Ha Minh Trung[1]

[1] National Institute of Plant Protection, Chem, Tuliem, Hanoi, Vietnam

The Blast disease of rice was first observed by French scientist in Ha Tinh province in 1921. The disease increased markedly after 1975 and now it is known to occur in the whole country. In 1995 leaf blast occurred over an area of 758,000ha, panicle blast developed on 183,000ha.

The use of resistant varieties has been considered as a safe and effective control against rice blast. From 1978, numerous field tests of IRBN collections have been made to determine varietal reaction to blast and to select for resistance. Some promising cultivars as IR 1820 (1979), IR 17494 (1982), C70 and C71 (1988) has been selected and released for mass production. However, after a few years, the resistance of these cultivars was broken down. These results showed that there is a big change in virulence of blast population.

To obtain cultivars with a stable resistance it is necessary to evaluate the pathogenicity and virulence of rice blast fungus at major rice growing areas. The molecular techniques are firstly used to study blast. By MGR-DNA finger print 254 isolates, collected from Red River Delta, Central Coast and Mekong Delta were grouped together in 4 lineage (A,B,C,D) with 50 haplotypes.

Preliminary results showed that there is a big difference between Pyricularia grisea population structure at major rice growing regions of Vietnam. For example, there are 4 lineage in the Red River Delta, where numerous genotypes of rice are grown. On the contrary, there is only one lineage A in the Central Coast, where rice cultivar IR 17494 is grown as a dominant genotype. It seems to give the answer on why the occurrence of blast in Central Coast is more often than that in Red River Delta. If so, the diversification of resistant genes in rice fields would show an excellent performance of genetic control against rice blast.

CLONING OF THE RICE BLAST RESISTANCE GENE *PI-B* USING A FUNCTIONAL GENOME LIBRARY

N-S. Jwa[1], Y. Tsunoda[1,2], S. Nakamura[1], K.Akiyama[1], T. Motomura[1], K. Kamihara[1], O. Kodama[2], S. Kawasaki[1]

[1]National Institute of Agrobiological Resources, Kannondai, Tsukuba, Ibaraki 305
[2]School of Agriculture, Ibaraki Univ., Ami, Ibaraki 300-03,

Rice blast resistance gene *Pi-b* was narrowed down in its target region to a single 150 kB BAC clone (BAC13) by RFLP analyses of its end or inner fragments. This was done by F2 analysis of only 194 susceptible individuals. Inserts of a cDNA library from the resistant cv. BL-1 with *Pi-b* were hybridized to the fragments of the BAC13 bound to latex beads, and this region specific cDNA library was constructed from the TE-eluting fraction. By screening and subtracting 12,000 clones. 17 classes of cDNAs were identified in this region. As the concentration ratio of the cDNAs in the region was 158 times, this screening is equivalent to that of 2 million clones. Among these cDNAs, some protein kinase and receptor protein motifs were found. The region also included a few kinds of resistance gene analogues. In parallel, this genome region was covered with a contig of clones of a functional genome library (average insert size about 40 kB) constructed from the resistant cv. BL-1 with a high capacity binary vector RAC (*Ri-ori* driven artificial chromosome; or pBIGRZ) developed by us. About 20-70 transformants were obtained from each contig component clone. Infection test was carried out for all the transformants of the components by spray inoculation of race 007. All the 68 transformants of one clone were completely resistant just as the same level as the original cv. BL-1, while all the rest-clone transformants were susceptible as the recipient cv. Only the single clone within the contig could confer the blast resistance.

ANALYSING GXE INTERACTION IN KRBN DATA USING PATTERN ANALYSIS

J.H.Roh*, S.W.Ahn**, C.I.Yang*, M.A.Ynalvez**, C.G.Mclaren**
National Crop Experiment Station,RDA,Suweon441-100,Korea*
International Rice Research Institute,P.O.Box933,Manila,Philippines**

Pattern Analysis provide powerful new methods for analyzing GxE interaction in variou agronomic traits in multi-environmental trials. In order to get basic information needed for researching rice blast resistance and developing rice blast resistant cultivars, korea rice blast nursery (KRBN) data of 44 test varieties at 21 sites in korea over 15 years was analyzed. Seven location groups (LGPs) and eight genotype groups (GGPs) were identified for rice blast reaction patterns. Among the LGPs, LGP15 (Jaecheon), LGP19 (Naju) and LGP2 (Icheon) showed distinct reaction pattern thus they formed individual location group. GGP80 was broadly susceptible across diverse environments but GGP65 anf GGP75 were broadly resistant. While GGP77 has a high differential ability to differentiate location groups, LGP15 and LGP19 have a high differential ability to differentiate genotype groups.

Ahn.S.W.,M.F.Koch. 1988. A conceptual model of disease resistance in rice pathosystems and ts implications for evaluating resistance. Int.Rice Res.Newsl.13(6):12-13.
IRRI,1988. Standard Evaluation System for Rice. 3rded.pp14.Philippines. Kwak,Tae-Soon.1986.
Interpretation of testing sites and classicfication of rice cultivars based on varietal reaction patterns to rice blast (Pyricularia oryzae). Res. Rept. RDA (Crops) 28:1-26.

TESTING A "LINEAGE-EXCLUSION" STRATEGY FOR BREEDING RICE RESISTANT TO BLAST DISEASE

A.L. Tapiero and M. Levy
Department of Biological Sciences, Purdue University, West Lafayette, IN USA 47907-1392

The rice blast fungus, *Pyricularia* (Magnaporthe) *grisea*, typically has a definable virulence spectrum on rice cultivars that is limited by avirulences shared, historically, by all isolate members of each distinct DNA-fingerprint (*MGR*) lineage. Combining resistance to exclude all lineages in the pathogen population can provide a more durable resistance to the disease in contrast to the conventional strategy that combines resistance to exclude observed virulence types. In Colombia we evaluated the field performance of rice lines constructed to pyramid two dominant blast resistance resistance genes after crossing the near-isogenic lines C101A51 (bearing *Pi-2(t)*) and C101LAC (bearing *Pi-1(t)*). The presence of the genes was confirmed in the resistant *F-3* and *F-4* progeny with RFLP polymorphisms for linked rice loci (*RG64* for *Pi-2(t)* and *RG303* and *G181* for *Pi-1(t)*). Pathogenicity tests had detected no isolate capable of overcoming both genes in Colombia, although isolates from lineages SRL-1 and SRL-2 commonly overcame *Pi-2(t)* and several others commonly overcame *Pi-1(t)*. The combination of genes excluded isolates from all ten recurrent Colombian lineages except SRL-6, where a minority of isolates overcame one or the other R-genes differentially. Field tests at three different sites in Colombia (including a "hot spot" blast nursery) in 1996-7 resulted in the infection of all R-gene pyramids. *MGR*-fingerprinting showed that infections on the pyramids were produced by isolates belonging to lineages SRL-6 and SRL-4. All SRL-6 isolates recovered from the pyramids were re-infective on their pyramid of origin as well as both R-gene parents, confirming that a shift in virulence type had been detected in the only lineage not excluded by the two R-genes used. Assays of the potential for similar virulence shifts among SRL-6 isolates and other lineages were conducted to evaluate the reliability of a lineage-exclusion strategy for rice blast resistance breeding in Colombia.

SELECTION FOR DURABLE RESISTANCE AGAINST RICE BLAST.

M.J. Vales
CIRAD-CA. CIAT AA 6713 Cali Colombia, mvales@cgnet.com

We went select for durable rice blast resistance. It is not easy to work with this notion of time. so we need other criterion... To increase the likelihood of durability, resistance against rice blast must be general, partial and polygenic. Why?

First of all why general? If a resitance is general, it avoids the specific adapation of the fungus and the breakdown of the resistance. But we are never sure that a resistance is general, and a not specific fungus adaptation is possible. So a blast populations monitoring is necessary.

Why partial? Because all the known complete rice blast resistances are specific. But warning, all the partial resistances are not general. So we have to select for partial resistance against various blast sources.

In the fied, with varying pathogen population, it is impossible to distinguish between complete and partial resistance because the both can reduce the lesion number and lesion size.
So to distinguish both mechanisms we have to work with controlled and virulent strains: strains with one or more virulence gene, if it is necessary.

Why polygenic? Because it is not easy for strains to aggregate numerous genes of pathogenicity. It is a question of probability. But this explanation requires the gene-for-gene relationship which is unknow between pathogenicity and polygenic resistance. Phenotypic attenuation is other reason for interest in polygenic resistance.

We have to study the polygenism the progenitors and the created new varieties. Recurrent selection is the better breeding method for polygenic traits.

The different qualities for a likely durable resistance, general, partial and polygenic, are independent and lead to independent methodological practice. Some risks come from the incertitude of certain theoretical basic hypothesis, but we need to act and it is possible to take those risks into consideration in the methodology.

BREEDING RICE VARIETIES RESITANT TO BLAST FOR THE MEKONG DELTA OF VIETNAM

B B. Bong, L. V. Quynh
Cuu Long Delta Rice Research Institute, Omon, Cantho, Vietnam

The Mekong Delta of Vietnam with rice area of 3 M ha contributes to 50% rice production and 90% rice export (3.7 M tons in 1997) of Vietnam. The most severe biotic constraints in rice production in this region are brown plant hopper and blast disease. Therefore, breeding effort for blast resistance has been made for the last fifteen years. As a result, rice varieties with durable resistance to blast have been released for production. The popular varieties like IR64, OM997, OM1706 and IR62032 were recorded resistant to blast at various locations in the Delta over cropping seasons. To help in selecting materials for breeding, we have tested 1209 rice varieties at four locations in the Delta. These varieties could be classified into four groups: 5.26% as high resistance (scale 0-3 over locations), 18.31% as intermediate resistance (scale 0-5 over locations), 46.91% as unstable resistance (scale varying from 0-9 over locations) and 29.52% as very susceptibility (scale 6-9 over locations). Disease severity index (DSI) were used to identify the varieties showing durable resistance to blast. They varieties including S969B, IR64, IR62032, OM1706, OM1570, OM723-11, Basmati 370, Tep Hanh, Sa Mo Van, NCM10-20 have been utilized in the breeding program for durable blast resistance.

DEVELOPMENT OF BLAST DISEASE RESISTANT TWO-LINE HYBRID RICE COMBINATIONS IN VIETNAM

Hoang Tuyet .Minh, Nghien Thi Nhan. Tran Duy Quy et al.
Institute of Agricultural Genetics, Hanoi Vietnam

Rice blast. caused by Pyricularia grisea (Cooke) Sacc. is one of most destructive factors in rice production. It occurs in all rice growing areas, especially in hybrid rice area and causes severe damage under favorable condition. In Vietnam about 150.000 hectares was coven by hybrid rice in 1997 and increasing year by year. The research and development on hybrid rice was given great priority by Vietnam Government. So for quickly development hybrid rice in our country, blast desease must be controlled by creating resistance varieties as well as parent lines in hybrid rice. During the period from 1990 to 1997, We carried out research on screening and selecting new TGMS lines and Two-line hybrid rice combinations that showed resistance to blast desease. Works have been focused on screening and identifying resistant resources of both female and male parental lines by using IRRI' methods. First step tried to collect blast desease isolates from the rice fields and isolation of monoconidia from infected leaves (Pyricular greases) and inoculation. Second step to make resistance reaction test of numerous materials, we could identify good resistance lines for making test-crosses. By the above mentioned way we obtained several combinations namely VN-01s/ D212 and VN-01s/ D18. In Farmer's field these combinations showed high yield (6-8 tons per hectare per season), short grow-duration (115-118 days), good cooking quality, long grain and resistance to blast desease. The VN-01s/ D212 has been released as a commercial combination fore ecological condition testing in 1997.

References:
- Trainees' Manual, Asian Rice Biotechnology Network, Boman, Mackill, Nelson McCouch, Zeigler and Others, (1993).
- Genetic Differentiation Among Isolates of Pyricularia Infecting Rice and Weed Hosts, E.S. Borromec, R.J. Nelson, J.M. Bonman, and H. Leung, Science Vol 252 p. 632.
-DNA Fingerprinting with a Dispersed Repeated Sequence Resolves Pathotype in the Rice blast fungus, Morris Levy a, Jose Romao a, Marco A. Marchetti b, and John E. Hamer a. a Department of Biological Science, Puredue University; b United States Department of Agriculture. Texas 77713, The Plant Cell, Vol. 3, 95-102, January 1991

SELECTION FOR RICE (*Oryza sativa*) BLAST (*Magnaporthe grisea*) RESISTANCE IN RECURRENT POPULATION WITH LARGE GENETIC BASE.

Michel J. Vales, Marc-Henri Chatel, Jaime Borrero and Yolima Ospina
CIRAD-CIAT-FLAR; CIAT AA 9713 Cali Colombia, mvales@cgnet.com

Introduction. Two non-exclusive strategies against rice blast are possible. The first one is the strategy of lineage exclusion using the complete resistance. The second one is the use of durable resistance. To increase the likelihood of durability, resistance against rice blast must be general, partial and polygenic. The polygenism leads to the use of recurent selection.

Material
The *Indica* tropical population PCT-6HB selected for resistance to the Hoja Blanca virus (RHBV). 12 strains from the 6 lineage observed in the cropping area. The virulence spectrum of each lineage is covered by two complementary strains.

Methods. The recurrent selection scheme has three parts with parallel execution and common genetic recombinations.

Selection for complete resistance. S1 lines with large complete resistance spectrum are used for the next genetic recombination. Complete resistance against the 6 lineage has to be selected.

Selection for partial resistance.. S1 plants showing no complete resistance against a given strain are multiplied. At field the S2 lines are selected for partial leaf and neck blast resistance using strain. S3 lines from the better S2 are used for the next genetic recombination.

Maintenance of population polymorphism and male sterile gene frequency. To avoid the risks of genetic derive and break down of the polymorphism, a population sample is conduced by successive harvest of male sterile plants. The S0 seeds are used for the genetic recombinations.

Prospects. When the concentration of complete resistance genes is considered sufficient, it is possible to fix the population for this type of resistance. Successive samples of the population will be breed by drastic recurrent solution. Then it is possible to develop breeding lines from the enhanced population through pedigree selection.

RECURRENT SELECTION FOR BLAST (*Magnaporthe grisea*) RESISTANCE USING RICE (*Oryza sativa*) POPULATIONS WITH NARROW GENETIC BASE.

Michel J. Vales, Marc-Henri Chatel, Jaime Borrero and Yolima Ospina
CIRAD-CIAT-FLAR; CIAT AA 9713 Cali Colombia, mvales@cgnet.com

Introduction. Recurrent selection is used to enhance populations with large genetic base (PLB) for polygenic traits like partial blast resistance. If the selection index is too high there is a risk of loosing useful genes not still in favorable genetic associations. The selection index has to be moderate, so recurrent selection breeding using PLB is a long-term process. The use of recurrent selection in populations with narrow genetic base (PNB) is a mean to avoid the risk of loss of useful genes and speed-up the genetic progress at medium-term.

Material and methods. To develop PNB the first step is to select few parents with the highest known level of expression of a specific trait, for example IRAT 13 for partial blast resistance. At the end of the recurrent selection process the fixed lines would present each trait at the corresponding parent level, for example the partial resistance level of IRAT 13.

Discussion

Advantages In comparison with the use of F2 populations the potential genetic progress is better because the recurrent cycles permit the use of the best parent for each individual trait, without warring for its other qualities. In comparison with the use of PLB the genes are less diluted so the use of PNB requires less time for the development of good lines. The risk of genetic derive is also reduced and it is possible to fix the PNB for oligogenic traits before the starting of the recurrent selection breeding. The live of PNB is short, so it is easier make new populations than adapt old ones to follow objective changes.

Disadvantages. In comparison with the use of F2 populations more time is needed to develop fixed lines because the first step is to enhance the PNB before the pedigree selection. The recurrent selection method is not adapted to the breeding for oligogenic traits. In comparison with the use of PLB the potential genetic progress is reduce.

Conclusion. The use of PNB is a technical compromise between the use of F2 populations and PLB use. Three types of population are complementary parts of a same general breeding scheme.

POPULATION BIOLOGY AND GENETIC ORGANIZATION OF VIRULENCE DIVERSITY IN THE RICE BLAST FUNGUS

Morris Levy
Department of Biological Sciences, Purdue University, West Lafayette, IN USA 47907-1392

The population structure of the rice blast fungus worldwide is typically organized into distinct genetic lineages, as defined by MGR-DNA fingerprints, that remain distinguishable over decades of sampling. Coupled with the generalized sterility of field isolates, the fingerprints indicate a genealogy that is predominantly or exclusively asexual; at this writing there is no reliable evidence that even rare recombination is a significant influence on rice blast diversity. MGR-signatures, as well as those for other retrotransposons, distinguish rice blast fungi from various conspecific grass-infecting forms, further indicating no genetic exchange between rice blasts and other conspecifics. Each country (or agroecological zone within a country) contains tens of lineages, the majority of which are indigenous. Lineage diversity in Asian countries, where rice agriculture is ancient and numerous traditional rice varieties are grown along with modern cultivars, is 2-7 fold greater than in the Americas, where several lineages are widely distributed. Each MGR-lineage has a limited cultivar range, although most express multiple pathotypes that are permutations of that lineage's compatibilities. This limitation is most obviously marked in assays using near-isogenic testers with single resistance genes. Nearly always, one or more blast resistance genes exclude all members of a lineage. Lineages, thus, express particular, historically conserved, "avirulences" as well as a defined spectrum of compatibilities. Accordingly, we can now identify blast resistance genes that, in combination, may durably exclude all of the lineages in a particular region. Complementary evidence of the evolution of rice blast fungus structure is becoming available from RFLP variation associated with two *P. grisea* avirulence (Avr) genes, *Avr2-Yamo* (governing incompatibility with the rice cultivar, Yashiro-mochi) and *Avr-Pwl2* (governing incompatibility with the grass species, *Eragrostis curvula*). For each Avr-gene, the RFLPs and associated pathogenicity are generally modal within lineages and often lineage-specific; sequencing of *Avr2-Yamo* alleles in Colombian lineages demonstrates strong conservation of the coding sequence in all alleles but with distinctive amino acid differences between lineage-specific "avirulent" and lineage-specific "virulent" alleles.

PHYLOGENETYIC AND ECOGEOGRAPHIC DIVERSITY IN MAGNAPORTHE GRISEA.

Biju-Duval1, J.L. Notteghem2, M.H. Lebrun1,3
IIGM, Bat 400, Université Paris Sud, 91405 Orsay, France; 2CIRAD-CA, BP 5035, 34 032 Montpellier, France; 3UMR41 CNRS-RPA, Rhône-Poulenc Agrochimie, 69009 Lyon, France.

Isolates from Magnaporthe grisea species are frequently isolated from diseased leaves of a large number of wild and cultivated grasses, including rice. The genetic diversity of isolates from M. grisea species was estimated using RFLP markers and rDNA sequences. A collection of 75 M. grisea isolates from diverse geographic origins and 20 host plants was classified into ten distinct groups differing by their rDNA restriction site maps. The phylogenetic tree deduced from sequences comparisons of rDNA intergenic transcribed spacers (ITS1 and ITS2) revealed four highly divergent populations within M. grisea species. Despite lack of morphological differences, these populations could be considered as subspecies using either molecular criteria (> 7% nucleotidic divergence for ITS) or biological criteria (crosses without viable progeny). Three of these populations were monomorphic for their ITS1/2 sequences and had a very limited and specific host range (Digitaria spp. or Pennisetum spp.). The fourth population was subdivided into 3 related sub-groups that differed by few bp in their ITS1/2 sequences. Each of these sub-groups corresponded to isolates from a limited number of host plants, such as rice and Panicum repens (sub-group 1), Eleusine spp. and wheat (sub-group 2) or Eleusine spp. and Eragrotis (sub-group 3). Cluster analysis using single-copy gene RFLPs confirmed this classification. The analysis of the distribution of RFLP alleles showed that M. grisea populations were structured according to their host plants and not to their geographic origins. Groups of isolates strongly related by this multilocus analysis were pathogenic to the same host plant and distributed worldwide. Transposon pot3/MGR586 or retrotransposons Grasshopper and Maggy were detected in high copy number only in specific groups of isolates sharing the same or related alleles at all loci analyzed. This discontinuous distribution of repeated sequences strongly supported the absence of genetically exchange between such populations. Overall, M. grisea species is composed of genetically isolated populations restricted to one or few host plants that are dispersed worldwide.

RACE DISTRIBUTION AND VIRULENT GENE FREQUENCY OF RICE BLAST FUNGUS IN THE SOUTHERN CHINA

B.T. Chen, H. L. Chen, D. P. Zhang, Y. F. Xie, Q. F. Zhang
State Key Laboratory of Genetic Improvement of Crops, Huazhong Agricultural University, 430070 Wuhan, P. R. China

About eight hundred isolates, which came from 200 rice cultivars growing in 12 provinces of the South of China in 1996-1997, were evaluated by spraying on 11 near-isogenic lines (NILs) of rice blast-resistance genes (Mackill et al, 1992). The data about race distribution and virulence gene frequency of rice blast pathogen population in different rice-growing regions and administration provinces, three rice blast severely-developed areas, and the data on the same hybrid rice cross, Shuangyou 63, which was planted in different regions, has obtained. The results are as follows. Race 120 or J041 named after Gilmor's octal notation designation (Gilmor, 1973), is the indica or japonica preponderance races of rice blast pathogen population in the South of China, but there is difference in race distribution and virulence gene frequency of pathogen among different pathogen sub-populations from different rice-growing regions and provinces, and among these from different lesions of the same cultivar from different collection sites. NILs C101A51and F145-2, which carry with blast-resistance gene $Pi-2(t)$ and $Pi-ta2$ respectively, had the widest disease resistance spectra to fungus, with virulent gene frequency of 7.64% and 10.37%, whereas 38.47% and 37.79% of isolates expressed virulence to NILs carrying $Pi-3(t)$ and $Pi-kp$ respectively. It is important to pay much attention to application of rice blast resistance gene $Pi-2(t)$ and $Pi-ta2$ in rice breeding program in the South of China.

Reference
Gilmor J. (1973) Octal notation for designating physiologic races of plant pathogen. Nature 242, 620.
Mackill, D.J. and Bonman, J.M. (1992) Inheritance of near-isogenic lines of rice. Phytopathology 82, 746-749.

GENETIC DIVERSITY OF THE RICE BLAST PATHOGEN POPULATIONS IN WEST AFRICA

J.Chipili[1], S.Sreenivasaprasad[1] and N.J.Talbot[2]
[1]Horticulture Research International, Wellesbourne, Warwickshire, CV35 9EF, UK
[2]University of Exeter, Perry Road, Exeter, EX4 4QG, UK

Blast, caused by *Magnaporthe grisea*, is one of the most widespread and damaging diseases of rice in west Africa. Assessing the genetic diversity of *M. grisea* populations is important in understanding the distribution and evolution of races pathogenic to rice. Whilst considerable work has been done in different parts of the world on the genetic diversity and complexity of this important pathogen of rice, not much information is available from w. Africa.

More than three hundred blast samples were obtained from various rice screening sites in some w.African countries during the period 1994-97. Up to 100 isolates have been isolated so far and are being characterised using a variety of molecular markers.

MGR586 fingerprints generated for ten monoconidial isolates each, obtained from a limited number of field isolates showed no variation within a field isolate. This suggests that individual field isolates of *M. grisea* are genetically homogenous and stable.

MGR586 fingerprint profiles (30 - 40 bands) were generated for *M. grisea* isolates from different sites in Ivory Coast, Ghana, Nigeria and Togo. The isolates were divided into at least 4 distinct lineages with the overall similarity ranging from as low as 13 to 100%. Two isolates obtained from rice blast samples from Nigeria gave atypical patterns with 7-9 bands only. These isolates did not produce typical blast symptoms on rice in preliminary pathogenicity tests and their epidemiological significance is unknown.

RAPD-PCR analysis of the atypical isolates along with some typical rice and non-rice isolates yielded 4 distinct groups, but did not reveal any close relationship between the atypical isolates and typical rice or non-rice isolates.

Comparison of the ITS sequence data generated from *M. grisea* isolates with those from a number of different hosts (kindly provided by M.H. Lebrun, University of Paris, France) revealed a close relationship between the atypical isolates from this study and *M. grisea* isolates from *Pennisetum*, Ginger and *Digitaria*.

The poster will give an assessment of the genetic diversity of *M. grisea* isolates examined so far from different sites and the distribution pattern of various genotypes.

VIRULENCE PATTERN IN FIELD POPULATIONS OF *PYRICULARIA GRISEA* FROM RICE CULTIVAR METICA-1 IN BRAZIL

A.S.Prabhu, M.C. Filippi
Embrapa Arroz e Feijão, C.P. 179, 74.001-970, Goiania-GO, Brazil

Rice blast (*Pyricularia grisea*) is the most destructive disease in irrigated rice in the State of Tocantins where Metica-1 is planted annually in extensive areas covering over 30.000 hectares since 1984. The variation in phenotypic virulence within the pathogen population was studied in samples collected from four individual farms under epidemic conditions of leaf blast occurrence on Metica -1. A set of 90 isolates were tested on 28 rice genotypes including eight international rice differentials, and five near isogenic lines of CO 39. The degree of similarity in the reaction type among test isolates and the composition of pathotypes were examined. Ninety nine percent of the isolates pertain to race ID-14. The frequency of virulence to cultivars IR 50, IR 5 and Colombia was high. Isolates with virulence to isogenic lines C 101 LAC (Pi-1) and C 101 A 51 (Pi -2) were rare. Also, isolates from Metica-1 were not virulent to Tetep, IR- 8, CICA- 8, CICA- 9, BR-IRGA 409 and Aliança. Virulence to the newly released cultivar Rio Formoso was not present in the population. The diversity of virulence in the sample population was very narrow. The little diversity of pathogen population of Metica-1 can possibly be attributed to continuous directional selection of the pathogen to the matching resistance in the cultivar over years.

GENETIC DIVERSITY AND VIRULENCE OF *PYRICULARIA GRISEA* ISOLATES FROM BRAZIL

M.C.Filippi [1], A.S.Prabhu[1], M. Levy[2]
[1]Embrapa Arroz e Feijão, Caixa Postal 179, 74001-970 Goiania-GO, Brazil.
[2]Department of Biological Sciences, Purdue University, West Lafayette, IN., U.S.A.

Studies were conducted on genetic diversity and virulence pattern of 73 isolates of *P. grisea* collected from 10 upland rice cultivars at three breeding sites in Brazil. MGR-DNA fingerprint analysis of these isolates was done to determine the lineage structure. The phenotypic virulence was tested on 27 rice genotypes including eight from upland and six from irrigated commercial rice cultivars, five near isogenic lines of CO39 and eight international rice differentials. Fourteen races were identified based on the reaction type on eight standard international differentials, the most predominant races being IB-9 and IB-41. Eighty two percent of isolates belonged to the race IB-9 which was recovered from all ten upland rice cultivars including the improved ones such as Carajas and Rio Paranaiba. Distinct differences in virulence pattern ware observed in upland and irrigated rice cultivars. The virulence frequency of isolates on upland and irrigated rice cultivars was 56.6% and 17.8%, respectively. None of the isolates were virulent on Cica-9 and Oryzica Llanos 5. Both virulent and avirulent isolates were detected in the sample population on the known genes in the near isogenic lines. Virulence frequencies were relatively low, in descending order, on Pi-1, Pi-4b and Pi-2. The recurrent parent CO-39 may also possess a major resistance gene because 6% of the isolates were avirulent. The coefficient of similarity for pathotype varied from 0.28 to 1.00 among the 73 isolates. Eighteen MGR-DNA fingerprint lineages were identified , the most predominant ones being 3Z-1, BZ-2, BZ-4 and BZ-12.

VARIABILITY AND DIVERSITY OF BLAST IN BHUTAN

Thinlay[1], R.S. Zeigler[2], A.C. Bordeos[2] and M.R. Finckh[3]
[1]National Plant Protection Centre Semtokha, Thimphu, Bhutan
[2] International Rice Research Institute, PO Box 933 Los Banos, Manila 1099, Philippines
[3]Swiss Federal Institute of Technology, ETH Zentrum, Zurich, CH-8092

Leaf blast (caused by the fungus *Pyricularia grisea*) is one of the important seedling diseases in the mid altitude (1000-1800m) and neck and node blast often affects rice in the high altitude (1800-2600m) rice growing areas in Bhutan. A serious blast epidemic in 1995 affected traditional local varieties mostly in the high altitude. During the epidemic, 71 isolates were collected from several areas in Bhutan and analysed for their variability with MGR finger printing (Hamer et al., 1989). Thirty of these isolates and 80 collected in 1996 were tested for virulence on CO39 and LTH near isogenic lines (NILs) and 12 Bhutanese varieties. Nurseries consisting of CO39 and LTH and five NILs of each, respectively, 12 varieties from the International Rice Blast Nursery set (Ahn, 1994) and 20 varieties from Bhutan were also set up in ten different sites in the high and mid altitudes to study virulence diversity in 1996 and 1997. Fingerprinting of the 71 isolates from the 1995 collection resulted in 13 different lineage groups and 58 haplotypes. Twenty six and fifteen isolates belonged to lineage 1 and 4 respectively. The remaining lineages comprised between 1 to 6 isolates per lineage. Virulence analysis of the 110 isolates resulted in 51 pathotypes when including only the CO39 and LTH and their NILs. When considering in addition the Bhutanese varieties, 87 pathotypes could be differentiated. CO39 was resistant to seven of the 110 isolates while its isoline C104 PKT (resistance gene *Pi-3(t)*) was susceptible. Overall, trap nurseries detected blast throughout the season (July to September) in the mid altitude while blast was mostly present during August in the high altitude sites. Many more trap nursery entries were infected in the mid altitude (mean=9.3) than in the high altitude (mean=3.9) indicating that the blast population in the mid altitude sites was more diverse than in high altitude sites. However, during the 1995 blast epidemic and in the nurseries in 1996 and 1997 mid altitude varieties were not severely affected while high altitude varieties were much more susceptible. This suggests strong selection for resistance due to diverse and ever present blast inoculum in the mid altitude areas. Thus, these materials should present good sources for resistance breeding.

Ahn, S. W., (1994). In: Rice Blast Disease. Edition: Zeigler et al., 1995. pp:137-54.
Hamer, J.E., Farall, L., Orbach, M. J., Valent, B., and Chumley, F. G. (1989). Proc. Nat. Acad. Sci 86; 9981-85.

PYRICULARIA ORYZAE IS THE PROPER NAME FOR THE RICE BLAST FUNGUS

H. K. Manandhar, Plant Pathology Division, Nepal Agricultural Research Council, Khumaltar, Lalitpur, Nepal

The blast fungus on rice was first described by F. Cavara in 1891 who called it *Pyricularia oryzae*. Identification of this species became controversial later with *P. grisea* which was described earlier by P. A. Saccardo on crab grass in 1880. However, the name *P. oryzae* has been maintained for rice-infecting isolates and *P. grisea* for grass-infecting isolates (3). Based on morphological similarity, cross infectivity and mating ability between *P. oryzae* and *P. grisea* Rossman et al. (4) published an article in 1990 stating that *P. grisea* is the correct name for the rice blast fungus. Since then, the name *P. oryzae* has been used less frequently. In my opinion the dispute is not settled yet since the distinction is evident in terms of host specificity despite the evidence of some cross-infectivity. Positive results obtained in cross inoculation experiments do not necessarily mean that the inoculants will always result in producing disease in the field. It is also my experience in Nepal that *Pyricularia* isolates from *Echinochloa colona*, *Eleusine indica* and *E. coracana* never infected rice plants when inoculated artificially on several cultivars susceptible to blast in 1994 and 1995. Molecular studies also reveal that the rice- and the grass-infecting isolates are generally host-limited (2) and genetic isolation exists between them (1). Therefore, their names should be maintained as published originally. *P. oryzae* has been used in hundreds of research articles and a number of books. The books published even after the Rossman et al.'s article, such as Ainsworth & Bisby's Dictionary of Fungi of 1995 edition (International Mycological Institute, CAB International) and Rice Blast Disease by K. Manibhushan Rao of 1994 edition (Daya Publishing House, Delhi, India) have still maintained the name *P. oryzae*. I plead to scientific community that the name *P. oryzae*, which has been used for more than a century, deserves to be retained.

Literature cited
1. Borromeo, E.S., Nelson, R.J., Bonman, J.M., and Leung, H. 1993. Genetic differentiation among isolates of *Pyricularia* infecting rice and weed hosts. Phytopathology 83:393-399.
2. Leong, S.A., Farman, M., Smith, J., Budde, A., Tosa, Y., and Nitta, N. 1994. Molecular genetic approach to the study of cultivar specificity in the rice blast fungus. Pages 87-110 in: Rice Blast Disease. R.S. Zeigler, S.A. Leong and P.S. Teng, eds. CAB International, Oxon, UK, and International Rice Research Institute, Manila, Philippines.
3. Ou, S.H. 1985. Rice Diseases. 2nd ed. Commonwealth Mycological Institute, Kew, UK.
4. Rossman, A.Y., Howard, R.J., and Valent, B. 1990. *Pyricularia grisea*, the correct name for the rice blast disease fungus. Mycologia 82:509-512.

ESTIMATION OF GENETIC DIVERSITY AMONG *PYRICULARIA GRISEA* ISOLATES FROM URUGUAYAN RICE CULTIVARS

F. Capdevielle [1], A. Branda [1], E. Avila [2]
[1] INIA Biotechnology Unit, P.O. Box 33085, Canelones (90200), Uruguay
[2] INIA Rice Program, P.O. Box 42, Treinta y Tres, Uruguay

Uruguay grows 130.000 hectares of irrigated rice and 85 % of the production is exported. Grain quality is one of the basic characters for the local breeding program, as well as high yield potential, cold tolerance and short growth duration. The incidence of rice blast is often limited to some introduced germplasm (Indica type) under particular environments; however an archival collection of isolates representing geographically sampling collected over the last 4 years under episodic disease conditions is presently available for estimation of genetic diversity among isolates of *P. grisea* from Uruguayan rice cultivars. Fourteen monosporic isolates DNA's from 7 cultivars have been compared based on RAPD pattern, using arbitrary 10-mer primers for DNA amplification. DNA preparation was according with Graham et al (1) and RAPD protocols are based on Williams et al (2). *P. grisea* isolates from diverse cultivars, classified according geographic origin (9 locations) and year of collection (1993-1997), were clustered based on pairwise distances (SM coefficient) between isolates using UPGMA cluster analysis. Three main clusters with 3 to 5 isolates each one were statistically tested using the AMOVA (Analysis of Molecular Variance) program, Excoffier (3), for computation of molecular variance components at different hierarchical levels: I among isolates, II among isolates within cluster groups, and III among cluster groups. All three levels were highly significant, with DNA differences among isolates explaining more than 70 % of the total variation. Differentiation among isolates from traditional and recently released cultivars suggests selective process in *P.grisea* occuring in the past 5-6 years. From this perspective it is therefore highly advisable to intensify research in such areas as physiological specialization, inheritance of resistance and search for effective sources of resistance in order to control this important disease.

References: (1) G.Graham, P.Mayers, R.Henry, 1994. Biotechniques Vol 16, N° 1
(2) J. Williams et al, 1990. Nucleic Acids Res. 18:6531-6535
(3) L.Excoffier, P.Smouse, J.Quattro, 1992. Genetics 131:479-491

POPULATION STRUCTURE OF THE RICE BLAST FUNGUS IN JAPAN REVEALED BY DNA FINGERPRINTING

L.D. Don, M. Kusaba, A.S. Urashima, Y. Tosa, H. Nakayashiki, and S. Mayama
Laboratory of Plant Pathology, Faculty of Agriculture, Nada, Kobe 657-8501, JAPAN

In 1993 devastating epidemics of the rice blast disease occurred in Japan. To reveal the structure of the current population of the rice blast fungus, we collected 280 isolates in 1993-1997 from various prefectures in Japan, and analyzed their haplotypes by DNA fingerprinting with MAGGY and MGR586 probes. A UPGMA dendrogram showed that all isolates were grouped into two distinct lineages. One (tentatively designated as Lf1) was a major lineage accounting for 97% of the total isolates and present throughout Japan, while the other (Lf2) was a minor lineage accounting for only 3%. Twenty-two isolates collected in 1976 at various prefectures were also grouped into two distinct lineages, a major one (Ls1) accounting for 77% and a minor one (Ls2) accounting for 23%. A dendrogram constructed from pooled data of these two populations showed that Lf1 and Lf2 corresponded to Ls1 and Ls2, respectively. These two lineages were finally designated as JL1 and JL2. These results shows that there are relatively few lineages of M. grisea in Japan in contrast to ones in USA and South Asia. It was clearly found that these lineages have a very similar virulence spectrum and that there is no relationship between lineages and races.
On the other hand, "standard isolates" collected before 1960 showed higher diversity. Out of the seven isolates tested, one was sorted into JL1, one or two into JL2, but one produced a distinct fingerprint strikingly different from those of JL1 or JL2. The other isolates showed some similarity to JL1 or JL2, but could not be sorted into them. These results suggest that the rice blast population in Japan was composed of diverse haplotypes before 1960, but have drastically changed thereafter, resulting in the predominant distribution of JL1.

THE USE OF VEGETATIVE COMPATIBILITY TO CHARACTERIZE POPULATION DIVERSITY IN PYRICULARIA GRISEA

J. C. Correll1, T. L. Harp1, F. N. Lee1, and R. S. Zeigler2
1Department of Plant Pathology, University of Arkansas, Fayetteville, AR 72701 USA (jcorrell@comp.uark.edu)
2Division of Entomology and Plant Pathology, IRRI, P. O. Box 933, Manila, Philippines

Spontaneous nitrate nonutilizing (nit) and sulfate non-utilizing (sul) mutants were recovered from P. grisea and used to characterize a collection of archived and contemporary rice-infecting field isolates into vegetative compatibility groups (VCGs). Pairing of certain phenotypically distinct nit mutants yielded robust heterokaryons as did certain nit x sul pairings. However, some cross-feeding was evident in some nit x sul pairings indicating that caution should be exercised in the interpretation of such reactions. Four distinct VCGs were identified among archival and contemporary rice-infecting isolates of P. grisea collected from Arkansas, Texas, Louisiana, and California. Each VCG corresponded with a distinct MGR586 group (or lineage) among the contemporary isolates in all but one case; VCG US-004 contained contemporary isolates collected from Arkansas belonging to MGR586 group D and contemporary isolates recovered from California in 1997 which belonged to group H. VCG US-004 also contained two isolates recovered from Texas and Arkansas in the mid 1970's which belonged to MGR586 group H. The data indicate that distinct barriers to vegetative compatibility are present in P. grisea and that certain VCGs may represent genetically distinct populations in Arkansas. However, the impact vegetative compatibility barriers have on horizontal gene transfer in P. grisea is unknown. A preliminary genetic analysis of vegetative compatibility with the sexually compatible strains Guy11 and 2539 of P. grisea indicated that multiple vegetative incompatibility (vic) loci were present. The use of VCGs may be very helpful in characterizing the population structure of this important pathogen worldwide.

REGIONAL POPULATION DIVERSITY OF THE RICE BLAST PATHOGEN, PYRICULARIA GRISEA

J. C. Correll, J. C. Guerber, F. N. Lee, and J. Q. Xia
Department of Plant Pathology, University of Arkansas, Fayetteville, AR 72701 USA (jcorrell@comp.uark.edu)

MGR586 DNA fingerprinting, mitochondrial DNA (mtDNA)RFLPs, and virulence phenotyping were used to examine the regional genetic diversity of the rice blast pathogen, Pyricularia grisea, in contemporary collections in Arkansas. Isolates were collected from commercial fields in nine counties in Arkansas. All isolates were examined for nuclear DNA RFLPs with the MGR586 DNA fingerprint probe. Four distinct MGR586 DNA fingerprint groups, or lineages, (designated A, B, C, and D) were identified among the contemporary regional collection of 470 field isolates. All four lineages were found in nine of the 18 locations; three lineages were found in four locations, two lineages in three locations, and one lineage in two locations. Ten, 19, 16, and 13 haplotypes (isolates which had MGR586 DNA fingerprints which differed by 1-20%) were identified within lineages A, B, C, and D, respectively, among the 470 field isolates examined. However, within each lineage, a single haplotype predominated representing 51-71% of the isolates collected for each of the four lineages. Overall, 60% of the 470 isolates belonged to one of only four haplotypes (A1, B1, C1, and D1) and these four predominant haplotypes were recovered from between 7 and 14 of the 18 locations sampled indicating a widespread distribution of these four apparent clones. These data indicate that an exceptionally low level of genetic diversity in the regional rice blast pathogen population in Arkansas relative to several other populations of P. grisea studied. Furthermore, no mtDNA RFLPs were detected among representative haplotypes within each of the lineages examined. Examination of virulence indicated that two races predominated in the regional collection. All 30 isolates in lineages A and C tested had an IB-49 virulence phenotype. Twenty nine out of 30 isolates in lineages B and D had an IC-17 virulence phenotype. One isolate in lineage B, collected isolated from a highly susceptible cultivar (L201), had an IG-1 virulence phenotype. The frequencies of the four lineages varied among the locations sampled and may have been due, in part, to the cultivar from which isolates were recovered. A single lineage was recovered from two cultivars, Mars and Millie. Although only a single field of each of these cultivars was sampled, the data indicate that certain cultivars grown in Arkansas may serve as a "bottleneck" selecting out specific lineages in the regional population.

MOLECULAR CHARACTERIZATION OF RICE BLAST FUNGUS (*MAGNAPORTHE GRISEA*) BY RAPD ANALYSIS AND REACTION OF RICE GENOTYPES TO INDIAN ISOLATES

Srinivasachary, Shivayogi. S, Shailaja Hittalmani, Girish Kumar, K and Shashidhar, H.E.
Department of Genetics and Plant Breeding, College of Agriculture, University of Agricultural Sciences, GKVK, Bangalore-560 065, India.

Severely blasted rice samples were collected from four locations in Southern Karnataka, India and 500 single spore cultures were isolated. The DNA of 27 isolates selected across three different locations comprising of different agro-climatic zones were fingerprinted using 10-mer Random Amplified Polymorphic DNA primers. The polymorphic DNA banding patterns specific to each isolate was observed and polymorphic bands generated by 176 markers levels by 30 primers were analyzed for genetic variation and grouping of isolates using cluster analysis (NTSYS). Three distinct groups were observed at 70% similarity levels. Most isolates form one of the location (Ponnampet), a hot spot for rice blast had representatives in all the three clusters identified. This hot spot is known to have different pathotypes whose genetic variation may now be related to the genetic differences identified by DNA fingerprinting. The isolates from the other two locations, RRS, Mandya and MRS, Hebbal clustered into a single group. This study reveals that random primers can be an efficient tool to identify genetic differences in the fungal races which would be useful in understanding the pathogen dynamics and evolution of new races.
Sixteen rice genotypes including land races, improved varieties and pyramided lines with known genes were inoculated with the selected isolates. Of the genotypes tested, IRAT177, Moroberekan, Apura, Doddi (land race), and the isolines $Pi1$, $Pi1+Pi4$ and $Pi2+Pi1$ showed high degree of resistance at seedling stage. Rice varieties HR12, CO39, Azucena and Irylon short showed susceptible reaction. We are in the process of identifying differentials using different isolates.

GENETIC ISOLATION AMONG MAGNAPORTHE GRISEA FROM DIFFERENT HOSTS IN INDIAN HIMALAYAS

J. Kumar[1], M. Ramos[2], H. Leung[2], and R.S. Zeigler[2]

[1]Hill Campus, GBPUA&T, Ranichauri 249199, U.P., India

[2]Intenational RiceResearch Institute, P. O., Box 933, Manila, Philippines

Magnaporthe grisea, the blast pathogen of cover 50 monocots has been, considered to comprise of host-limited forms. In Indian Himalayas, rice-derived *M. grisea* forms from the region are highly diverse, and frequent blast epidemics in rice and several millet hosts co-cultivated in traditional rice system are commonly encountered. The pathogen population in this region offers a unique opportunity to test if host-delimited forms of pathogen remain genetically isolated. We obtained *M. grisea* isolates from rice, and two millets, *Eleusine coracana* (ragi) and *Setaria italica* (foxtail millet) from near-by or inter sown plots from one location in Indian Himalayas. *Eco*RI digests of genomic DNA of each host-derived form were hybridized to multilocus and low-copy probes. Repeated sequences such as MGR586, MAGGY and *grasshopper* were present in high copy in rice-, setara-or *Eleusine*-derived isolates, respectively. While all rice-derived isolates showed high copy number MAGGY, about half *Setaria*-derived isolates had no MAGGY but high-copy *grasshopper*. None of the rice-derived isolates hybridized to *grasshopper*, and none of the *Eleusine*-derived isolates hybridized to MAGGY. Thus, while multilocus DNA profiles of the two host-derived forms suggested genetic isolation, *Setaria*-derived forms appeared to be a bridge between rice and *Eleusine* pathogens. Single-copy probes revealed a similar pattern as the multilocus probes; including the dimorphic pattern of *Setaria* isolates. However within each host-derived form occasional atypical alleles, also found in one of the other host-derived forms, were encountered. The existing data suggests strong host limitation and genetic isolation in *M. grisea* in Indian Himalayas. We have used AFIP and RFLP markers to generate linkage map based on segregation among progeny of a cross between rice-derived and *Setaria*-derived pathogens. Using mapped single-locus markers, we are determining if there is limited gene flow among different host-derived pathogen populations in nature.

EVIDENCE OF PARASEXUAL RECOMBINATION OF *MAGNAPORTHE GRISEA*

M.Tsujimoto, N.Yasuda and Y.Fujita
Laboratory of Plant Disease Control, Hokuriku National Agricultural Experiments Station, Inada, 1-2-1, Jouetsu, 943-0193, Niigata, Japan.

Hyphal fusion is one of pathogencity variation factor. Auxotrophic complementation and segregation of nutritional marker have been used to demonstrate parasexual recombination, but such marker can not eliminate the possibility that variants is a result of mutation. Then we attempted to determine whether parasexual exchange of DNA in rice blast fungus, M. grisea may occur. Two blast strains Y90-71BI introduced Ignite/basta-resisterance (bar) gene and 3514-R-2BS introduced blastcidin S deaminase gene (BSD) were cocultured in liquid yeast extract medium for 7 days and then transferred to the selective PDA medium containg bialaphos and blastcidin S (at consentration of 800 µg/ml bialaphos and 100 µg/ml blastcidine S). After 7days , the strains resistant to both drugs isolated By southern blot analysis of these bialaphos and blastcydine S resistant strains genomic DNA ,it was showed that they had both resistance gene. This finding suggests that Y90-71BI and 3514-R-2BS occurred hyphal fusion in coculture and acquired resistance to these drugs by parasexual exchanges of DNA . These lawn on oatmeal agar plates were different. BIS1 was gray lawn and similar to 3514-R-2BS, BIS3 was white lawn and similar to Y90-71BI. But BIS2 was black lawn and different from each other. Mating type of BIS1, BIS2 were same as 3514-R-2BS(1-2) and BIS3 was same as Y90-71BI(1-2). Pathogencity of BIS1 and BIS2 were same as Y90-71BI (race 102). It seems that exchanges of mating types and pathogencities supports evidence of parasexual recombination. By staining conidia collected from the variants with 4µ,6-diamond -2-phenylindole (DAPI) , all conidia were found to be only one nucleus per cell. It seems that variants form uninuclear after formation of heterokaryon by parasexual recombination.

Genetic Diversity in *Magnaporthe grisea* & Pathogenic Variation of its Progeny by Hyphal Fusion

Y Shen[1], XP.Yuan[1], CY.Li[2], J.Manry[3], XH.Zhao[1], PL.Zhu[1], ZX.Luo[2], YL.Wang[1], M.Levy[3]
1 China National Rice Research Institute, Hangzhou 310006, China
2 Yunnan Academy of Agricultural Sciences, Kunmin 650205, China
3 Purdue University, West Lafayette, IN 47907, USA

To develop improved breeding strategies for control the rice blast disease. MGR-DNA fingerprinting and pathotyping were combine to determine the population structure of the pathogen and the relationship between this virulence diversity of the pathogen in China. The MGR-DNA fingerprints of 475 isolates of *Magnaporthe grisea* from 146 sites were collected. Forty-five pathotypes of *M.grisea* were distinguished into 56 separate lineages. Each lineage had a limited host range and these relationship were maintained several years over broad geographic range. A set of isolates with different virulence spectra to select suitable donors for more durable resistance breeding were also selected.

The MGR-DNA fingerprintings of 40 isolates, including 11 parents isolates(asexual / sexual) with hyphal fusion, 22 recultural isolates of hyphal fusion spots and the 5th generational isolates of 7 paired isolates in *M.grisea* have been studied with a conversed repetitive sequence probe MGR586/EcoR1 digest combination. Most of the isolates with hyphal fusion were distinguished into different lineages based on fingerprinting similarity. Compared with the results of pathotype identification, we concluded that hyphal fusion is one of the important causes to pathogenic variation, being as important as sexual hybridization, heterokaryosis, and natural mutation in rice blast fungus.

POPULATION GENETIC STUDIES ON THE RICE BLAST PATHOGEN, MAGNAPORTHE GRISEA, IN THAILAND

Roumen, E.; Luangsa-ard, J.; Sirithunya1, P.; Na-Lampang2, A.; Saranun, S.; Pimpisit-thavorn, S.; Boonchitsirikul, C. and Pan, Q. BIOTEC, National Science & Technology Development Agency, 73/1 Rama VI rd, Rajdhevee, Bangkok 10400.
1 Lampang Agricultural Research Center, PO Box 89, Lampang. 2 Phitsanulok
Rice Research Center, amphur Wang Thong, Phitsanulok 65130. E-mail : Edroumen @biotec.or.th

Recent DNA finger-printing study (Meeka-ta-na-karn, et al. 1996) showed an unusually high degree of genetic diversity within the Thailand blast pathogen population compared to other countries. Before effective resistance breeding can be carried out, more information on the pathogen population structure is needed. A possible explanation for the high degree of genetic variation in Thailand is the occurrence of sexual recombination among rice pathogenic Magnaporthe isolates. Therefore, at BIOTEC, we set out to investigate the mating type and fertility traits of the pathogen population. Among the 300 monospore isolates obtained from rice leaves and panicles collected in the central plain and Northern provinces of Thailand in 1996 and 1997, most isolates (58%) were unable to form perithecia. Nor did they induce perithecia formation in the testers. The ratio between 'MAT1-2' and 'MAT1-1' genotype among the remaining isolates was 8:1, and 3% of the isolates appeared herma-phroditic. Inoculation tests confirmed pathogenicity of these fertile isolates to rice and showed variation in virulence. Molecular analysis also revealed the fertile isolates to be distinct from each other, even within the same mating type. Interestingly, in certain isolate combinations, hyphae in bordering zones dis-inte-grated, leaving a transparant, hyphae-free band between the opposing isolates. Symptoms of this highly repeatable finding were typical of a 'vegetative incompatibility' system.

The possibility that the high degree of genetic diversity of the blast pathogen population in Thailand is explained by sexual recombination cannot be ruled out, since highly fertile, herma-phroditic isolates of opposite mating type were found in the same field. On the other hand, the vast majority of the isolates are highly sterile, and the structure of the pathogen population does not seem essentially different from those in other countries. If sexual re-combination occurs, it is most likely a rare event.

Meekatanakarn, P., et al . (1996). Characterization of the Pyricularia grisea population by lineage using MGR-DNA fingerprinting in Thailand. In "Third Asia pacific Conference on Agricultural Biotechnology: issues and Choices", pp. 66, Hua-Hin, Prachuapkirikhan, Thailand.

LINES AND SINES IN MAGNAPORTHE GRISEA.

M. Anthony Meyn1, Lennie Farrall2, Forrest G. Chumley2, Barbara Valent2 and Marc J. Orbach1
1Dept. of Plant Pathology, University of Arizona, Tucson, AZ, USA
2 Dupont Company, Wilmington, DE, USA

Several transposable elements have been identified in the rice blast pathogen, Magnaporthe grisea, in attempts to determine their role in the genetic instability observed in this fungus. We report the characterization of the first LINE retrotransposon found in M. grisea, that we have designated MGL, which is homologous to MGR583. This 6 kb element is highly conserved both in rice pathogens and pathogens of other grasses, although there is variability in copy number among the strains. Surprisingly, although full length transcripts have been observed, movement of MGL has not. MGL contains two open reading frames (ORFs) which overlap by a single base pair. The first ORF contains homology to retroviral gag proteins and the second to reverse transcriptase (RT) polypeptides. Most MGL copies are flanked by short, variable-length, direct repeats as is characteristic of this class of elements. The 3' end of MGL is characterized by variable numbers of a TAC repeat, a feature also observed in the SINE element, MG-SINE, of M. grisea (Kachroo et al. 1995). Sequence comparisons reveal that the last 240 nt of MG-SINE are 90% identical to MGL. This homology suggests a model for the origin of MG-SINE, and a mechanism for MG-SINE transposition that relies on the RT of MGL, and utilizes the end of MGL for initiation of reverse transcription. A revised distribution of these two elements will be presented.

Kachroo, P., Leong, S.A., Chattoo, B.B. (1995) Mg-SINE: a short interspersed nuclear element from the rice blast fungus, Magnaporthe grisea. Proc. Natl. Acad. Sci. USA 92:11125-11129

CHROMOSOME NUMBER AND GENETIC SIMILARITY IN *MAGNAPORTHE GRISEA*

M. C. Chung[1], Y. J. Chen[1], H. K. Wu[1]
[1] Institute of Botany, Academia Sinica, Nankang, Taipei, 11529, Taiwan.

The chromosome number of *M. grisea* is nine. It was recently determined in a fertile cross between Guy11 and 2539 at meiotic pachytene, diakinesis as well as at metaphase using aceto-lactic orcein technique. This finding is contradictory to those data obtained from mitotic division but gives fully support to our previous report in the crosses between Ken-82-13 and the local field isolates. It can be predicated with good confidence that a new linkage group would be soon added to the recently constructed genetic map that has already consisted of eight linkage groups.

To test the genetic similarity among the *M. grisea* isolates, RAPD markers were amplified using the 60 oligomers from OPA, OPF and OPM sets. Re-amplifying of the primary markers gave 73 markers that were reproducible and recognizable. Calculation of the similarity value of the isolates was based on the 25 randomly chosen but recognizable RAPD markers. It varies from 0.40 to 0.48 between Guy11 and the four local field isolates, but is higher i.e. 0.52 between Guy11 and 2539. This higher genetic similarity is faithfully reflected on the high fertility of their hybrids.

Out of the 15 RAPD markers tested, only one, M08-M2, can be used to map. It hybridized to a certain band of each lane on the blotted filter. It seems that this marker is chromosome specific. Much more of such markers are expected. The remaining 14 markers tested look like repetitive sequences that may be useful in fingerprinting to distinguish pathotypes among isolates.

TOWARDS A PHYSICAL MAP OF THE RICE BLAST FUNGUS, *MAGNAPORTHE GRISEA*

Heng Zhu, Barbara Blackmon, Maciek Sasinowski and Ralph-Dean (Dept. of Plant Pathology and Physiology, and Clemson University Genomics Institute, Clemson University, Clemson, SC 29634, USA)

The rice blast fungus, *Magnaporthe grisea*, is one of the most important plant pathogens for studying various aspects of host-plant interactions. It has been widely adopted as a model organism because it is ideally suited for genetic and biological studies. To facilitate map based cloning, chromosome walking, and genome organization studies of *M. grisea*, a complete physical map of chromosome 7 in *M. grisea* has been constructed utilizing a bacterial artificial chromosome (BAC) library. Using 147 chromosome 7 specific single-copy BAC clones and 19 RFLP markers on chromosome 7, 624 BAC clones were identified by hybridization. Hybridization contigs were constructed using a random cost algorithm, while fingerprinting contigs were constructed using the software package FPC. Results from both methods were generally in agreement, but numerous anomalies were observed. The combined data produced 2 robust anchored contigs covering over 95% of the 4.2 Mbp chromosome 7. The genetic and physical maps closely agreed, with the exception of a single chromosome inversion. Based on the contig maps, the BAC clone coverage of chromosome 7 was also investigated. The distribution of repetitive elements on chromosome 7 was investigated by hybridization with different *M. grisea* repeats against the minimum BAC tile containing 41 BAC clones. Similarly, the locations of genes expressed at different developmental stages were also determined. Initial cDNA sequence analysis and progress towards constructing a physical map of the entire genome will be presented.

DIFFERENTIAL CLONING OF APPRESSORIUM FORMATION STAGE-SPECIFIC GENES OF *MAGNAPORTHE GRISEA*

T. Kamakura[1], S. Yamaguchi[2], K. Saitoh[2], T. Teraoka[2], I. Yamaguchi[1]

[1] RIKEN Institute, Wako, Saitama 351-0198, Japan

[2] Tokyo University of Agriculture and Technology, Fucyu, Tokyo 183-8509, Japan

The conidial germ tube of *Magnaporthe grisea* differentiates a specific infection structure, an appressorium, for penetration into the host. Formation of appressorium was observed not only on rice leaves but also on synthetic solid substara such as polycarbonate. We found that germ tubes of *M. grisea* were potentially anti-hydrotactic and that increase in frequency of appressorium formation was well correlated with the substratum hardness. Several proteolytic and glycolytic enzymes were found to inhibit germling adhesion and appressorium formation, but not conidial germination. Furthermore, we observed that a plant lectin, concanavalin A, specifically suppressed the appressorium formation without affecting the germling adhesion if it was applied within 2-3 hours after germination. From the result, we have constructed a cDNA library which represents the appressorium differentiation stage from the 2.5 hours-old germ tubes. For the construction of the library, a subtractive cDNA cloning strategy using combination of biotin labeled driver DNA method and adapter-primed PCR method was performed to condense the differential cDNAs that were presumably expressed during early stage of the development of conidial germ tube and/or appressorium formation. Out of 686 colonies of the library, about 150 distinct clones' nucleotide sequence were determined partially. We have further analyzed some of those candidates which showed differential expression by RT-PCR analysis. cDNA and genomic DNA structure and presumable character of those genes will be discussed.

cAMP SIGNALINGS IN FUNGAL PATHOGENESIS: DIVERGENT PATHWAYS REGULATE GROWTH AND PATHOGENESIS IN THE RICE BLAST FUNGUS.

K. Adachi, J.E. Hamer
Department of Biological Sciences, Purdue University, West Lafayette IN 47907, USA

cAMP is involved in signaling appressorium formation in the rice blast fungus, *Magnaporthe grisea*. However null mutations in a protein kinase A catalytic subunit gene, *CPKA*, do not block appressorium formation (1, 2), and mutations in adenylate cyclase gene have pleiotropic effects on growth, conidiation, sexual development and appressorium formation (3). Thus cAMP signaling plays roles in both growth and morphogenesis, as well as appressorium formation. To further clarify cAMP signaling in *M.grisea* we identify strains in which a null mutation in adenylate cyclase gene (*MAC1*) has an unstable phenotype, such that by-pass suppressors of the Mac1⁻ phenotype (*sum*) can be identified. *sum* mutations completely restore growth, sexual and asexual morphogenesis and lead to an ability to form appressoria under conditions inhibitory to wildtype. Protein kinase A (PKA) assays and molecular cloning show that one suppressor mutation (*sum1-99*) alters a conserved amino acid in the cAMP binding domain A of the regulatory subunit gene of PKA (*SUM1*), while other suppressor mutations act independently of PKA activity. PKA assays demonstrate that the catalytic subunit gene, *CPKA*, encodes the only detectable PKA activity in *M. grisea*. Because *CPKA* is dispensable for growth, morphogenesis and appressorium formation divergent catalytic subunit genes must play roles in these processes. These results suggest a model where both saprophytic and pathogenic growth of *M. grisea* are regulated by adenylate cyclase but that different effectors of cAMP mediate downstream effects specific for either cell morphogenesis or pathogenesis.

1. Mitchell, T.K. and Dean, R.A. (1995). Plant Cell 7: 1869-1878.
2. Xu, J.-R. et al. (1997). Mol. Plant-Microbe Interact. 10: 187-194.
3. Choi, W. and Dean, R.A. (1997). Plant Cell 9: 1973-1983.

EARLY SIGNALING EVENTS DURING INFECTION STRUCTURE FORMATION BY THE RICE BLAST FUNGUS.

J.A. Sweigard, T.M. DeZwaan, A. Carroll, L. Farrall and B. Valent. Central Research and Development Department, E.I. duPont de Nemours and Co., Inc. P.O. Box 80402, Wilmington, DE 19880-0402

The filamentous ascomycete *Magnaporthe grisea* causes a devastating disease of rice and other grasses. Plant infection requires the fungus to form a specialized infection structure called an appressorium. From the appressorium the fungus penetrates the plant cuticle and invades the underlying epidermal cells. The signaling pathways required for appressoria formation appear to be complex and include a cAMP-regulated pathway and a MAP kinase pathway. We have discovered two new genes involved in appressorium formation. *PTH11* encodes a novel protein with seven to nine transmembrane domains. *pth11* mutants form very long germ tubes on the plant surface and few of these germ tubes form appressoria. Consequently, these mutants cause very limited disease. Current data suggests that Pth11p may be the receptor for plant wax monomers and/or have a role in general surface sensing. Surprisingly, not all *M. grisea* strains require *PTH11* for appressorium formation. *PTH12* encodes a homeodomain transcription factor homolog. *pth12* mutants appear to differentiate appressoria but these appressoria do not mature. These mutants cause no disease symptoms. Progress on these two genes will be discussed and integrated into the current model of the signaling pathways required for appressorium formation.

INTERACTION BETWEEN RICE BACTERIAL BLIGHT RESISTANCE GENE PRODUCT XA21 AND PATHOGENIC PROTEIN FROM *XATHOMONAS* ORYZA PV. ORYZA (Xoo)

· Y. Shen, M.-S. Chen, P. C. Ronald
Department of Plant Pathology, University of California, Davis, CA 95616, USA

Some disease resistance gene products act as receptors of pathogen molecules. Upon binding downstream responses are triggered which ultimately lead to restriction of pathogen growth. Rice bacterial blight resistance gene *Xa21* is a unique type of resistance gene which encodes a receptor-like kinase consisting of a presumed extracellular leucine rich repeat (LRR) domain and a cytoplasmic kinase domain. In the present study, the possible interactions between XA21 and Xoo proteins were explored using yeast two-hybrid system.

A genomic DNA library of Xoo (Philippine race 6) was constructed in activation domain vector pPC86, which contains 1×10^7 Xoo genomic DNA clones at a size range from 500bp to 5kb. The library was first screened with the LRR domain as a bait in two yeast strains, FH7C and PJ69-4A. No single Xoo clone from 1×10^8 transformants was found to encode a protein which interacts with XA21 LRR domain. However, when the library was screened with kinase domain as a bait, a unique 3kb Xoo clone was isolated from 4 yeast colonies in two separate screening experiments in two different host strains. Further analysis of the clone showed that a 1.7kb ORF (ORF1) encode a protein which interacts with XA21 kinase domain. The interaction was verified in the third yeast strain Y190.

Structure analysis found that the kinase interacting Xoo protein is a GTP binding protein with well conserved GTP binding motives. Conjugation of a virulent Korean Xoo strain J18 with ORF1 containing Xoo genomic clone pURF027kx (5.5kb) was made. Inoculation of *Xa21* possessing rice plants PBR221 with the transconjugants did not display avirulence which might be conferred by the Xoo protein. It is therefore postulated that the Xoo protein may be a virulence factor which may be secreted from bacterial cell and enter into the rice cell where it interacts with the kinase domain of XA21 and interfere with defence signal transduction pathway. Studies on the biological function of the gene through unmarked mutation or marker exchange is now underway.

IDENTIFICATION OF PATHOGENICITY MUTANTS OF THE RICE BLAST FUNGUS MAGNAPORTHE GRISEA BY INSERTIONAL MUTAGENESIS

P. V. Balhadère, A. J. Foster and N. J. Talbot
School of Biological Sciences, University of Exeter, Exeter EX4 4QG, UK

Restriction enzyme-mediated insertion was used to characterise mutants of the rice blast fungus Magnaporthe grisea impaired in pathogenicity. Three REMI protocols were critically evaluated and the frequency of REMI insertions determined for each. A REMI library of 3527 transformants was generated and 1150 transformants have so far been screened for defects in pathogenicity. Five mutants have been identified and characterised phenotypically. Genetic analysis showed that each mutant was the result of a single gene mutation, but that only two of the mutations had arisen due to REMI. Two mutants (2029 and 2050) were impaired in appressorium function, showing reduced penetration of intact barley cuticle, but were unaltered in appressorial turgor generation. Two further mutants, 125 and 130, were altered in conidial morphology, conidial development and appressorium function. Significantly, mutant 130 was also a methionine auxotroph and methionine auxotrophy co-segregated with the reduction in pathogenicity. An additional mutant, 80, was non-pathogenic on susceptible rice cultivars but fully pathogenic on barley. The reduction in pathogenicity in mutant 80 was associated with a delay in conidial germination and appressorium development. The genetic loci in mutants 2029, 2050, 125, 130 and 80 were termed PDE1, PDE2, IGD1, MET1 and GDE1 respectively. PDE1 and PDE2 were non-allelic to CPKA, indicating that each mutation has occurred at a previously undescribed locus in M. grisea. The mutations at PDE1 and GDE1 were generated by insertional mutagenesis, while the remaining mutations were spontaneous, or induced by the REMI procedure. The results indicate the utility of REMI for studying fungal pathogenicity, but also highlight the requirement for rigorous genetic and phenotypic analysis.

REGULATION OF THE PATHOGENICITY GENE MPG1 IN THE RICE BLAST FUNGUS MAGNAPORTHE GRISEA

D.M. Soanes, M.J. Kershaw and N.J. Talbot
School of Biological Sciences, University of Exeter, Exeter EX4 4QG, UK

MPG1 encodes a fungal hydrophobin which is highly expressed during appressorium development and disease symptom outbreak[1]. Targeted replacement of MPG1 leads to mutants which are unable to carry out efficient appressorium development and are also impaired in conidiogenesis. Conidiating cultures of Dmpg1 mutants are reduced in cell surface hydrophobicity due to the absence of the MPG1-encoded hydrophobin rodlet layer which surrounds conidia. During appressorium development MPG1 appears to be secreted onto the rice surface where interfacial self-assmebly of the hydrophobin plays a role in hyphal attachment to the surface. Critically, this process also acts as a developmental signal for appressorium differentiation. We have found that a number of class I hydrophobins are able to complement the mpg1 mutant phenotype if they are expressed under control of the MPG1 promoter[2]. This suggests that the regulation of MPG1 is fundamental to its role in appressorium formation. To investigate this we have fused a 1.2kb MPG1 promoter fragment to a synthetic allele of the GFP gene encoding green fluorescent protein. This has confirmed that MPG1 is highly expressed during both conidiogenesis and appressorium formation. MPG1 expression becomes loaclaied in the developing appressorium and appears to continue after cuticle penetration has occurred. We are carrying out a functional analysis of the MPG1 promoter and wil report progress toward determining the likely cis-acting elements and trans-acting factors which regulate MPG1 expression during pathogenesis in M. grisea.

1. Talbot et al. (1996) Plant Cell 8, 985-999
2. Kershaw et al. (1998) EMBO J. (in press)

AZOXYSTROBIN: A FLEXIBLE FUNGICIDE FOR RICE

D.W.Bartlett[1], Toshio Enoyoshi[2], C.F.N. PIJLS[1]
[1] Zeneca Agrochemicals, Jealott's Hill Research Station, Bracknell, Berkshire RG42 6ET, UK
[2] Zeneca K.K. Agrochemicals, JARS, 780 Kuno-cho, Ushiku, Ibaraki 300-11, Japan

Azoxystrobin, inspired by the natural product strobilurin A, is a broad spectrum fungicide showing excellent activity across a wide range of Ascomycete, Basidiomycete, Deuteromycete and Oomycete pathogens (Godwin et al., 1992). It inhibits mitochondrial respiration by blocking electron transfer at ubiquinol: cytochrome c oxidoreductase (cytochrome bc1 complex), a mode of action with no cross-resistance to other classes of agricultural fungicide.

Azoxystrobin controls leaf and panicle blast caused by Pyricularia grisea and all other major fungal diseases of rice including sheath blight (Rhizoctonia solani), black sheath rot (Gaeumannomyces graminis var. graminis), stem rot (Sclerotium oryzae) and brown spot (Cochliobolus miyabeanus).

Azoxystrobin is systemic in rice. Therefore, it can be applied either as a nursery box granule or foliar application for control of rice blast. Foliar uptake of azoxystrobin is a gradual process. Absorbed fungicide moves in the xylem and results in an even distribution throughout treated foliage with no accumulation at leaf tips. There is no phloem translocation or vapour activity. Granules containing 14C-azoxystrobin have been used to demonstrate that it moves rapidly following application to the soil surface of nursery boxes and can be traced throughout the foliar tissue of rice seedlings by 3 hours after application.

The dominant effect of azoxystrobin on the development of Pyricularia grisea is inhibiton of spore germination and appressorium formation resulting in excellent preventative activity when applied either as a nursery box granule or as a foliar spray.

The unique properties of azoxystrobin result in significant benefits to yield and milling quality in comparison to conventional rice fungicides.

Godwin, J.R.; Anthony, V.M.; Clough, J.M.; Godfrey, C.R.A. (1992) ICIA5504: a novel broad spectrum, systemic β-methoxyacrylate fungicide. Proceedings 1992 Brighton Crop Protection Conference - Pests and Diseases, 1, 435-42.

CLONING OF PATHOGENICITY GENES BY INSERTIONAL MUTAGENESIS IN MAGNAPORTHE GRISEA.

Clergeot Pierre-Henri1, Firon Arnaud1, Otero Sarah1,2, Latorse Marie-Pascale2, Tharreau Didier3, Notteghem Jean-Loup3 and Lebrun Marc-Henri1

1 IGM, Université Paris-Sud, 914105 Orsay and UMR 41 CNRS-RPA, Rhône-Poulenc Agrochimie;
2 Fongicides, Rhône-Poulenc Agrochimie, 69009 Lyon, France ;
3 CIRAD-CA, 34032 Montpellier, France.

Unraveling functions implicated in the infection process of plant pathogenic fungi is an important challenge for crop protection in the future. We are searching for pathogenicity genes of the rice blast fungus Magnaporthe grisea using a REMI-based plasmid insertional mutagenesis strategy (1). We analyzed 1400 transformants for their pathogenicity by spore inoculation on detached rice and barley leaves as a preliminary screen and on rice plants to confirm putative mutants. Seven non-pathogenic mutants were identified (0,5%). Among four mutants crossed with a compatible wild type strain, two (M421 and M763) had their pathogenicity gene tagged by one copy of the plasmid. The number of lesions caused by mutant M763 was dramatically reduced compared to wild type (-95%). The non-pathogenic mutant M421 (no lesions) was able to differenciate appressoria on plastic coverslips, but slightly earlier than wild type. Its appressoria were more susceptible to collapse at high glycerol concentration than wild type, suggesting a defect in a critical step controlling osmotic pressure levels required for appressorium fonction (2). Flanking regions to the plasmid were cloned by inverse-PCR for M421 and by plasmid rescue for M763. Two cosmids hybridizing with the M421 flanking region were identified from which a genomic subclone of 1,8 kb complemented the mutation. Localization of transcribed regions by reverse Northern within this subclone and their sequencing highlighted a small 0,85 kb ORF interrupted two short introns. The protein deduced from this ORF shared no homology with proteins from. Isolation of M421 cDNA confirmed the position of the two introns. Characterisation of M763 flanking regions is in progress.

(1) Sweigard et al. (1998), Mol. Plant Micr. Interact., 5: 404-412.
(2) de Jong et al. (1997), Nature, 389: 244-245.

AVIRULENCE GENES CLONING IN THE RICE BLAST FUNGUS MAGNAPORTHE GRISEA.

K. Percet3, S. Sibuet3, W. Dioh1, D. Tharreau2, J.L. Notteghem2, M.H. Lebrun1,3
1IGM, Bat 400, Universit* Paris Sud, 91405 Orsay, France
2CIRAD-CA. BP 5035, 34 032 Montpellier, France
3Biotechnologies et UMR41 CNRS-RPA, Rhône-Poulenc Agrochimie, 69009 Lyon, France.

Genetic studies with M. grisea isolates pathogenic on rice and fertile in crosses revealed three genetically independent avirulence genes [1] that are likely to interact with so far undescribed rice resistance genes. We initiated the positional cloning of these three genes using RAPD markers screened by bulk segregant analysis and RFLP markers. Two avirulence genes mapped near chromosome ends (avr1-MedNo* and avr1-Ku86). The avirulence gene avr1-Irat7 mapped on chromosome one at 30 cM from avr1Co39. We identified 16 RAPD markers closely linked to these avirulence genes (0-10 cM). Most RAPD markers corresponded to junction fragments between M. grisea genome and known transposons [2]. Positional cloning was started by the screening of a cosmid genomic library from an avirulent progeny by hybridisation with single copy sequence RAPD markers. For RAPD markers containing repeated sequence, we performed a RAPD analysis on cosmid pools and clones [3]. Cosmids contigs were constructed for these three loci. Starting from RAPD marker OPE-Y13 completly linked to avr1-Irat7, we constructed a contig of 100 kb. Two cosmids from this contig conferred avirulence on rice cultivar Irat7 when introduced by transformation in a virulent recipient strain, defining a 30 kb region where avr1-Irat7 is located. Up to know, the smallest subclone from this region able to complement virulence is a 20 kb fragment. The two other contigs are in the process of extension towards avirulence genes.

1. Silue D. et al. 1992. Phytopathology 82: 1462-1467.
2. Dioh W. et al. 1996. In Rice Genetics 3, IRRI, pp. 916-920.
3. Dioh W. et al. 1997. Nucleic Acid Research 25: 5130-5131.

ISOLATION OF THE CULTIVAR SPECIFICITY GENE AVR1-MARA OF MAGNAPORTHE GRISEA

M.A. Mandel, U. P. Gunawardena, T. M. Harper, M. J. Orbach
Dept. Plant Pathology, University of Arizona, Tucson, AZ, USA

Molecular and genetic approaches are being used to clone the *AVR1-MARA* gene of *Magnaporthe grisea*, and address its apparent genetic stability. We have reported the cloning of the virulent allele, *avr1-MARA*, and the mapping of the avirulent allele. The two alleles differ by the presence of two regions of 40 kb and 14 kb only in the avirulent locus. The avirulent locus is approximately 65 kb with major portions unclonable in *E. coli* (Mandel et al., 1997). To localize the avirulent gene within the locus, we have used transformation-mediated gene disruption methods to delete all, or part of the locus. By this method, the gene has been localized to the region of the locus that contains the 40 kb *AVR*-associated sequences. Two approaches are being taken to isolate this region; one, a combination of Long Distance and Inverse PCR method, has resulted in the isolation of portions of this region. Sequence analysis of these segments has shown them to be unusual in *M. grisea*, with the DNA being 70% AT. Analyses of these sequences will be presented. The second approach to clone this region is Transformation-Associated-Recombination, an *in vivo* ligation method using *Saccharomyces cerevisiae* as a cloning host.
Analyses of virulent mutants of *AVR1-MARA* and the distribution of the *AVR1-MARA* locus in populations of *M. grisea* will be presented.

Mandel, M.A., Crouch, V.W., Gunawardena, U.P., Harper, T.M., and Orbach M.J. 1997. Mol. Plant-Microbe Interact. 10:1102-1105.

MOLECULAR CHARACTERIZATION OF THE *Magnaporthe grisea* AVIRULENCE GENE *AVR2-YAMO* AND CORRESPONDING RESISTANCE GENE *Pi-ta*

B. Valent1, G.T. Bryan1, Y. Jia1, L. Farrall1, K. Wu2, M.J. Orbach3, G.K. Donaldson1, H.P. Hershey1 and S.A. McAdams1
1 Dupont Agricultural Products, Post Office Box 80402, Wilmington DE19880-0402, USA
2 Current Address: Ceregen, 800 Lindbergh Boulevard, St. Louis, MO 63167, USA
3 Current Address: Dept. of Plant Pathology, University of Arizona, Tucson, AZ 85721, USA

Genetic analysis has demonstrated a classical gene-for-gene interaction between races of the rice blast fungus and rice variety Yashiro-mochi, which contains resistance gene *Pi-ta*. Pathogen strains with the avirulence gene *AVR2-YAMO* fail to infect Yashiro-mochi, presumably due to the production of a signal molecule that interacts in some manner with *Pi-ta* and mediates recognition. The cloned avirulence gene encodes a protein with a predicted molecular weight of 25 kDa that has some sequence similarities to neutral zinc metalloproteases. Transformation of virulent strains of *M. grisea* with a functional copy of *AVR2-YAMO* confers avirulence on Yashiro-mochi. Mutation of specific amino acid residues in the putative zinc-binding region removes the ability to confer avirulence. Sequence comparisons with non-functional *avr2-yamo* homologs from rice pathogens also indicate that maintenance of the putative zinc binding region is essential. Reverse transcriptase PCR and *AVR2-YAMO* promoter fusions to reporter genes demonstrate that this avirulence gene, which appears not to be expressed in culture, is expressed during infection, suggesting a role in pathogenicity. Data will be presented on the cloning and characterization of the resistance gene *Pi-ta*. Current efforts to identify the signal molecule responsible for the *AVR2-YAMO Pi-ta*-induced resistance response will be described, including new data obtained using the yeast two-hybrid system that suggests a physical interaction between the *AVR2-YAMO* protein and the *Pi-ta* protein.

ALLELIC DIVERSITY OF AN AVIRULENCE GENE AMONG FIELD ISOLATES OF THE RICE BLAST FUNGUS IN COLOMBIA.

Montenegro, M. V (1), M. Levy (1) and B. Valent (2)
1. Dept. of Biological Sciences, Purdue University, West Lafayette, IN USA 47907-1392
2. DuPont Experimental Stn., E402/4244, Wilmington, DE USA 19880

To gain insight into the evolution of pathogenic change, allelic variation for *Avr2-Yamo*, the first cultivar-specificity avirulence gene cloned from *Magnaporthe grisea*, was analyzed among Colombian field isolates previously assigned to five genetic lineages based on MGR-DNA fingerprinting. Typically, each lineage expresses a uniform EcoRI-RFLP profile for *Avr2-Yamo* and an associated reaction on the cultivar Yashiro-mochi. Isolates of lineages SRL-1 and SRL-2 have a 3.2 kb RFLP and are avirulent. Isolates of lineages SRL-3, ALT-7 and SRL-6 are virulent but have RFLPs of 7.5 kb, 7.5 kb and 1.9 kb, and 1.9 kb or null, respectively. DNA-amplifications and sequencing confirm strong conservation of the coding sequence (based on the Chinese isolate O-137) in all alleles. Both avirulent and most virulent Colombian alleles putatively encode a protein with 224 amino acids, i.e., the same length as the functional protein of O-137 plus a one leucine insertion. Deletions in upstream regions relative to O-137 are observed for all alleles but the majority of allelic differences involve single base mutations rather than sequence length changes, indicating that avr-gene instability is not a common feature in Colombian field isolates. A group of seven putative amino acid changes distinguish the avirulent (O-137, and SRL-1 and SRL-2) alleles from the virulent (SRL-3 and the two ALT-7) alleles. Phylogenetic analysis indicates that the ALT-7 genotype evolved by gene duplication, where the original locus (7.5kb) remained largely conserved and the duplicate locus (1.9kb) diverged significantly. The 1.9 kb "virulent" allele present in some SRL-6 isolates is closest in sequence to 0-137, although it contains two frameshift mutations that alter completely its inferred protein sequence. The SRL-6 allele also conserves intact the upstream sequence of the gene that is absent in both ALT-7 alleles, ruling out recombination as a source of the ALT-7 genotype. Molecular and pathogenic characterization of transformants (e.g., SRL-1 allele into SRL-3 background) is underway to test avirulence functions of alleles. The 0-137 upstream region contains two repetitive fragments, MGR 608 and MGR 619, sequences previously found in association with the species-specificity avirulence genes, *Avr-Pwl1* and *Avr-Pwl2*. Further sequence analysis of *Avr2-Yamo* alleles and associated flanking regions should also reveal major historical patterns of global pathogen migration.

PHYSICAL AND GENETIC PARAMETERS OF THE *MAGNAPORTHE GRISEA* GENOME

S. Leong

Over the last decade, my research program has focused on developing a fundamental understanding of the genome organization of the ascomycete phytopathogenic fungus *Magnaporthe grisea*. A genetic map containing over 200 markers was developed and linked to the electrophoretic karyotype of the fungus. This map was also integrated with other genetic maps of the organism thus increasing the available markers to over 400. Several classes of transposable elements were discovered and found in clusters throughout the genome. Within the context of this framework analysis of the *M. grisea* genome, we have recently employed a map-based cloning method to clone *AVR1-CO39*, a gene which controls cultivar-specific interactions of *M. grisea* with rice plants. Achilles' cleavage was used to both facilitate walking and to overcome several difficulties encountered in the chromosome walk. Interestingly, the ratio of genetic to physical distance observed in the region analyzed during the walk to *AVR1-CO39* was found to vary by 14 fold. Our work provides important lessons and tools for map-based cloning approaches in fungi.

RICEBLASTDB: A DATABASE FOR MAGNAPORTHE GRISEA

Yap (1), S. Gnanamanickam (2), W. Wong (3), A. Amante-Bordeos (4), R. Zeigler (4), M. Levy (5), S. Leong (3), and S. McCouch (1)
1 Cornell Univ, Dept of Plant Breeding and Biometrics, 252 Emerson,
Ithaca, NY 14853, USA
2 Univ of Madras, Adv Studies in Botany Center, Madras 600025,
Tamilnadu, INDIA
3 Univ of Wisconsin, Dept of Plant Pathology and USDA/ARS, 1630 Linden
Dr, Madison, WI 53706, USA
4 IRRI, EPPD, PO Box 933, 1099 Manila, PHILIPPINES
5 Purdue Univ, Dept of Biological Sciences, G-420 Lilly Hall, West
Lafayette, IN 47907, USA

The database RiceBlastDB aims to integrate the diverse types of information about the causal agent of rice blast, Magnaporthe grisea (anamorph Pyricularia grisea). The database currently includes an extensive bibliography about the fungus; a genetic map based on a cross between isolates Guy11 and 2539; a physical map based on a BAC contig library derived from strain 70-15; and data on the rice blast populations in Europe and the Philippines, including DNA fingerprint variation, lineage structure, and pathotype information; work is currently underway to incorporate information on additional blast populations worldwide. It is publicly accessible and may be browsed on the World Wide Web through the USDA AgriculturalGenomeInformation System at <http://probe.nalusda.gov:8000/plant/aboutRiceBlastDB.html>. We are working closely with the RiceGenes curators to integrate data between host and pathogen. When more fully developed, we hope that this database will facilitate greater understanding of the genetics of pathogenesis and host-plant resistance, and serve as a guide to breeders for resistance gene deployment.

SIGNALING PATHWAYS AND APPRESSORIUM FORMATION IN *MAGNAPORTHE GRISEA.*

Ralph A. Dean, Department of Plant Pathology & Physiology, and Clemson University Genomics Institute, Clemson University, USA.

It remains largely a mystery how an emerging germ-tube from a conidium recognizes it is on a suitable host and sets in motion the process of appressorium formation that leads directly to infection of the underlying tissues. The first step in the infection process occurs when the 3-celled teardrop shaped conidium lands on and attaches to the leaf surface. A germ-tube emerges and after a few hours growth, polar elongation ceases and the tip hooks and swells to form the appressorial initial. This process can be induced on a hydrophobic surface, such as GelBond. Other contact surfaces, such as cleaned glass and the hydrophilic surface of GelBond are typically non-inductive. On non-inductive surfaces, the germ-tube continues to extend and develops into typical vegetative mycelia.

As the appressorium continues to mature, a dense layer of cross-linked fibrillar materials and melanin are laid over the pre-existing wall, except across a pore at the plant interface. The nucleus during this time undergoes a mitotic division and one daughter enters the appressorium. The appressorium eventually appears to be partially embedded in the plant cuticle and is firmly anchored by an adhesive material. Infection ultimately takes place when a narrow infection hypha emerges through the pore and penetrates directly into the plant tissues. The penetration process is driven by turgor pressure generated by the mobilization of carbohydrate reserves into glycerol.

Our present understanding of the signal pathways regulating appressorium formation will be discussed. In the model, I propose a dual signaling pathway involving both thigmotropic and chemical stimuli from the environment being transduced into the germ-tube by separate and distinct pathways. There are several lines of evidence supporting this model. *App2*- mutants, which do not form appressoria in response to a hydrophobic surface, form appressoria in response to 1,16 hexadecanediol. Furthermore, mutants lacking *CPKA*, the catalytic subunit gene of protein kinase A, do not form normal appressoria in response to 1,16 hexadecanediol or a hydrophobic surface. Thus, the hydrophobic and 1,16 hexadecanediol signals were proposed to merge at or before protein kinase activation. Recent work with G protein α subunit mutants and an adenylate cyclase mutant has allowed the model to be refined. *Mag*B- mutants, which are defective in appressorium formation on hydrophobic surfaces, respond to 1,16 hexadecanediol. However, mutants lacking the adenylate cyclase gene, *MAC1*, are unable to form appressoria in response to 1,16 hexadecanediol. These data suggest convergence of these 2 pathways at or before the activation of adenylate cyclase.

In most recent work, the intrinsic GTPase activity of the G α subunit and the ability of the GTP-bound α subunit to dissociate from the G βγ subunits was affected by changing certain amino acids. Constitutive activation of G α was accomplished by substituting arginine for glycine at position 42. A dominant interfering mutation was created by replacing glycine 203 with arginine. The results of these studies will be discussed and integrated into our emerging model.

DEFENSE OF RICE PLANT TO RESPIRATORY INHIBITION BY A BLASTICIDE, METOMINOSTROBIN (SSF126)

A. Mizutani, N. Miki, M. Masuko
Aburahi Laboratories, Shionogi & Co., Ltd., 1405 Gotanda, Koka, Shiga 520-3423, Japan

Metominostrobin (SSF126) has been developed as a systemic fungicide for control of rice blast caused by *Pyricularia oryzae* (anamorgh of *Magnaporthe grizea*). It inhibits fungal respiration by blockage of electron flux through the cytochrome bc_1 segment in the mitochondrial respiratory chain (1). Superoxide anion (O_2^-) generated by blockage of the cytochrome pathway mediated de novo synthesis of the alternative oxidase protein responsible for the cyanide-resistant alternative respiration. However, flavonoids in plants inhibits the metominostrobin-dependent induction of the alternative respiration by scavenging the oxygen radical, indicating that metominostrobin controls rice blast in conjunction with rice plant components (2). In rice plants, both the cytochrome and the alternative pathways contribute to the respiration and the capacity of the alternative pathway is enhanced in response to inhibition of the cytochrome pathway. To determine whether expression of the alternative oxidase protein correlates with the increased capacity of the alternative pathway, rice plant mitochondrial proteins were immunoblotted with a monoclonal antibody against the alternative oxidase protein. The immunoblot analysis revealed that the alternative oxidase proteins are constitutive in rice plants, unlike *P. oryzae*. Furthermore, high molecular mass species (72-88KDa) detected were converted to the corresponding low molecular mass species (36-40Kda) at inhibition of electron flux through the cytochrome pathway. The conversion closely correlated with enhancement of the alternative respiration, indicating that the capacity of the alternative pathway is regulated by interconversion of the alternative oxidase proteins in rice plants. Possibly, the regulatory mechanism of respiration enables rice plants to survive exposure to metominostrobin.
References
1. Mizutani A, Yukioka H, Tamura H, Miki N, Masuko M, Takeda R. 1995. *Phytopathology* 85, 306-311.
2. Mizutani A, Miki N, Yukioka H, Tamura H, Masuko M, 1996. *Phytopathology* 86, 295-300.

THE INVESTIGATION AND INTERPRETATION OF GENOTYPE BY ENVIRONMENT INTERACTION FOR RICE BLAST SUSCEPTIBILITY IN REPUBLIC OF KOREA

C.I.Yang1, S.W.Ahn2, J.H.Roh1, M.A.Ynalvez2, C.G.Mclaren2
1 National Crop Experiment Station, RDA, Suwon 441-100, Korea
2 International Rice Research Institute, P.O.Box 933, Manila, Philippines

Addititive Main Effect and Multiplicative Interaction Model (AMMI) with a mean polished transformation is useful for analyzing G x E interaction in various agronomic traits in multi-environment variety trials and provide insight into the environmental factors accounting for G x E interactions. Data from Korea Rice Blast Nursery (KRBN) was analyzed with AMMI model for the past 15 years (1981-1996) over 21 test sites for 44 common test varieties with the exclusion of the 1983. It was found that genotype variability (89%) was highest and G x E interaction (8%) was larger than that of environment (3%). These results that much of the variability in mean blast scores is largely accounted for by variations in genotypes. Major sources of variation and environmental factors affecting G x E in blast nursery test in korea were temperature and precipitation during the testing periods. From the results of the correlation analysis, meteorological factors were important than soil factors in the mentioned periods.

M. Cooper and I.H.Delacy. 1994. Relationships among analytical methods used to study genotypic variation and genotype-by-environment interaction in plant breeding multienvironment experiments. Theor Appl Genet 88 : 561-572
H.G.Gauch,Jr and R W.Zobel. 1990. Inputing missing yield trial data Theor Appl Genet 79 : 753 – 761
Dan Brabu and K. Ruben Gabriel. 1978. The Biplot as a Diagnostic Tool for Models of Two-Way Tables TECHNOMETRIC, VOL.20, NO.1, FEB.

INTEGRATED RICE BLAST MANAGEMENT IN GILAN PROVINCE

M. Izadyar
Plant Pests & Diseases Research Institute
P.O.Box 1454, Tehran 19395, Iran

Rice blast caused by Magnaporthae grisea (Pyricularia oryzae Cav.) is one of the serious diseases of local rice cultivars with high cooking quality in Caspian sea Area in north of Iran. All local cultivars are susceptible to all physiological races prevalent throughout the rice growing fields in Gilan and Mazandaran which are the main rice growing area in Iran. So cultivation of resistant breeded lines are the most important method to control rice blast in north of Iran and a few improved resistant varieties such as Khazar, Sepeed-Roud, Bejar and etc. have been introduced to farmers and nearly 30% of fields are under cultivation of these varieties. So other fields are under cultivation of local cultivars which are very susceptible to much prevalent physiological races such as IA-81 and IA-89. Since there are much fluctuation in disease severity in year to year, region to region, cultivar to cultivar and even field to field. So chemical control has not been successful without forecasting criteria.
Some forecasting criteria such as host plant predisposition by inoculation of leaf sheath of different local cultivars with prevalent rice blast strains in the region and incubated at 26oC for 40 hours and the rate of mycelial growth in epidermis cells were evaluated into 6 grades under microscope and the severity of leaf blast were estimated by Y=19.33X- 9.8 (Izadyar, 1980), meteorological factors (weather temperature, humidity and precipitation) and spores intensity in the air and the relationship between number of days from the date of transplanting time until the date of first appearance of leaf blast lesion in the field with severity of leaf blast in the field for rice blast predicting and application of effective fungicides: Edifenphos and Tricyclazole were studied and introduced to execute by plant protection experts and farmers .

Izadyar, M. 1980.The Iranian Phytopathological Society,Vol. 16 No.(1-4) 15-23.

PREDICTING LOSSES FROM RICE BLAST IN INDIA

N.V.L. R.Rajeswari, K. Muralidharan
Directorate of Rice Research, Hyderabad 500 030, India

Rice is the most important food crop in India both in acreage and productivity. Rice requirement is estimated at 94 m t by 2002, and at 104 mt by 2007. Production growth would necessitate increased use of agricultural inputs. In all rice ecosystems, blast disease outbreaks limit yields. Increased use of nitrogen fertilizer would naturally escalate the blast outbreaks. The degree of occurrence of blast disease on rice is well documented in production oriented surveys made every year. The magnitude of losses caused by *Pyricularia grisea* becomes apparent only when the grain yield harvested in fungicide treated plot is compared with unprotected plot. We pooled 30 years data (1968-1997) from experiments under different ecosystems, on rice blast control. For each ecosystem, we derived empirical estimates on blast induced yield losses. In farmers' fields often blast occurs with one or more pests. Published data from such experiments made (1951-1966) at the Central Rice Research Institute, Cuttack, before the introduction of high-yieding rice cultivars were also analysed to build a general regression model. Farmers have been experiencing losses to Basmati rice in the northern states. We made attempts to predict the yield losses in farmers' fields in some states and verify actual yield losses. Using systematic sampling methods, data were collected on blast incidence in large areas. A mere 10% neck blast incidence in irrigated fields resulted in 0.40t/ha yield loss In Basmati rice besides direct grain yield loss, the quality of the grain harvested was found affected by the blast pathogen. This paper discusses the usefulness of the empirical approach in arriving at loss estimates in fields under different rice ecosystems.

CONTROL OF RICE BLAST BY PLANT MONOTERPENES

S. Narasimhan and S. Masilamani
Centre for Agrochemical Research, SPIC Science Foundation, 110, Mount Road
Guindy, Chennai, INDIA.

The use of synthetic fungicides has come in for increasing criticisms in recent years, with aspects of public concern including both the possible accumulation of fungicide residues in soil following repeated applications to crops and deleterious effects on soil micro - organisms with adverse long term consequences for soil fertility. In our laboratory, we are working on natural product chemistry for very long time and screened many plant extracts and pure compounds of plant origin for antifungal property. Recently we have isolated a monoterpene which has proved to be very good antifungal against various plant pathogens. The monoterpene and its derivatives were tested for antifungal activity against rice blast (*Pyricularia oryzae*). Most of the derivatives were having inhibition of fungal mycelium at 1000 ppm. The biologically active monoterpene is developed into commercial product by mixing it with suitable surfactant and sprayed in the field against rice blast. The product is as good as commercial synthetic fungicides, and it is having all the necessary characteristics to compete with the best synthetic agrochemical. The source of the compound is easily available and economically viable. Structure activity relationship studies of this and related monoterpenes is also in progress. The results will be discussed in detail.

MANAGEMENT OF PADDY BLAST THROUGH CULTURAL PRACTICES, HOST RESISTANCE AND FUNGICIDES

Ram Singh, S. Sunder, D.P. Nandal and M. Ram
CCS HAU Rice Research Station, Kaul-136 021, Haryana, India

Field trials conducted at CCS HAU Rice Research Station, Kaul revealed that the incidence of leaf, neck and node blast was significantly higher in biweekly irrigated plots with a corresponding decrease in grain yield over submerged treatment. Among six basmati cvs. planted on 1, 15 and 30 July, Ranbir êBasmati had maximum neck blast while Type 3 and HKR 90-416 the least. Significantly higher disease incidence and lower grain yield was recorded in 30 July plantings over other two dates in all the varieties except neck blast incidence in Type 3 and Basmati 370. Amongst 219 scented genotypes evaluated in UBSN, 13 genotypes were found consistently resistant during three years. Out of four fungitoxicants evaluated, tricyclazole proved most effective and reduced neck blast incidence by 74.6 per cent along with 23.2 per cent increase in grain yield followed by epoxiconazole.

REFERENCES
Chaudhary, R.G. and Vishwadhar, 1988. Indian Phytopathol.41: 552-557.
Gill, M.A; Bonman, J.M; Khan, M.G. and Butt, M.A. 1993. Pak. J. Phytopathol. 5(1-2): 10-17.
Kumar, S; Mandakini Nayak and Sridhar, R. 1997. J. Mycol. Pl. Pathol. 27: 1-5.
Singh, R. and Dodan, D.S. 1996. Intern. Rice Res. Notes. 21 (2&3): 45.
Woloshuk, C.P. and Sisler, H.D. 1982. J. Pestic. Sci. 7: 161-166.

Key words index

Developments in Plant Pathology

1. R. Johnson and G.J. Jellis (eds.): *Breeding for Disease Resistance.* 1993
   ISBN 0-7923-1607-X
2. B. Fritig and M. Legrand (eds.): *Mechanisms of Plant Defense Responses.* 1993
   ISBN 0-7923-2154-5
3. C.I. Kado and J.H. Crosa (eds.): *Molecular Mechanisms of Bacterial Virulence.* 1994
   ISBN 0-7923-1901-X
4. R. Hammerschmidt and J. Kuć (eds.), *Induced Resistance to Disease in Plants.* 1995
   ISBN 0-7923-3215-6
5. C. Oropeza, F.W. Howard, G. R. Ashburner (eds.): *Lethal Yellowing: Research and Practical Aspects.* 1995 ISBN 0-7923-3723-9
6. W. Decraemer: *The Family Trichodoridae: Stubby Root and Virus Vector Nematodes.* 1995 ISBN 0-7923-3773-5
7. M. Nicole and V. Gianinazzi-Pearson (eds.): *Histology, Ultrestructure and Molecular Cytology of Plant-Microorganism Interaction.* 1996 ISBN 0-7923-3886-3
8. D.F. Jensen, H.-B. Jansson and A. Tronsmo (eds.): *Monitoring Antagonistic Fungi Deliberately Released into the Environment.* 1996 ISBN 0-7923-4077-9
9. K. Rudolph, T.J. Burr, J.W. Mansfield, D. Stead, A. Vivian and J. von Kietzell (eds.): *Pseudomonas Syringae Pathovars and Related Pathogens.* 1997
   ISBN 0-7923-4601-7
10. C. Fenoll, F.M.W. Grundler and S.A. Ohl (eds.): *Cellular and Molecular Aspects of Plant-Nematode Interactions.* 1997 ISBN 0-7923-4637-8
11. H.-W. Dehne, G. Adam, M. Diekmann, J. Frahm, A. Mauler-Machnik and P. van Halteren (eds.): *Diagnosis and Identification of Plant Pathogens.* 1997
   ISBN 0-7923-4771-4
12. A.C. Cassells (ed.): *Pathogen and Microbial Contamination Management in Micro-propagation.* 1997 ISBN 0-7923-4784-6
13. K. Kohmoto and O.C. Yoder (eds.): *Molecular Genetics of Host-Specific Toxins in Plant Disease.* 1998 ISBN 0-7923-4981-4
14. R. Albajes, M. Lodovica Gullino, J.C. van Lenteren and Y. Elad (eds.): *Integrated Pest and Disease Management in Greenhouse Crops.* 1999 ISBN 0-7923-6257-8
15. D. Tharreau, M.H. Lebrun, N.J. Talbot and J.L. Notteghem (eds.): *Advances in Rice Blast Research.* Proceedings of the 2nd International Rice Blast Conference 4-8 August 1998, Montpellier, France. 2000 ISBN 0-7923-6257-8

KLUWER ACADEMIC PUBLISHERS – DORDRECHT / BOSTON / LONDON